A WATER RESOURCES TECHNICAL PUBLICATION

Engineering Monograph No. 25

Hydraulic Design of Stilling Basins and Energy Dissipators

By A. J. PETERKA

Denver, Colorado

United States Department of the Interior

BUREAU OF RECLAMATION

As the Nation's principal conservation agency, the Department of the Interior has responsibility for most of our nationally owned public lands and natural resources. This includes fostering the wisest use of our land and water resources, protecting our fish and wildlife, preserving the environmental and cultural values of our national parks and historical places, and providing for the enjoyment of life through outdoor recreation. The Department assesses our energy and mineral resources and works to assure that their development is in the best interests of all our people. The Department also has a major responsibility for American Indian reservation communities and for people who live in Island Territories under U.S. administration.

ENGINEERING MONOGRAPHS are published in limited editions for the technical staff of the Bureau of Reclamation and interested technical circles in Government and private agencies. Their purpose is to record developments, innovations, and progress in the engineering and scientific techniques and practices that are employed in the planning, design, construction, and operation of Reclamation structures and equipment.

First Printing: September 1958
Second Printing—Revised: July 1963
Third Printing: March 1974
Fourth Printing—Revised: January 1978
Fifth Printing: May 1979
Sixth Printing: October 1980
Seventh Printing: May 1983
Eighth Printing: May 1984

Preface

THIS MONOGRAPH generalizes the design of stilling basins, energy dissipators of several kinds and associated appurtenances. General design rules are presented so that the necessary dimensions for a particular structure may be easily and quickly determined, and the selected values checked by others without the need for exceptional judgment or extensive previous experience.

Proper use of the material in this monograph will eliminate the need for hydraulic model tests on many individual structures, particularly the smaller ones. Designs of structures obtained by following the recommendations presented here will be conservative in that they will provide a desirable factor of safety. However, model studies will still prove beneficial to reduce structure sizes further, to account for nonsymmetrical conditions of approach or getaway, or to evaluate other unusual conditions not described herein.

In most instances design rules and procedures are clearly stated in simple terms and limits are fixed in a definite range. However, it is occasionally necessary to set procedures and limits in broader terms, making it necessary that the accompanying text be carefully read.

At the end of this monograph is a graphic summary, giving some of the essential material covered, and a nomograph which may be used as a computation aid. These sheets are particularly useful when making preliminary or rough estimates of basin sizes and dimensions.

The monograph contains essentially the information contained in the following Bureau of Reclamation's Hydraulic Laboratory Reports: Hyd-399 dated June 1, 1955, by J. N. Bradley and A. J. Peterka; Hyd-409 dated February 23, 1956, by A. J. Peterka; Hyd-415 dated July 1, 1956, by G. L. Beichley and A. J. Peterka; Hyd-445 dated April 28, 1961, by A. J. Peterka; Hyd-446 dated April 18, 1960, by G. L. Beichley and A. J. Peterka; and Hyd PAP-125 dated July 1959, by T. J. Rhone and A. J. Peterka.

A previous edition of this monograph dated September 1958 contained material from Hyd-399 and Hyd-415 only.

Hyd-399 was published in the October 1957 Journal of the Hydraulics Division, American Society of Civil Engineers, in a series of six papers under the title of "The Hydraulic Design of Stilling Basins." Hyd-415 was published in the Journal of the Hydraulics Division, ASCE, October 1959, under the title "The Hydraulic Design of Slotted Spillway Buckets." Hyd-446 was published in the Journal of the Hydraulics Division, ASCE, September 1961, under the title "Hydraulic Design of Hollow-Jet Valve Stilling Basins," and later in Transactions for 1962, ASCE, Vol. 127, Part 1, Paper No. 3296. Hyd PAP-125 was published in the Journal of the Hydraulics Division, ASCE, December 1959, under the title, "Improved Tunnel Spillway Flip Buckets," and later in Transactions for 1961, ASCE, Vol. 126, Part 1, Paper No. 3236.

Hyd-409 was rewritten for inclusion in this monograph, and new data and more extensive conclusions and recommendations have been added. Hyd-445 was also modified for inclusion in this monograph and contains additional information for chute slopes flatter than 2:1.

Contents

Page

Section 3.—Short Stilling Basin for Canal Structures, Small Outlet Works, and Small Spillways (Basin III)

Section 4.—Stilling Basin Design and Wave Suppressors for Canal Structures, Outlet Works and Diversion Dams (Basin IV)

Section 5.—Stilling Basin With Sloping Apron (Basin V)

Section 6.—Stilling Basin for Pipe or Open Channel Outlets (Basin VI)

Section 7.—Slotted and Solid Buckets for High, Medium, and Low Dam Spillways (Basin VII)

Section 8.—Hydraulic Design of Hollow-Jet Valve Stilling Basins (Basin VIII)

Section 9.—Baffled Apron for Canal or Spillway Drops (Basin IX)

Section 10.—Improved Tunnel Spillway Flip Buckets (Basin X)

Section 11.—Size of Riprap To Be Used Downstream From Stilling Basins

LIST OF FIGURES

LIST OF TABLES

Introduction

ALTHOUGH HUNDREDS of stilling basins and energy-dissipating devices have been designed in conjunction with spillways, outlet works, and canal structures, it is often necessary to make model studies of individual structures to be certain that these will operate as anticipated. The reason for these repetitive tests is that a factor of uncertainty exists regarding the overall performance characteristics of energy dissipators.

The many laboratory studies made on individual structures over a period of years have been made by different personnel, for different groups of designers, each structure having different allowable design limitations. Since no two structures were exactly alike, attempts to generalize the assembled data resulted in sketchy and, at times, inconsistent results having only vague connecting links. Extensive library research into the works of others revealed the fact that the necessary correlation factors are nonexistent.

To fill the need for up-to-date hydraulic design information on stilling basins and energy dissipators, a research program on this general subject was begun with a study of the hydraulic jump, observing all phases as it occurs in open channel flow. With a broader understanding of this phenomenon it was then possible to proceed to the more practical aspects of stilling basin design.

Existing knowledge, including laboratory and field tests collected from Bureau of Reclamation records and experiences over a 23-year period, was used to establish a direct approach to the practical problems encountered in hydraulic design. Hundreds of tests were also performed on both available and specially constructed equipment to obtain a fuller understanding of the data at hand. Testing and analysis were coordinated to establish valid curves in critical regimes to provide sufficient understanding of energy dissipators in their many forms, and to establish workable design criteria. Since all the test points were obtained by the same personnel, using standards established before testing began, and since results and conclusions were evaluated from the same datum of reference, the data presented are believed to be consistent and reliable.

Six test flumes were used at one time or another to obtain the experimental data required on Hydraulic Jump Basins I through V—Flumes A and B, Figure 1; Flumes C and D, Figure 2; and Flume F, Figure 3. The arrangement shown as Flume E, Figure 3, actually occupied a portion of Flume D during one stage of the testing, but it is designated as a separate flume for ease of reference. Flumes A through E contained overflow sections so that the jet entered the stilling basin at an

A—Test flume A. *Width of basin 5 feet, drop 3 feet, discharge 6 c.f.s.*

B—Test flume B. *Width 2 feet, drop 5.5 feet, discharge 12 c.f.s.*

FIGURE 1.—*Test flumes.*

angle to the horizontal. The degree of the angle varied in each test flume. In Flume F, the entering jet was horizontal, since it emerged from under a vertical slide gate.

Each flume served a useful purpose either in verifying the similarity of flow patterns of different physical size or in extending the range of the experiments started in one flume and completed in others. The different flume sizes and arrangements also made it possible to determine the effect of flume width and angle of entry of the flow.

The experiments were started in an existing model of a flat-chute spillway, Figure 1A, having a small discharge and low velocity. This was not an ideal piece of equipment for general experiments as the training walls on the chute were diverging. The rapid expansion caused the distribution of flow entering the stilling basin to shift with each change in discharge; however, this piece of equipment served a purpose in that it aided in establishing the procedures used in the research program.

Tests were then continued in a glass-sided laboratory flume 2 feet wide and 40 feet long in which an overflow section was installed, Flume B, Figure 1B. The crest of the overflow section was 5.5 feet above the floor, and the downstream face was on a slope of 0.7:1. The discharge capacity was about 12 c.f.s.

Later, the work was carried on at the base of a chute 18 inches wide having a slope of 2 horizontal to 1 vertical and a drop of approximately 10 feet, Flume C, Figure 2A. The stilling basin had a glass wall on one side. The discharge capacity was 5 c.f.s.

The largest scale experiments were made on a glass-sided laboratory flume 4 feet wide and 80 feet long, in which an overfall crest having a slope of 0.8:1 was installed, Flume D, Figure 2B. The drop from headwater to tail water in this case was approximately 12 feet, and the maximum discharge capacity was 28 c.f.s.

The downstream end of the above flume was also utilized for testing small overflow sections 0.5 to 1.5 feet in height. The maximum discharge used was 10 c.f.s. As stated above, this piece of equipment is designated as Flume E, and is shown in Figure 3A.

The sixth testing device was a tilting flume which could be adjusted to provide slopes up to 12°, Flume F, Figure 3B. This flume was 1 foot wide by 20 feet long; the head available was 2.5 feet, and the flow was controlled by a slide gate. The discharge capacity was about 3 c.f.s.

Each flume contained a head gage, a tail gage, a scale for measuring the length of the jump, a point gage for measuring the average depth of flow entering the jump, and a means of regulating the tail water depth. The discharge in all cases was measured through the laboratory venturi meters or portable venturi-orifice meters. The tail water depth was measured by a point gage operating in a stilling well. The tail water depth was regulated by an adjustable weir at the end of each flume.

Flume B was also used for the tests to develop the slotted-bucket energy dissipator described in Section 7, Basin VII. Other test setups used to develop the impact basin, the wave suppressors, the baffled chutes, the flip buckets, the hollow-jet valve stilling basin, and the riprap size data, are described in appropriate sections.

FIGURE 2. *Test flumes.*

A—Test flume C. *Width 1.5 feet, drop 10 feet, discharge 5 c.f.s., slope 2:1.*

B—Test flume D. *Width 4 feet, drop 12 feet, discharge 28 c.f.s., slope 0.8:1.*

A—Test flume E. *Width 4 feet, drop 0.5–1.5 feet, discharge 10 c.f.s.*

B—Test flume F. *Adjustable tilting type, maximum slope 12 degrees, width 1 foot, discharge 5 c.f.s.*

FIGURE 3.—*Test flumes.*

Section 1

General investigation of the hydraulic jump on horizontal aprons (Basin I)

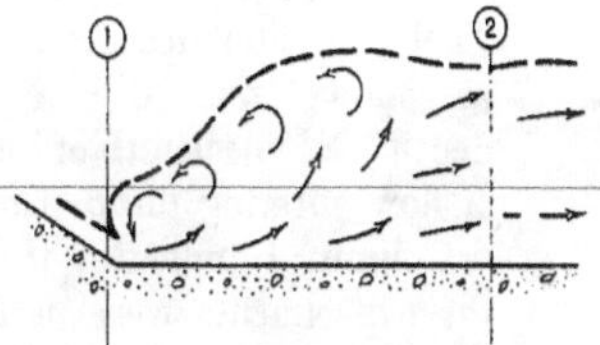

A TREMENDOUS amount of experimental, as well as theoretical, work has been performed in connection with the hydraulic jump on a horizontal apron. To mention a few of the experimenters who contributed basic information, there are: Bakhmeteff and Matzke (*1, 9*),[1] Safranez (*3*), Woycicki (*4*), Chertonosov (*10*), Einwachter (*11*), Ellms (*12*), Hinds (*14*), Forchheimer (*21*), Kennison (*22*), Kozeny (*23*), Rehbock (*24*), Schoklitsch (*25*), Woodward (*26*), and others. There is probably no phase of hydraulics that has received more attention; however, from a practical viewpoint, there is still much to be learned.

The first phase of this study consisted of observing and measuring the hydraulic jump in its various forms. The results were then correlated with those of others, the primary purpose being to become better acquainted with the overall jump phenomenon. The objectives of the study were: (1) to determine the applicability of the hydraulic jump formula for the entire range of conditions experienced in design; (2) to determine the length of the jump over the entire practical range and to correlate the findings with results of other experimenters where possible; and (3) to observe, catalog, and evaluate the various forms of the jump.

Hydraulic Jump Experiments

Observation of the hydraulic jump throughout its entire range required tests in all six test flumes. As indicated in Table 1, this involved about 125 tests for discharges of 1 to 28 c.f.s. The number of flumes used enhanced the value of the results in that it was possible to observe the degree of similitude obtained for the various sizes of jumps. Greatest reliance was placed on the results from the larger flumes, since the action in small jumps is too rapid for the eye to follow and, also, friction and viscosity become a measurable factor. This was demonstrated by the fact that the length of jump obtained from the two smaller flumes, A and F, was consistently shorter than that observed

[1] Numbers refer to references in "Bibliography."

for the larger flumes. Out-of-scale frictional resistance on the floor and side walls produced a short jump. As testing advanced and this deficiency became better understood, some allowance was made for this effect in the observations.

Experimental Results

Definitions of the symbols used in connection with the hydraulic jump on a horizontal floor are shown in Figure 4. The procedure followed in each test of this series was to establish a flow and then gradually increase the tail water depth until the front of the jump moved upstream to Section 1, indicated in Figure 4. The tail water depth was then measured, the length of the jump recorded, and the depth of flow entering the jump, D_1, was obtained by averaging a generous number of point gage measurements taken immediately upstream from Section 1. The results of the measurements and succeeding computations are tabulated in Table 1. The measured quantities are tabulated as follows: total discharge (Col. 3); tail water depth (Col. 6); length of jump (Col. 11), and depth of flow entering jump (Col. 8).

Column 1 indicates the test flumes in which the experiments were performed, and Column 4 shows the width of each flume. All computations are based on discharge per foot width of flume; unit discharges (q) are shown in Column 5.

The velocity entering the jump V_1, Column 7, was computed by dividing q (Col. 5) by D_1 (Col. 8).

The Froude Number

The Froude number, Column 10, Table 1, is:

$$F_1=\frac{V_1}{\sqrt{gD_1}} \quad (1)$$

where F_1 is a dimensionless parameter, V_1 and D_1 are velocity and depth of flow, respectively, entering the jump, and g is the acceleration of gravity. The law of similitude states that where gravitational forces predominate, as they do in open channel phenomena, the Froude number should have the same value in model and prototype. Therefore, a model jump in a test flume

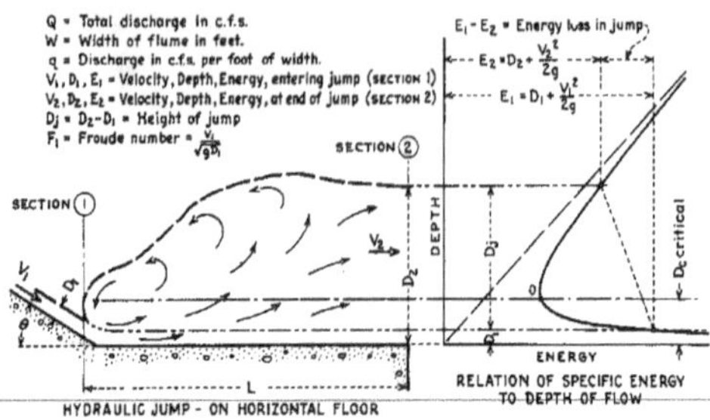

Figure 4.—*Definition of symbols (Basin I).*

will have the identical characteristics of a prototype jump in a stilling basin, if the Froude numbers of the incoming flows are the same. Although energy conversions in a hydraulic jump bear some relation to the Reynolds number, gravity forces predominate, and the Froude number becomes most useful in plotting stilling basin characteristics. Bakhmeteff and Matzke (*1*) demonstrated this application in 1936 when they related stilling basin characteristics to the square of the Froude number, $\frac{V^2}{gD_1}$, which they termed the kinetic flow factor.

The Froude number, equation (1), is used throughout this monograph. As the acceleration of gravity is a constant, the term g could be omitted. However, its inclusion makes the expression dimensionless, and the form shown as equation (1) is preferred.

Applicability of Hydraulic Jump Formula

The theory of the hydraulic jump in horizontal channels has been treated thoroughly by others (see "Bibliography"), and will not be repeated here. The expression for the hydraulic jump, based on pressure-momentum may be written (*15*):

$$D_2=-\frac{D_1}{2}+\sqrt{\frac{D_1^2}{4}+\frac{2V_1^2D_1}{g}}$$

or (2)

$$D_2=-\frac{D_1}{2}+\sqrt{\frac{D_1^2}{4}+\frac{2V_1^2D_1^2}{gD_1}}$$

where D_1 and D_2 are the depths before and after the jump, Figure 4. These depths are often called conjugate or sequent depths.

Transposing D_1 to the left side of the equation and substituting F_1^2 for $\frac{V_1^2}{gD_1}$.

$$\frac{D_2}{D_1}=-1/2+\sqrt{1/4+2F_1^2}$$

or (3)

$$\frac{D_2}{D_1}=1/2\left(\sqrt{1+8F_1^2}-1\right)$$

Equation (3) shows that the ratio of depths is a function of the Froude number. The ratio $\frac{D_2}{D_1}$ is plotted with respect to the Froude number on Figure 5. The line, which is virtually straight except for the lower end, represents the above expression for the hydraulic jump; the points, which are experimental, are from Columns 9 and 10, Table 1. The agreement is excellent over the entire range, indicating that equation (3) is applicable when the flow enters the jump at an appreciable angle to the horizontal.

There is an unsuspected characteristic in the curve, however, which is mentioned here but will be enlarged on later. Although the tail water depth, recorded in Column 6 of Table 1, was sufficient to bring the front of the jump to Section 1 (Fig. 4) in each test, the ability of the jump to remain at Section 1 for a slight lowering of tail water depth became more difficult for the higher and lower values of the Froude number. The jump was least sensitive to variation in tail water depth in the middle range, or values of F_1 from 4.5 to 9.

Length of Jump

The length of the jump measurement, Column 11, Table 1, was the most difficult to determine. Special care was therefore given to this measurement. Where chutes or overfalls were used, the front of the jump was held at the intersection of the chute and the horizontal floor, as shown in Figure 4. The length of jump was measured from this point to a point downstream where either the high-velocity jet began to leave the floor or to a point on the surface immediately downstream from the roller. whichever was the longer. In the case of Flume F, where the flow discharged from a gate onto a horizontal floor, the front of the jump was maintained just downstream from the completed contraction of the entering jet. In both cases the point at which the high-velocity jet begins to rise from the floor is not fixed, but tends to shift upstream and downstream. This is also true of the roller on the surface. It was at first difficult to repeat length observations within 5 percent by either criterion, but with practice satisfactory measurements became possible. It was the intention to judge the length of the jump from a practical standpoint; in other words, the end of the jump, as chosen, would represent the end of the concrete floor and side walls of a conventional stilling basin.

The length of jump has been plotted in two ways. Although the first method is perhaps the better method, the second is the more common and useful. The first method is shown in Figure 6 where the ratio, length of jump to D_1 (Col. 13, Table 1), is plotted with respect to the Froude number (Col. 10) for results from the six test flumes. The resulting curve is of fairly uniform curvature, which is the principal advantage of these coordinates. The second method of plotting, where the ratio, length of jump to the conjugate tail water depth D_2 (Col. 12) is plotted with respect to the Froude number, is presented in Figure 7. This latter method of plotting will be used throughout the study. The points represent the experimental values.

In addition to the curve established by the test points, curves representing the results of three other experimenters are shown in Figure 7. The best known and most widely accepted curve for length of jump is that of Bakhmeteff and Matzke (*1*) which was determined from experiments made at Columbia University. The greater portion of this curve, labeled "1," is at variance with the present experimental results. Because of the wide use this curve has experienced, a rather complete explanation is presented regarding this disagreement.

The experiments of Bakhmeteff and Matzke were performed in a flume 6 inches wide, having a limited testing head. The depth of flow entering the jump was adjusted by a vertical slide gate. The maximum discharge was approximately 0.7 c.f.s., and the thickness of the jet entering the jump, D_1, was 0.25 foot for a Froude number of 1.94. The results up to a Froude number of 2.5 are in agreement with the present experiments. To increase the Froude number, it was necessary for Bakhmeteff and Matzke to decrease the gate

TABLE 1.—*Natural stilling basin with horizontal floor (Basin I)*

Test flume	S=tan Slope of floor	Q c.f.s.	W Width of stilling basin ft.	q per ft. of W	TW= D_2 ft.	V_1 ft. per sec.	D_1 ft.	$\frac{D_2}{D_1}$	$F_1=\frac{V_1}{\sqrt{gD_1}}$	L Length of jump ft.	$\frac{L}{D_2}$	$\frac{L}{D_1}$	$E_1=d_1+\frac{V_1^2}{2g}$ ft.	$E_2=d_2+\frac{V_2^2}{2g}$ ft.	$E_L=E_1-E_2$ ft.	$\frac{E_L}{D_1}$	$\frac{E_L}{E_1}$ %	Gate opening ft.
(1)	(2)	(3)	(4)	(5)	(6)	(7)	(8)	(9)	(10)	(11)	(12)	(13)	(14)	(15)	(16)	(17)	(18)	(19)
A	0	3.000	4.915	0.610	0.564	8.47	0.072	7.83	5.56	3.0	5.32	42	1.187	0.582	0.605	8.40	51.0	
		3.500		0.712	0.612	8.79	.081	7.56	5.44	3.3	5.39	41	1.282	0.633	0.649	8.01	50.6	
		4.000		0.814	0.651	8.95	.091	7.15	5.23	3.5	5.38	38	1.336	0.675	0.661	7.26	49.5	
		4.500		0.916	0.694	8.98	.102	6.80	4.96	3.8	5.48	37	1.354	0.721	0.633	6.21	46.8	
		5.000		1.017	0.730	9.08	.112	6.52	4.78	4.1	5.62	37	1.392	0.760	0.632	5.64	45.4	
B	0	4.300	2.000	2.150	1.480	17.48	.123	12.03	8.78	9.0	6.08	73	4.868	1.513	3.355	27.28	68.9	
		5.000		2.500	1.600	17.48	.143	11.19	8.15	10.1	6.31	71	4.888	1.638	3.250	22.73	66.5	
		3.000		1.500	1.236	17.24	.087	14.20	10.30	7.5	6.07	86	4.702	1.259	3.443	39.57	73.2	
		6.000		3.000	1.754	17.54	.171	10.26	7.47	10.6	6.04	62	4.948	1.799	3.149	18.42	63.6	
		7.000		3.500	1.908	17.59	.199	9.59	6.95	11.5	6.03	58	5.003	1.960	3.043	15.29	60.8	
		8.000		4.000	2.016	17.47	.229	8.80	6.43	12.3	6.10	54	4.968	2.077	2.891	12.62	58.2	
		7.000		3.500	1.900	17.41	.201	9.45	6.84	11.5	6.05	57	4.908	1.953	2.955	14.70	60.2	
		3.110		1.550	1.240	17.61	.088	14.09	10.46	7.9	6.37	90	4.903	1.264	3.639	41.35	74.2	
		4.100		2.050	1.440	17.52	.117	12.31	9.03	8.7	6.04	74	4.883	1.471	3.412	29.16	69.9	
		5.975		2.988	1.760	17.58	.170	10.35	7.51	10.2	5.80	60	4.969	1.805	3.164	18.61	63.7	
		7.010		3.505	1.900	17.52	.200	9.50	6.90	11.3	5.95	56	4.967	1.953	3.014	15.07	60.7	
		8.000		4.000	2.030	17.54	.228	8.90	6.48	12.6	6.21	55	5.006	2.090	2.916	12.79	58.3	
		4.150		2.075	1.450	17.44	.119	12.18	8.91	8.8	6.07	74	4.842	1.482	3.360	28.24	69.4	
		5.500		2.750	1.691	17.40	.158	10.70	7.72	10.7	6.33	68	4.860	1.732	3.128	19.80	64.4	
		6.000		3.000	1.764	17.44	.172	10.26	7.41	11.0	6.24	64	4.895	1.809	3.086	17.94	63.0	
		6.500		3.250	1.827	17.38	.187	9.77	7.08	11.3	6.18	60	4.878	1.876	3.002	16.05	61.5	
		2.000		1.000	1.000	17.24	.058	17.24	12.62	6.4	6.40	110	4.674	1.016	3.658	63.07	78.3	
		2.500		1.250	1.104	17.36	.072	15.33	11.40	6.7	6.07	93	4.752	1.124	3.628	50.39	76.3	
		3.000		1.500	1.235	17.44	.086	14.36	10.48	7.4	5.99	86	4.809	1.258	3.551	41.29	73.8	
		3.500		1.750	1.325	17.50	.100	13.25	9.75	7.8	5.89	78	4.856	1.352	3.504	35.04	72.2	
		4.000		2.000	1.433	17.39	.115	12.46	9.04	8.5	5.93	74	4.811	1.463	3.348	29.11	69.6	
		4.500		2.250	1.517	17.44	.129	11.76	8.56	9.4	6.20	73	4.852	1.551	3.301	25.59	68.0	
		5.000		2.500	1.599	17.24	.145	11.03	7.98	10.0	6.25	69	4.761	1.637	3.124	21.54	65.6	
		5.500		2.750	1.691	17.40	.158	10.70	7.72	10.7	6.33	68	4.860	1.732	3.128	19.80	64.4	
C	0	1.000	1.500	0.667	0.910	20.21	.033	27.58	19.55	5.0	5.49	152	6.375	0.918	5.457	165.36	85.6	
		1.500		1.000	1.125	20.43	.048	23.44	16.76	6.4	5.69	133	6.785	1.137	5.648	117.07	83.2	
		2.000		1.333	1.300	21.16	.063	20.63	14.86	7.7	5.92	122	7.016	1.316	5.700	90.48	81.2	
		2.500		1.667	1.465	21.39	.078	18.78	13.48	8.9	6.08	114	7.169	1.485	5.684	72.87	79.3	
		3.000		2.000	1.615	21.74	.092	17.55	12.63	9.7	6.01	105	7.431	1.639	5.792	62.96	77.9	
		3.380		2.253	1.730	22.09	.102	16.96	12.19	10.8	6.24	106	7.679	1.756	5.923	58.07	77.1	
		4.000		2.667	1.890	22.79	.117	16.15	11.74	11.1	5.87	95	8.152	1.921	6.261	53.51	76.5	
		4.440		2.960	2.010	23.13	.128	15.70	11.39	12.3	6.11	96	8.435	2.044	6.391	49.93	75.8	

		1.250		0.833	0.914	17.35	.048	19.04	13.96	5.4	5.91	112	4.723	0.927	3.796	79.08	80.4
		1.750		1.167	1.135	18.82	.062	18.30	13.32	6.5	5.73	105	5.563	1.151	4.412	71.16	79.3
		2.250		1.500	1.320	19.48	.077	17.14	12.37	7.8	5.91	101	5.969	1.340	4.629	60.12	77.6
		2.750		1.833	1.468	20.37	.090	16.31	11.97	9.1	6.20	101	6.533	1.492	5.041	56.01	77.1
		3.250		2.167	1.616	20.84	.104	15.54	11.39	10.0	6.19	96	6.849	1.644	5.205	50.05	76.0
		3.750		2.500	1.736	21.19	.118	14.71	10.87	11.0	6.34	93	7.091	1.768	5.323	45.11	75.1
		4.250		2.833	1.870	21.14	.134	13.96	10.18	11.6	6.20	87	7.074	1.905	5.169	38.57	73.1
D	0	4.000	3.970	1.008	1.110	20.16	.050	22.20	15.89	6.5	5.86	130	6.361	1.123	5.238	104.76	82.3
		5.000		1.259	1.220	20.31	.062	19.68	14.37	7.5	6.15	121	6.467	1.236	5.231	84.37	80.9
		6.000		1.511	1.376	20.43	.074	18.19	13.23	8.4	6.24	114	6.549	1.365	5.184	70.05	79.2
		7.000		1.763	1.460	20.50	.086	16.98	12.32	9.0	6.16	105	6.612	1.483	5.129	59.64	77.6
		8.000		2.015	1.570	20.56	.098	16.02	11.55	9.7	6.18	99	6.662	1.595	5.067	51.70	76.1
		9.000		2.267	1.670	20.80	.109	15.32	11.11	10.0	5.99	92	6.827	1.699	5.128	47.05	75.1
		8.080		2.035	1.600	20.56	.099	16.16	11.52	9.5	5.94	96	6.663	1.625	5.038	50.89	75.6
		11.730		2.955	1.962	21.41	.138	14.22	10.16	12.4	6.32	90	7.256	1.997	5.259	38.11	72.5
		10.000		2.519	1.752	20.99	.120	14.60	10.68	10.4	5.94	87	6.961	1.784	5.177	43.14	74.4
		3.000		0.756	0.954	19.89	.038	25.11	12.98	5.4	5.66	142	6.181	0.964	5.217	137.29	84.4
		5.000		1.259	1.250	20.31	.062	20.16	14.37	7.4	5.92	119	6.467	1.266	5.201	83.89	80.4
		7.000		1.763	1.452	20.50	.086	16.88	12.12	8.7	5.99	101	6.612	1.475	5.137	59.73	77.7
		9.000		2.267	1.693	20.80	.109	15.53	11.11	10.5	6.20	96	6.827	1.721	5.106	46.84	74.8
		11.720		2.952	1.922	21.39	.138	13.93	10.15	11.4	5.93	83	7.243	1.959	5.284	38.29	73.0
		10.000		2.519	1.780	20.99	.120	14.83	10.68	11.2	6.29	93	6.961	1.811	5.150	42.92	74.0
		12.000		3.023	1.953	21.29	.142	13.75	9.96	12.3	6.30	87	7.180	1.990	5.190	36.55	72.3
		14.000		3.526	2.163	21.63	.163	13.27	9.44	13.0	6.01	80	7.428	2.204	5.224	32.05	70.3
		16.000		4.030	2.330	22.02	.183	12.73	9.07	13.8	5.92	75	7.712	2.376	5.336	29.16	69.2
		18.000		4.534	2.495	22.56	.201	12.41	8.87	15.4	6.17	77	8.104	2.546	5.558	27.65	68.6
		4.980		1.254	1.220	20.23	.062	19.68	14.32	7.0	5.74	113	6.417	1.236	5.181	83.56	80.7
		10.000		2.519	1.792	20.99	.120	14.93	10.68	11.0	6.14	92	6.961	1.823	5.138	42.82	73.8
		11.000		2.771	1.867	21.15	.131	14.25	10.30	11.3	6.05	86	7.077	1.901	5.176	39.51	73.1
		13.000		3.275	2.009	21.55	.152	13.22	9.74	12.4	6.17	82	7.363	2.050	5.313	34.95	72.2
		15.000		3.778	2.180	21.84	.173	12.60	9.25	13.3	6.10	77	7.580	2.226	5.354	30.95	70.6
		6.500		1.637	1.412	20.46	.080	17.65	12.75	8.9	6.30	111	6.580	1.433	5.147	64.34	78.2
		4.980		1.254	1.220	20.23	.062	19.68	14.32	7.0	5.74	113	6.417	1.236	5.181	83.56	80.7
		17.000		4.282	2.410	22.30	.192	12.56	8.77	14.6	6.05	76	7.914	2.461	5.453	28.40	68.9
		19.000		4.786	2.560	22.79	.210	12.19	8.77	15.3	5.98	73	8.275	2.614	5.661	26.96	68.4
		21.000		5.290	2.656	23.20	.228	11.65	8.56	16.0	6.02	70	8.586	2.717	5.869	25.74	68.4
		26.160		6.589	3.060	24.22	.272	11.25	8.19	19.4	6.34	71	9.381	3.132	6.249	22.97	66.6
		22.980		5.788	2.842	24.12	.240	11.84	8.68	18.7	6.58	78	9.274	2.907	6.367	26.53	68.7
		23.930		6.028	2.845	23.74	.254	11.20	8.30	18.3	6.43	72	8.998	2.915	6.083	23.95	67.6
		28.370		7.147	3.202	24.56	.291	11.00	8.02	21.0	6.56	72	9.657	3.279	6.378	21.92	66.0

TABLE 1.—*Natural stilling basin with horizontal floor (Basin I)*—Continued

Test flume	S=tan Slope of floor	Q c.f.s.	W Width of stilling basin ft.	q per ft. of W	T.W.= D_2 ft.	V_1 ft. per sec.	D_1 ft.	$\frac{D_2}{D_1}$	$F_1=\frac{V_1}{\sqrt{gD_1}}$	L Length of jump ft.	$\frac{L}{D_2}$	$\frac{L}{D_1}$	$E_1=d_1+\frac{V_1^2}{2g}$ ft.	$E_2=d_2+\frac{V_2^2}{2g}$ ft.	$E_L=E_1-E_2$ ft.	$\frac{E_L}{D_1}$	$\frac{E_L}{E_1}$ %	Gate opening ft.
(1)	(2)	(3)	(4)	(5)	(6)	(7)	(8)	(9)	(10)	(11)	(12)	(13)	(14)	(15)	(16)	(17)	(18)	(19)
E	0	5. 000	3. 970	1. 259	0. 840	10. 49	. 120	7. 00	5. 34	5. 0	5. 95	42	1. 831	0. 875	0. 956	7. 97	52. 2	Dam 1.5′ high.
		6. 000		1. 511	0. 940	10. 57	. 143	6. 57	4. 92	5. 6	5. 96	39	1. 880	0. 980	0. 900	6. 29	47. 9	
		7. 000		1. 763	0. 990	10. 75	. 164	6. 04	4. 67	5. 9	5. 96	36	1. 960	1. 039	0. 921	5. 62	47. 0	
		8. 000		2. 014	1. 080	10. 89	. 185	5. 84	4. 46	6. 3	5. 83	34	2. 029	1. 134	0. 895	4. 84	44. 1	
		9. 000		2. 266	1. 160	11. 05	. 205	5. 66	4. 30	6. 6	5. 69	28	2. 104	1. 219	0. 885	4. 32	42. 1	
		11. 000		2. 770	1. 260	11. 17	. 248	5. 08	3. 95	7. 1	5. 63	32	2. 188	1. 335	0. 853	3. 44	39. 0	
		4. 000		1. 008	0. 770	10. 28	. 098	7. 86	5. 79	4. 7	6. 10	48	1. 742	0. 796	0. 946	9. 65	54. 3	
		10. 000		2. 518	1. 220	11. 09	. 227	5. 37	4. 10	6. 9	5. 66	30	2. 139	1. 286	0. 853	3. 76	39. 9	
		10. 002		2. 518	1. 080	8. 99	. 280	3. 86	3. 00	6. 0	5. 56	21	1. 536	1. 164	0. 372	1. 33	24. 2	Dam 10″ high.
		9. 000		2. 266	1. 000	8. 78	. 258	3. 88	3. 05	5. 5	5. 50	21	1. 457	1. 080	0. 377	1. 46	25. 9	
		8. 000		2. 014	0. 960	8. 76	. 230	4. 17	3. 22	5. 0	5. 21	22	1. 413	1. 029	0. 384	1. 67	27. 2	
		7. 000		1. 763	0. 900	8. 24	. 214	4. 21	3. 13	4. 7	5. 22	22	1. 269	0. 961	0. 308	1. 44	24. 3	
		6. 000		1. 511	0. 820	8. 39	. 180	4. 56	3. 48	4. 3	5. 24	24	1. 274	0. 873	0. 401	2. 23	31. 5	
		5. 000		1. 259	0. 760	7. 77	. 162	4. 69	3. 40	4. 1	5. 39	25	1. 102	0. 803	0. 299	1. 85	27. 2	
		4. 000		1. 007	0. 660	7. 75	. 130	5. 08	3. 79	3. 7	5. 61	28	1. 064	0. 697	0. 367	2. 82	34. 5	
		3. 000		0. 755	0. 570	7. 95	. 100	5. 70	4. 21	3. 3	5. 79	33	1. 082	0. 597	0. 485	4. 85	44. 8	
		5. 084		1. 281	0. 620	5. 80	. 221	2. 81	2. 17	2. 6	4. 19	12	0. 744	0. 686	0. 058	0. 26		Dam 6″ high.
		3. 675		0. 926	0. 510	5. 12	. 168	3. 04	2. 20	2. 5	4. 90	15	0. 576	0. 561	0. 015	0. 09		
		2. 440		0. 615	0. 410	5. 44	. 113	3. 63	2. 85	2. 2	5. 36	19	0. 573	0. 445	0. 128	1. 13		
		7. 680		1. 934	0. 770	5. 69	. 340	2. 26	1. 72	3. 0	3. 90	9	0. 874	0. 866	0. 008	0. 02		
		6. 000		1. 511	0. 690	5. 68	. 266	2. 59	1. 93	2. 8	4. 06	10	0. 768	0. 765	0. 003	0. 01		
F	0	0. 960	1. 000	0. 960	0. 792	12. 15	. 079	10. 03	7. 62	4. 0	5. 05	51	2. 371	0. 815	1. 556	19. 70	65. 6	0. 125
		0. 815		0. 815	0. 653	9. 59	. 085	7. 68	5. 80	3. 0	4. 59	35	1. 513	0. 677	0. 836	9. 84	55. 3	
		0. 680		0. 680	0. 540	8. 61	. 079	6. 84	5. 40	2. 4	4. 44	30	1. 230	0. 565	0. 665	8. 42	54. 1	
		1. 580		1. 580	0. 992	11. 70	. 135	7. 35	5. 61	6. 2	6. 25	46	2. 261	1. 031	1. 230	9. 11	54. 4	. 208
		1. 200		1. 200	0. 740	8. 89	. 135	5. 48	4. 26	4. 3	5. 81	32	1. 362	0. 781	0. 581	4. 30	42. 7	
		1. 400		1. 400	0. 880	10. 37	. 135	6. 52	4. 97	5. 4	6. 14	40	1. 805	0. 919	0. 886	6. 56	49. 1	
		2. 230		2. 230	1. 220	12. 89	. 173	7. 05	5. 46	7. 3	5. 98	42	2. 753	1. 272	1. 481	8. 56	53. 8	. 281
		1. 730		1. 730	0. 927	10. 00	. 173	5. 36	4. 24	5. 2	5. 61	30	1. 726	0. 981	0. 745	4. 31	43. 2	
		1. 250		1. 250	0. 644	7. 23	. 173	3. 72	3. 06	3. 4	5. 28	20	0. 985	0. 702	0. 283	1. 64	28. 7	
		1. 150		1. 150	0. 581	6. 65	. 173	3. 36	2. 82	3. 1	5. 34	18	0. 860	0. 642	0. 218	1. 26	25. 3	
		1. 400		1. 400	0. 638	6. 69	. 210	3. 04	2. 57	3. 3	5. 17	16	0. 901	0. 712	0. 189	0. 90	21. 0	. 333

	1. 850		1. 850	0. 882	8. 81	. 210	4. 20	3. 39	5. 0	5. 67	24	1. 415	0. 950	0. 465	2. 21	32. 9	
				1. 075	10. 48	. 210	5. 12	4. 03	6. 2	5. 77	30	1. 915	1. 140	0. 775	3. 69	40. 5	
				1. 345	13. 05	. 210	6. 40	5. 02	8. 2	6. 10	39	2. 854	1. 410	1. 444	6. 88	50. 6	
				0. 753	7. 29	. 251	3. 00	2. 56	3. 5	4. 65	14	1. 076	0. 845	0. 231	0. 92	21. 5	0. 396
				1. 023	9. 36	. 251	4. 08	3. 29	5. 6	5. 47	22	1. 611	1. 105	0. 506	2. 02	31. 4	
				1. 235	11. 08	. 251	4. 92	3. 90	7. 2	5. 83	29	2. 157	1. 314	0. 843	3. 36	39. 1	
				1, 427	12. 61	. 251	5. 69	3. 44	8. 4	5. 89	33	2. 720	1. 504	1. 216	4. 84	44. 7	
				0. 704	6. 67	. 285	2. 47	2. 20	3. 2	4. 55	11	0. 976	0. 817	0. 159	0. 56	16. 3	. 458
				1. 016	8. 91	. 285	3. 56	2. 94	5. 3	5. 22	19	1. 518	1. 113	0. 405	1. 42	26. 7	
				1. 219	10. 53	. 285	4. 28	3. 48	7. 0	5. 74	25	2. 007	1. 313	0. 694	2. 44	34. 6	
				1. 435	12. 14	. 285	5. 04	4. 01	8. 3	5. 78	29	2. 574	1. 525	1. 049	3. 68	40. 8	

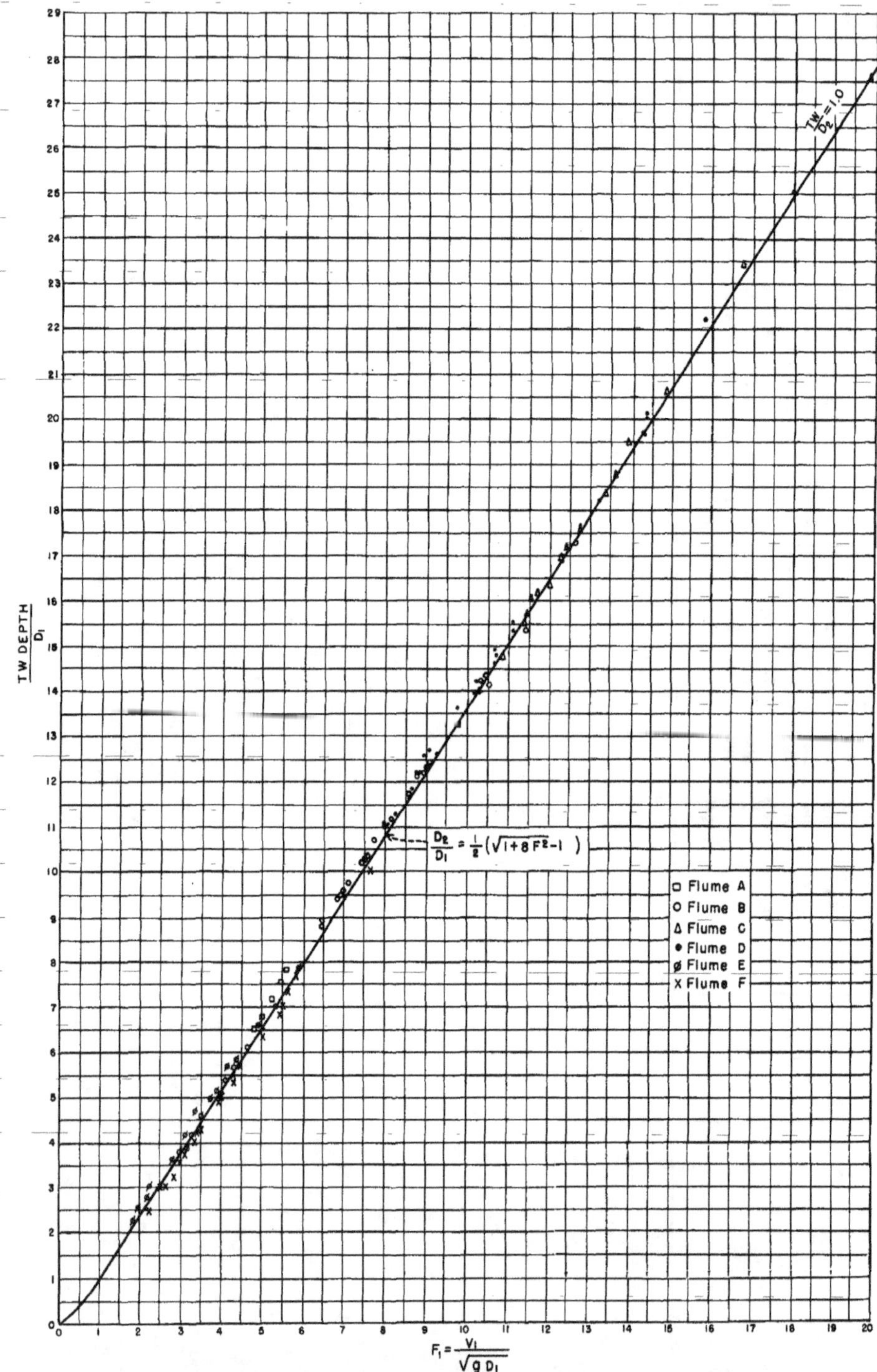

FIGURE 5.—*Ratio of tail water depth to D_1 (Basin I).*

opening. The extreme case involved a discharge of 0.14 c.f.s. and a value of D_1 of 0.032 foot, for $F_1=8.9$, which is much smaller than any discharge or value of D_1 used in the present experiments. Thus, it is reasoned that as the gate opening decreased, in the 6-inch-wide flume, frictional resistance in the channel downstream increased out of proportion to that which would have occurred in a larger flume or a prototype structure. Thus, the jump formed in a shorter length than it should. In laboratory language, this is known as "scale effect," and is construed to mean that prototype action is not faithfully reproduced. It is quite certain that this was the case for the major portion of curve 1. In fact, Bahkmeteff and Matzke were somewhat dubious concerning the small-scale experiments.

To confirm the above conclusion, it was found that results from Flume F, which was 1 foot wide, became erratic when the value of D_1 approached 0.10. Figures 6 and 7 show three points obtained with a value of D_1 of approximately 0.085. The three points are given the symbol ⊠ and fall short of the recommended curve.

The two remaining curves, labeled "3" and "4," on Figure 7, portray the same trend as the recommended curve. The criterion used by each experimenter for judging the length of the jump is undoubtedly responsible for the displacement. The curve labeled "3" was obtained at the Technical University of Berlin on a flume ½ meter wide by 10 meters long. The curve labeled "4" was determined from experiments performed at

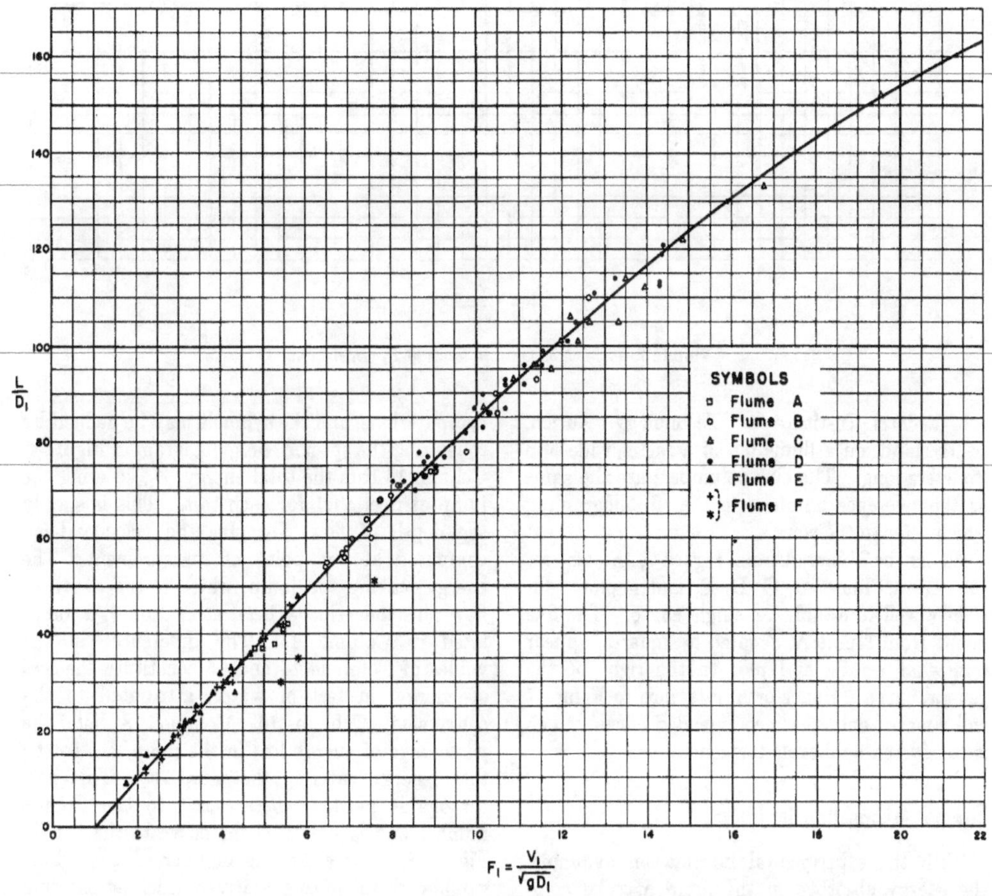

FIGURE 6.—*Length of jump in terms of D_1 (Basin I).*

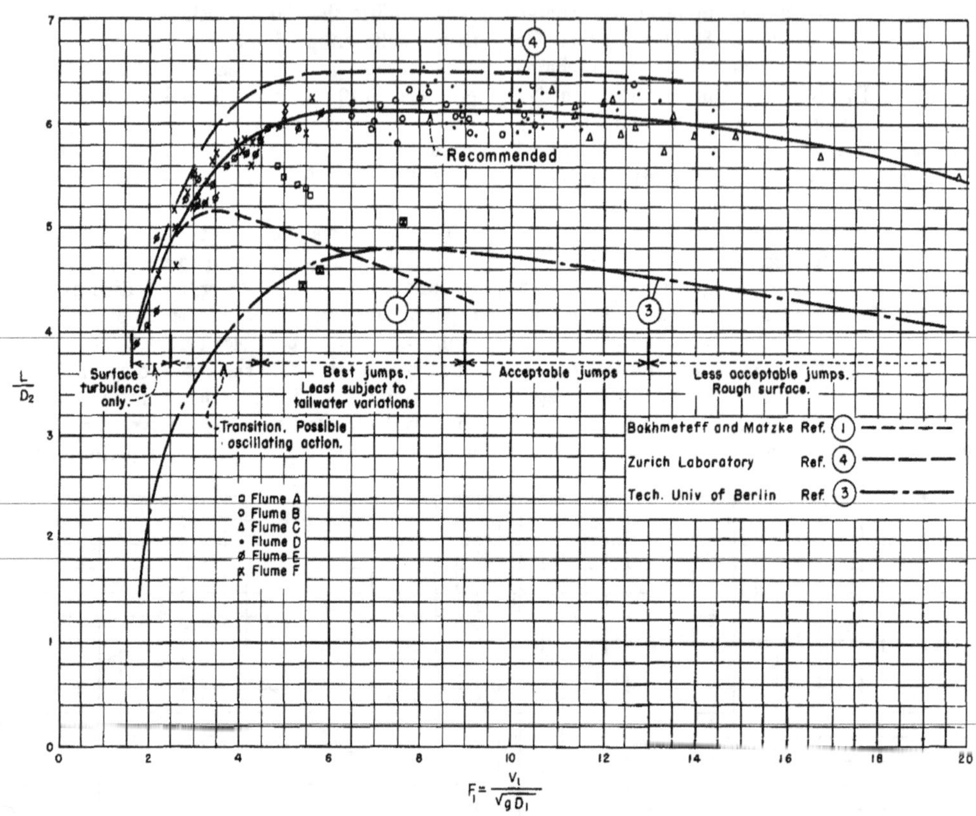

FIGURE 7.—*Length of jump in terms of D_2 (Basin I).*

the Federal Institute of Technology, Zurich, Switzerland, on a flume 0.6 of a meter wide and 7 meters long. The curve numbers are the same as the reference numbers in the "Bibliography" which refer to the work.

As can be observed from Figure 7, the test results from Flumes B, C, D, E, and F plot sufficiently well to establish a single curve. The five points from Flume A, denoted by squares, appear somewhat erratic and plot to the right of the general curve. Henceforth, reference to Figure 7 will concern only the recommended curve, which is considered applicable for general use.

Energy Absorption in Jump

With the experimental information available, the energy absorbed in the jump may be computed. Columns 14 through 18, Table 1, list the computations, and the symbols may be defined by consulting the specific energy diagram in Figure 4. Column 14 lists the total energy, E_1, entering the jump at Section 1 for each test. This is simply the depth of flow, D_1, plus the velocity head computed at the point of measurement. The energy leaving the jump, which is the depth of flow plus the velocity head at Section 2, is tabulated in Column 15. The differences in the values of Columns 14 and 15 constitute the loss of energy, in feet of water, attributed to the conversion, Column 16. Column 18 lists the percentage of energy lost in the jump, E_L, to the total energy entering the jump, E_1. This percentage is plotted with respect to the Froude number and is shown as the curve to the left on Figure 8. For a Froude number of 2.0, which would correspond to a relatively thick jet entering the jump at low velocity, the curve shows the energy absorbed in the jump to be about 7 percent of the total energy entering. Considering the other extreme, for a Froude number of 19, which would be produced by a relatively thin jet entering the jump at very high velocity, the absorption by the jump would amount to 85 percent of the energy entering. Thus, the hydraulic jump can perform over a wide range of conditions. There are poor jumps and good jumps, the most satisfactory occurring over the center portion of the curve.

Another method of expressing the energy absorption in a jump is to express the loss, E_L, in terms of D_1. The curve to the right on Figure 8 shows the ratio $\frac{E_L}{D_1}$ (Column 17, Table 1) plotted against the Froude number. Losses in feet of head are obtained from this method.

Forms of the Hydraulic Jump

The hydraulic jump may occur in at least four different distinct forms on a horizontal apron, as shown in Figure 9. All of these forms are encountered in practice. The internal characteristics of the jump and the energy absorption in the jump vary with each form. Fortunately these forms, some of which are desirable and some undesirable, can be cataloged conveniently with respect to the Froude number, as shown in Figure 9.

When the Froude number is unity, the water is flowing at critical depth; thus a jump cannot form. This corresponds to Point 0 on the specific energy diagram of Figure 4. For values of the Froude number between 1.0 and 1.7, there is only a slight difference in the conjugate depths D_1 and D_2. A slight ruffle on the water surface is the

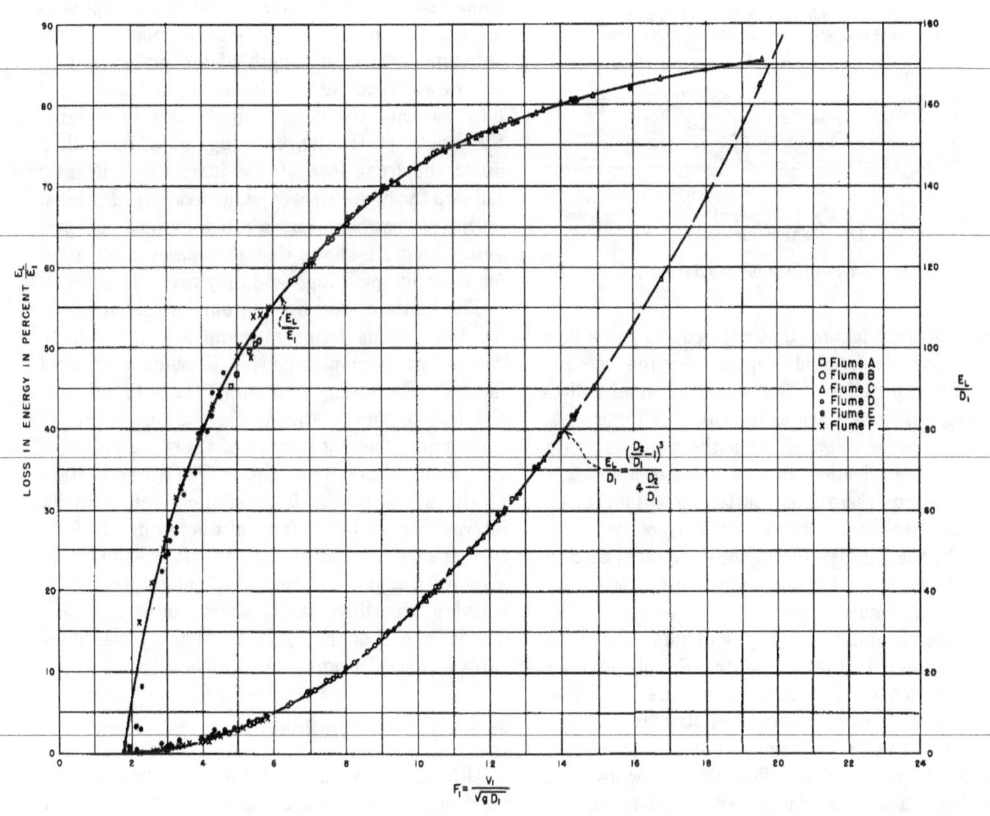

FIGURE 8.—*Loss of energy in jump on horizontal floor (Basin I).*

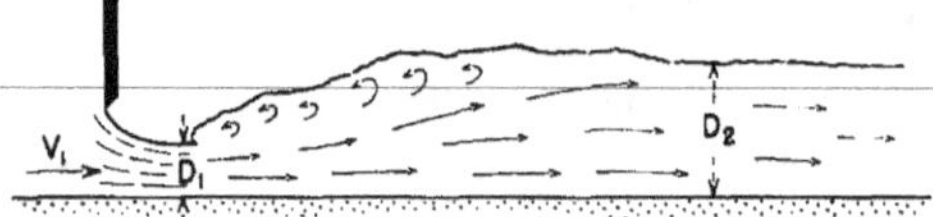

F_1=1.7 to 2.5
A—Pre-jump—very low energy loss

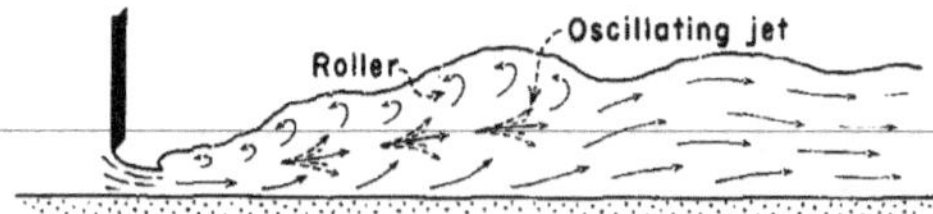

F_1=2.5 to 4.5
B—Transition—rough water surface

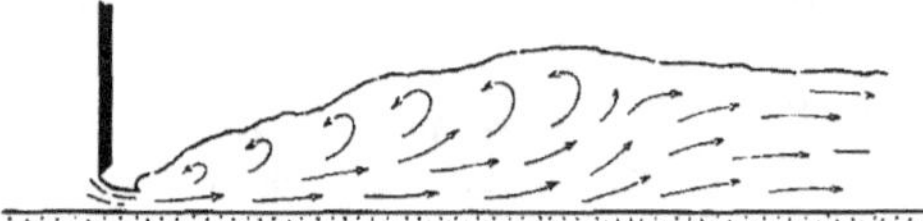

F_1=4.5 to 9.0—range of good jumps
C—Least affected by tail water variations

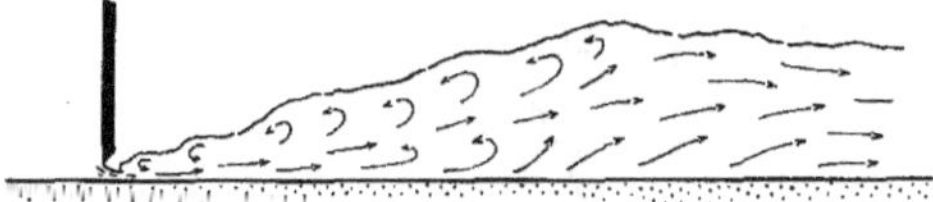

F_1=9.0 upward
D—effective but rough

FIGURE 9.—*Jump forms (Basin I).*

only apparent feature that differentiates this flow from flow at critical depth. As the Froude number approaches 1.7, a series of small rollers develop on the surface as indicated in Figure 9A, and this action remains much the same but with further intensification up to a value of about 2.5. In this range there is no particular stilling basin problem involved; the water surface is quite smooth, the velocity throughout the cross section is fairly uniform, and the energy loss is less than 20 percent, Figure 8.

Figure 9B indicates the type of jump that may be encountered at values of the Froude number from 2.5 to 4.5. This type has a pulsating action and is usually seen in low head structures. The entering jet oscillates from bottom to surface and has no regular period. Turbulence occurs near the bottom at one instant and entirely on the surface the next. Each oscillation produces a large wave of irregular period which in prototype structures has been observed to travel for miles causing damage to earth banks and riprap. This problem is of sufficient importance that a separate section, Section 4, has been devoted to the practical aspects of design.

A well-stabilized jump can be expected for the range of Froude numbers between 4.5 and 9, Figure 9C. In this range, the downstream extremity of the surface roller and the point at which the high-velocity jet tends to leave the floor occur in practically the same vertical plane. The jump is well balanced and the action is thus at its best. The energy absorption in the jump for Froude numbers from 4.5 to 9 ranges from 45 to 70 percent, Figure 8.

As the Froude number increases above 9, the form of the jump gradually changes to that shown in Figure 9D; V_1 is very high, D_1 is comparatively small, and the difference in conjugate depths is large. The high-velocity jet no longer carries through for the full length of the jump; that is, the downstream extremity of the surface roller now becomes the determining factor in judging the length of the jump. Slugs of water rolling down the front face of the jump intermittently fall into the high velocity jet, generating additional waves downstream, and a rough surface can prevail. Figure 8 shows that the energy dissipation for these jumps is high and may reach 85 percent.

The limits of the Froude number given above for the various forms of jump are not definite values but overlap somewhat depending on local factors. Returning to Figure 7, it is found that the length curve catalogs the various forms of the jump. The flat portion of the curve indicates the range of best operation. The steep portion of the curve to the left definitely indicates an internal change in the form of the jump. In fact, two changes are manifest, the form shown in Figure 9A and the form, which might better be called a transition stage, shown in Figure 9B. The right end of the curve on Figure 7 also indicates a change in form, but to less extent.

Practical Considerations

Although the academic rather than the practical viewpoint is stressed in this section, a few of the practical aspects of stilling basin design should be discussed. Considering the four forms of jump just discussed, the following are pertinent:

1. All jump forms shown in Figure 9 are encountered in stilling basin design.

2. The form in Figure 9A requires no baffles or special devices in the basin. The only requirement is to provide the proper length of pool, which is relatively short. This can be obtained from Figure 7.

3. The form in Figure 9B presents wave problems which are difficult to overcome. This jump is frequently encountered in the design of canal structures, diversion or low dam spillways, and even outlet works. Baffle piers or appurtenances in the basin are of little value. Waves are the main source of difficulty and methods for coping with them are discussed in Section 4. The present information may prove valuable in that it will help to restrict the use of jumps in the 2.5 to 4.5 Froude number range. In many cases the critical range cannot be avoided, but in others the jump may be brought into the desirable range by altering the dimensions of the structure.

4. No particular difficulty is encountered in the form shown in Figure 9C. Arrangements of baffles and sills will be found valuable as a means of shortening the length of basin. This is discussed in Sections 2 and 3.

5. As the Froude number increases, the jump becomes more sensitive to tail water depth. For numbers as low as 8, a tail water depth greater than the conjugate depth is advisable to be certain that the jump will stay on the apron. This phase is discussed in more detail in the following sections.

6. When the Froude number is greater than 10, the difference in conjugate depths is great, and, generally speaking, a very deep basin with high training walls is required. On high spillways the cost of the stilling basin may not be commensurate with the results obtained. A bucket-type dissipator may give comparable results at less cost. On lower head structures the action in the basin will be rugged in appearance with surface disturbances being of greatest concern.

7. High Froude numbers will always occur for flow through extremely small gate openings on even the smallest structures. Unless the discharge for these conditions represents an appreciable percentage of the design flow, the high Froude numbers have no significance.

Water-Surface Profiles and Pressures

Water-surface profiles for the jump on a horizontal floor were not measured as these have already been determined by Bakhmeteff and Matzke (*1*), Newman and LaBoon (*19*), and Moore (*27*, *18*). It has been shown by several experimenters that the vertical pressures on the floor of the stilling basin are virtually the same as the static head indicated by the water-surface profile.

Conclusions

The foregoing experiments and discussion serve to associate the Froude number with the hydraulic jump and stilling basin design. The ratio of conjugate depths, the length of jump, the type of jump to be expected, and the losses involved have all been related to this number. The principal advantage of this form of presentation is that one may analyze the problem, provide the solution, and determine the probable performance characteristics from relatively simple and rapid calculations.

Application of Results (Example 1)

Water flowing under a sluice gate discharges into a rectangular stilling basin the same width as the gate. The average velocity and the depth of flow after contraction of the jet is complete are: $V_1=85$ ft. per sec. and $D_1=5.6$ feet. Determine the conjugate tail water depth, the length of basin required to confine the jump, the effectiveness of the basin to dissipate energy, and the type of jump to be expected.

$$F_1=\frac{V_1}{\sqrt{gD_1}}=\frac{85}{\sqrt{32.2\times5.6}}=6.34$$

Entering Figure 5 with this value

$$\frac{D_2}{D_1}=8.5$$

The conjugate tail water depth

$$D_2=8.5\times5.6=47.6 \text{ feet}$$

Entering the recommended curve on Figure 7 with a Froude number of 6.34.

$$\frac{L}{D_2}=6.13$$

Length of basin necessary to confine the jump

$$L=6.13\times47.6=292 \text{ feet}$$

Entering Figure 8 with the above value of the Froude number, it is found that the energy absorbed in the jump is 58 percent of the energy entering.

By consulting Figure 9, it is apparent that a very satisfactory jump can be expected.

The following sections deal with the more practical aspects of stilling basin design, such as modifying the jump by baffles and sills to increase stability and shorten the length.

Section 2

Stilling basin for high dam and earth dam spillways and large canal structures (Basin II)

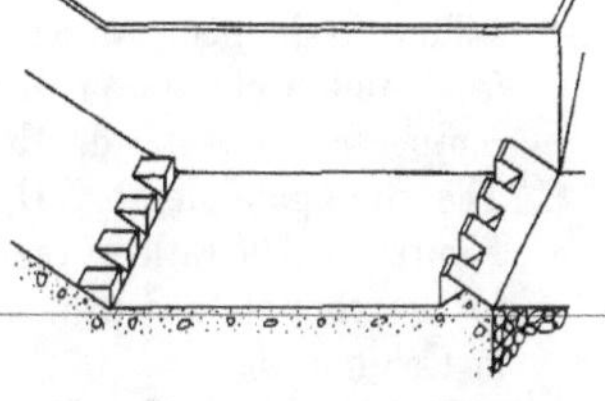

STILLING basins are seldom designed to confine the entire length of the hydraulic jump on the paved apron as was assumed in Section 1; first, for economic reasons, and second, because there are means for modifying the jump characteristics to obtain comparable or better performance in shorter lengths. It is possible to reduce the jump length by the installation of accessories such as baffles and sills in the stilling basin. In addition to shortening the jump, the accessories exert a stabilizing effect and in some cases increase the factor of safety.

Section 2 concerns stilling basins of the type which have been used on high dam and earth dam spillways, and large canal structures, and will be denoted as Basin II, Figure 10. The basin contains chute blocks at the upstream end and a dentated sill near the downstream end. No baffle piers are used in Basin II because of the relatively high velocities entering the jump. The object of these tests was to generalize the design, and determine the range of operating conditions for which this basin is best suited. Since many basins of this type have been designed, constructed, and operated, some of which were checked with models, the principal task in accomplishing the first objective was to tabulate and analyze the dimensions of existing structures. Only structures on which firsthand information was available were used.

Results of Compilation

With the aid of Figure 10, most of the symbols used in Table 2 are self-explanatory. The use of baffle piers is limited to Basin III. Column 1 lists the reference material used in compiling the table. Column 2 lists the maximum reservoir elevation, Column 3 the maximum tail water elevation, Column 5 the elevation of the stilling basin floor, and Column 6 the maximum discharge for each spillway. Column 4 indicates the height of the structure studied, showing a maximum fall from headwater to tail water of 179 feet, a mini-

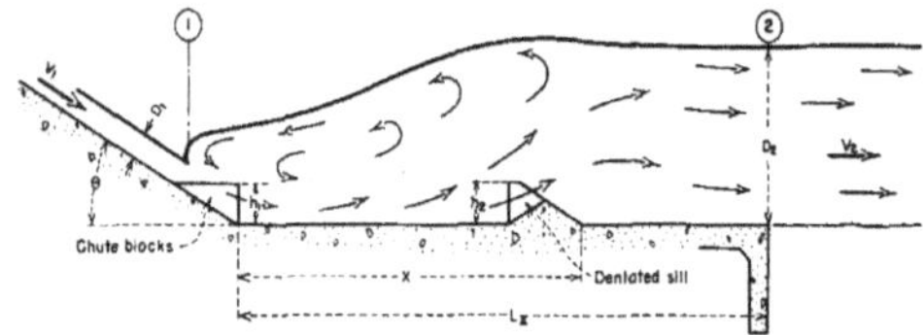

FIGURE 10.—*Definition of symbols (Basin II).*

mum of 14 feet, and an average of 85 feet. Column 7 shows that the width of the stilling basins varied from 1,197.5 to 20 feet. The discharge per foot of basin width, Column 8, varied from 760 to 52 c.f.s., with 265 as an average. The computed velocity, V_1 (hydraulic losses estimated in some cases), entering the stilling basin (Col. 9) varied from 108 to 38 feet per second, and the depth of flow, D_1, entering the basin (Col. 10) varied from 8.80 to 0.60 feet. The value of the Froude number (Col. 11) varied from 22.00 to 4.31. Column 12 shows the actual depth of tail water above the stilling basin floor, which varied from 60 to 12 feet, and Column 14 lists the computed, or conjugate, tail water depth for each stilling basin. The conjugate depths, D_2, were obtained from Figure 5. The ratio of the actual tail water depth to the conjugate depth is listed for each basin in Column 15.

Tail water depth. The ratio of actual tail water depth to conjugate depth shows a maximum of 1.67, a minimum of 0.73, and an average of 0.99. This means that, on the average, the basin floor was set to provide a tail water depth equal to the conjugate or necessary depth.

Chute blocks. The chute blocks used at the entrance to the stilling basin varied in size and spacing. Some basins contained nothing at this point, others a solid step, but in the majority of cases ordinary chute blocks were utilized. Chute blocks bear some resemblance to baffle piers but their function is altogether different. Chute blocks at the upstream end of a basin tend to corrugate the jet, lifting a portion of it from the floor to create a greater number of energy dissipating eddies, resulting in a shorter length of jump than would be possible without them. These blocks also reduce the tendency of the jump to sweep off the apron at tail water elevations below conjugate depths. The proportioning of chute blocks has been the subject of much discussion. The tabulation of Columns 19 through 24 of Table 2 shows the sizes which have been used. Column 20 shows the height of the chute blocks, while Column 21 gives the ratio of height of block to the depth, D_1. The ratios of height of block to D_1 indicate a maximum of 2.72, a minimum of 0.81, and an average of 1.35. This is somewhat higher than was shown to be necessary by the verification tests discussed later; a block equal to D_1 in height is sufficient.

The width of the blocks is shown in Column 2. Column 23 gives the ratio of width of the block to height, with a maximum of 1.67, a minimum of 0.44, and an average of 0.97. The ratio of width of block to spacing, tabulated in Column 24, shows a maximum of 1.91, a minimum of 0.95, and an average of 1.15. The three ratios indicate that the proportion: height equals width, equals spacing, equals D_1 should be a satisfactory standard for chute block design. The wide variation shows that these dimensions are not critical.

Dentated sill. The sill in or at the end of the basin was either solid or had some form of dentated arrangement, as designated in column 25. A dentated sill located at the end of the apron is recommended. The shape of the dentates and the angle of the sills varied considerably in the spillways tested, Columns 26 through 31. The position of the dentated sill also varied and this is indicated by the ratio $\frac{X}{L_{II}}$ in Column 26. The distance, X, is measured to the downstream edge of the sill, as illustrated in Figure 10. The ratio $\frac{X}{L_{II}}$ varied from 1 to 0.65; average 0.97.

The heights of the dentates are given in Column 27. The ratio of height of block to the conjugate tail water depth is shown in Column 28. These ratios show a maximum of 0.37, a minimum of 0.08, and an average of 0.20. The width to height ratio, Column 30, shows a maximum of 1.25, a minimum of 0.33, and average of 0.76. The ratio of width of block to spacing, Column 31, shows a maximum of 1.91, a minimum of 1.0, and an average of 1.13. For the purpose of generalization, the following proportions are recommended: (1) height of dentated sill$=0.2D_2$, (2) width of blocks$=0.15D_2$, and (3) spacing of blocks$=0.15D_2$, where D_2 is the conjugate tail water depth. It is recommended that the dentated sill be placed at the downstream end of the apron.

TABLE 2—*Model results on existing Type II basins*

Dam	Max res el	TW el	Fall HW to TW ft.	Basin floor el	Max Q c.f.s.	W Width of basin ft.	q per ft of W, cfs	V_1 ft./sec	D_1 ft.	$F_1=\frac{V_1}{\sqrt{gD_1}}$	TW depth ft.	$\frac{\text{TW depth}}{D_1}$	D_2 (Conj) ft.	$\frac{\text{TW depth}}{D_2\text{ (Conj)}}$	L Basin length, ft.	$\frac{L}{D_2\text{ (Conj)}}$	Slope of entry, °	Chute blocks Type	Chute blocks h_1 Ht ft.
(1)	(2)	(3)	(4)	(5)	(6)	(7)	(8)	(9)	(10)	(11)	(12)	(13)	(14)	(15)	(16)	(17)	(18)	(19)	(20)
Rye Patch	4, 123	4, 085. 5	37. 5	4, 062	20, 000	110	182	*53	3. 40	5. 15	23. 5	6. 81	22. 5	1. 04	46	2. 04	33	Solid	4. 0
Unity	3, 820	3, 771	49	3, 749	10, 000	55. 5	180	62	2. 90	6. 42	22	7. 59	25. 0	. 88	59	2. 36	26. 7	Teeth	3. 5
Alcova	5, 500	5, 354	146	5, 309	55, 000	150	367	98	3. 75	8. 88	45	12. 00	45. 5	. 97	125	2. 75	25. 0	Solid	4. 3
Shadow Mt	8, 367	8, 332	35	8, 313	10, 000	70	143	55	2. 60	6. 09	19	7. 31	21. 0	. 91	56	2. 67	28	T	3. 0
Boysen (Final)	4, 752	4, 628	124	4, 594	20, 000	66	303	85	3. 55	7. 98	34	9. 58	38. 2	. 89	151	3. 95	34	T	4. 0
Boysen (Prelim)	4, 752	4, 646. 5	105. 5	4, 600	62, 000	125	496	90	5. 70	6. 63	46. 5	8. 16	50. 7	. 92	140	2. 76	33. 7	T	6. 0
Scofield	7, 630	7, 583. 6	46. 4	7, 564	6, 200	40	155	61	2. 54	6. 74	19. 6	7. 72	23. 0	. 85	60	2. 61	33. 7	T	3. 5
Boca	5, 605	5, 508	97	5, 487	8, 000	75	107	73	1. 47	10. 53	21	14. 28	21. 2	. 99	58. 6	2. 66	26. 7	T	4. 0
Fresno	2, 591	2, 528	63	2, 499. 5	51, 000	190	268	70	3. 85	6. 27	28. 5	7. 40	32. 0	. 89	85	2. 66	33. 7	T	4. 0
Bull Lake	5, 805	5, 743. 5	61. 5	5, 725	10, 000	100	100	59	1. 70	7. 97	18. 5	10. 88	18. 3	1. 01	75	4. 10	14	None	---
Caballo	4, 182	4, 118	64	4, 086	33, 000	108	306	72	4. 22	6. 29	32	7. 58	35. 0	. 92	78. 2	2. 23	26. 7	T	4. 5
Moon Lake	8, 137	8, 028. 2	108. 8	8, 005	10, 000	75	133	74	1. 80	9. 74	23. 2	12. 90	23. 8	. 98	60	2. 52	26. 7	T	2. 6
Deer Creek	5, 417	5, 285	132	5, 260	12, 000	75	160	86	1. 87	11. 00	25	13. 37	28. 0	. 89	75	2. 68	26. 7	T	3. 0
Alamogordo	4, 275	4, 163	112	4, 118	56, 000	110	509	96	5. 30	7. 33	45	8. 49	52. 0	. 87	125	2. 40	26. 7	T	8. 0
Enders	3, 129. 5	3, 057	73	3, 016	200, 000	400	500	79	6. 33	5. 54	41	6. 38	46. 5	. 88	115	2. 47	26. 7	T	6. 0
Medicine Creek	2, 408. 9	2, 328	81	2, 287	97, 800	262	373	80	6. 32	5. 55	41	7. 31	47. 0	. 87	125	2. 66	26. 7	T	6. 75
Cedar Bluff	2, 192	2, 074. 3	118	2, 035. 5	87, 400	200	437	93	4. 68	7. 60	39	8. 33	48. 0	. 81	141	2. 94	18. 5	T	7. 0
Falcon	314. 2	235	79	175	456, 000	600	760	86	8. 80	5. 13	60	6. 82	60. 0	1. 00	180	3. 00	26. 7	T	8. 0
Trenton	2, 785	2, 700. 6	84	2, 653	133, 000	266	500	81	6. 20	5. 76	48	7. 74	47. 6	1. 01	125	2. 63	18. 5	T	5. 0
Cachuma	757. 6	578. 8	179	523	161, 000	322	500	108	4. 63	8. 84	56	12. 09	55. 8	1. 00	153	2. 74	26. 7	T	5. 5
Tiber	3, 014. 9	2, 835. 8	179	2, 797	54, 250	200	271	97	2. 79	10. 26	39	13. 98	39. 0	1. 00	117	3. 25	18	T	7. 0
Imperial Spillway	191	168	23	150	150, 000	1, 197. 5	125	*49	2. 60	5. 44	16	6. 16	18. 5	. 87	41	2. 22	14	T	2. 33
Imperial Sluiceway	181	155	26	140	24, 000	248. 3	97	*38	2. 50	4. 31	15	6. 00	14. 0	1. 07	69	4. 93	14	T	3. 33
Grassy Lake	7, 210	7, 100	110	7, 086	1, 200	20	60	63	. 95	11. 41	14	14. 73	14. 8	. 95	45	3. 04	26. 7	T	1. 0
Box Butte	4, 014	3, 961	53	3, 946. 5	2, 500	40	62	56	1. 60	7. 84	15	9. 37	16. 9	. 89	50	2. 96	22	T	3. 3
Siphon Drop	169. 7	150. 7	19	136	2, 000	33. 5	60	*40	1. 50	5. 80	15	10. 0	11. 5	1. 30	36	3. 13	22	T	2. 25
Pilot Knob	170. 26	124	46	94. 5	13, 155	140	94	*59	1. 60	8. 34	30	18. 7	18. 0	1. 67	60	3. 33	18. 5	T	2. 5
AA Canal Drp 1	43. 6	29. 2	14	13. 3	8, 700	118	74	*40	1. 80	5. 32	16	8. 89	12. 5	1. 28	27	2. 16	22	T	1. 75
Wasteway #2	1, 185. 75	1, 027. 4	159	1, 014. 9	2, 100	40	52	*92	. 60	22. 00	12	21. 4	16. 5	. 73	45	2. 73	24. 7	Vanes	---
Big Sandy #2	6, 761. 3	6, 702	59	6, 679	7, 500	50	150	66	2. 24	7. 70	23	10. 26	23. 3	. 99	75	3. 57	33. 7	T	2. 5
Cherry Creek	5, 632. 4	5, 558	74	5, 518	45, 000	116	388	*80	4. 90	6. 46	40	8. 16	42. 0	. 95	120	2. 86	25	None	---
Pine View	4, 870	4, 817	53	4, 785	9, 000	40	225	*59	3. 80	5. 42	32	8. 42	27. 0	1. 19	96	3. 55	33. 7	None	---
Agency Valley	3, 340	3, 266. 5	74	3, 234	10, 000	50	200	66	3. 00	6. 82	32	10. 68	27. 0	1. 19	110	4. 07	33. 7	Solid	2. 5
Davis	647	515. 5	131	460	175, 000	246	711	*97	7. 30	6. 41	56	7. 64	62. 0	. 90	100	1. 61	14	T	14. 3
Bonny	3, 737. 6	3, 623	114. 6	3, 589	64, 700	215	301	84	3. 60	7. 76	34	9. 44	38. 0	. 89	102	2. 68	20	T	7. 0
Cle Elum	2, 240	2, 130	110	2, 097	40, 000	200	200	*82	2. 40	9. 46	---	13. 74	30. 0	1. 10	108	3. 60	33. 7	None	---
Maximum	---	---	179	---	---	1, 197. 5	760	108	8. 80	22. 00	60	21. 40	62. 0	1. 67	180	4. 93	34	---	---
Minimum	---	---	14	---	---	20	52	38	0. 60	4. 31	12	6. 00	11. 5	0. 73	27	1. 61	14	---	---
Average	---	---	85	---	---	---	265	---	---	---	---	---	---	0. 99	---	2. 90	---	---	---

*Estimated hydraulic losses.

TABLE 2.—*Model results on existing Type II basins*—Continued

Dam	Chute blocks—Con.				End sill							Intermediate baffle blocks							Type of basin	Wing walls at end of basin
	$\frac{h_1}{D_1}$	W_1 Width ft.	$\frac{W_1}{h_1}$	$\frac{W_1}{s_1}$	Type	Pos on apron $\frac{X}{L}$	h_2 Ht ft.	$\frac{h_2}{D_2}$	W_2 Width ft.	$\frac{W_2}{h_2}$	$\frac{W_2}{s_2}$	Type	Pos on apron $\frac{X}{L}$	Ht ft.	$\frac{Ht}{D_2}$	Width ft.	$\frac{Width}{Ht}$	$\frac{Width}{Spacing}$		
	(21)	(22)	(23)	(24)	(25)	(26)	(27)	(28)	(29)	(30)	(31)	(32)	(33)	(34)	(35)	(36)	(37)	(38)	(39)	(40)
Rye Patch	1. 18				Solid	1. 0	2. 0	0. 09											R	45° warp.
Unity	1. 21	1. 83	0. 52	1. 0	Teeth	. 93	5. 5	. 22	2. 17	0. 39	1. 0								R	Normal.
Alcova	1. 15				do	1. 0	10. 0	. 22	5. 0	. 50	1. 0								R	None.
Shadow Mt	1. 16	4. 0	1. 33	1. 78	do	1. 0	3. 5	. 17	4. 0	1. 14	1. 78								R	Normal.
Boysen (Final)	1. 13	5. 25	1. 31	1. 75	do	. 68	8. 75	. 23	5. 25	. 60	1. 75								R	None.
Boysen (Prelim)	1. 05	7. 5	1. 25	1. 8	do	1. 0	8. 0	. 16	7. 5	. 94	1. 8								R	None.
Scofield	1. 38	3. 5	1. 0	1. 91	do	1. 0	4. 0	. 17	3. 5	. 88	1. 91								R	Normal.
Boca	2. 72	3. 0	. 75	1. 0	do	1. 0	4. 0	. 19	3. 0	. 75	1. 0								R	Normal.
Fresno	1. 04	4. 0	1. 0	1. 0	do	1. 0	6. 0	. 19	4. 0	. 67	1. 0								R	45° warp.
Bull Lake					do	1. 0	4. 0	. 22	5. 0	1. 25	1. 0								R	Normal.
Caballo	1. 07	4. 0	. 89	1. 0	do	. 90	6. 5	. 19	4. 0	. 62	1. 0								R	Normal.
Moon Lake	1. 44	1. 875	. 72	1. 0	do	. 85	5. 0	. 21	3. 75	. 75	1. 0								R	Normal.
Deer Creek	1. 60	3. 0	1. 0	1. 0	do	1. 0	5. 0	. 18	3. 0	. 60	1. 0								R	Normal.
Alamogordo	1. 51	3. 5	. 44	1. 0	do	1. 0	9. 0	. 17	4. 0	. 44	1. 0								R	Normal.
Enders	. 95	5. 0	. 83	1. 0	do	1. 0	12. 0	. 26	5. 0	. 42	1. 0								R	Normal.
Medicine Creek	1. 07	6. 0	. 89	1. 0	do	1. 0	8. 0	. 17	6. 0	. 75	1. 0								R	Normal.
Cedar Bluff	1. 49	6. 0	. 86	1. 0	do	1. 0	9. 0	. 19	6. 0	. 67	1. 0								R	Normal.
Falcon	. 91	10. 0	1. 25	1. 0	do	1. 0	12. 0	. 20	10. 0	. 83	1. 0								R	45° vert.
Trenton	. 81	5. 0	1. 0	1. 0	do	1. 0	9. 0	. 19	5. 0	. 56	1. 0								R	Normal.
Cachuma	1. 19	5. 08	. 92	1. 0	do	1. 0	12. 0	. 21	5. 08	. 42	1. 0								R	Normal.
Tiber	2. 51	5. 0	. 71	1. 0	do	1. 0	8. 0	. 21	5. 0	. 62	1. 0								R	Normal.
Imperial Spillway	. 90	2. 49	1. 02	1. 03	do	1. 0	3. 75	. 20	3. 27	1. 01	1. 0								R	20° warp.
Imperial Sluiceway	1. 33	3. 0	. 90	. 95	do	. 65	5. 0	. 36	5. 67	1. 13	1. 1								R	20° warp.
Grassy Lake	1. 05	1. 67	1. 67	1. 0	do	1. 0	2. 0	. 14	1. 33	. 66	1. 0								R	Normal.
Box Butte	2. 06	3. 0	. 91	1. 0	do	1. 0	3. 0	. 18	3. 0	1. 0	1. 0								R	Normal.
Siphon Drop	1. 50	2. 25	1. 0	1. 0	do	1. 0	3. 25	. 29	2. 25	. 69	1. 0	T'th	0. 72	3. 25	0. 28	2. 25	1. 44	1. 0	R	45° warp.
Pilot Knob	1. 56	2. 5	1. 0	1. 0	do	1. 0	5. 0	. 28	5. 0	1. 0	1. 0								R	45° warp.
AA Canal Drp 1	. 97	1. 75	1. 0	1. 0	Solid	1. 0	1. 0	. 08											R	30° warp.
Wasteway #2					Solid	1. 0	2. 0	. 12				T'th	. 71	2. 41	. 15	2. 41	1. 0	1. 0	R	30° warp.
Big Sandy #2	1. 12	3. 25	1. 30	1. 0	Teeth	1. 0	3. 0	. 12	3. 25	1. 08	1. 0								T	Normal.
Cherry Creek					Solid	1. 0	5. 0	. 12				T'th	. 73	5. 0	. 12	3. 75	. 75	1. 0	R	None.
Pine View					Teeth	1. 0	5. 0	. 13	6. 0	1. 20	1. 0								T	Warp curve.
Agency Valley	. 83				do	. 91	10. 0	. 37	3. 33	. 33	1. 0								T	50° warp.
Davis	1. 96	13. 0	. 91	1. 86	do	1. 0	14. 3	. 23	13. 0	. 91	1. 86								R	Normal.
Bonny	1. 94	5. 0	. 72	1. 0	do	1. 0	8. 0	. 21	5. 0	. 62	1. 0								R	Normal.
Cle Elum					do	1. 0	10. 0	. 33	10. 0	1. 0	1. 0								R	None.
Maximum	2. 72		1. 67	1. 91		1. 0		. 37		1. 25	1. 91		. 73		. 28		1. 44	1. 0		
Minimum	. 81		. 44	. 95		. 65		. 08		. 33	1. 00		. 71		. 12		. 75	1. 0		
Average	1. 35		. 97	1. 15		. 97		. 20		. 76	1. 13							1. 0		

R = Rectangular. T = Trapezoidal.

Columns 32 through 38 show the proportions of additional baffle piers used on three of the stilling basins. These are not necessary and are not recommended for this type of basin.

Additional details. Column 18 indicates the angle with the horizontal at which the high-velocity jet enters the stilling basin for each of the spillways. The maximum angle was 34° and the minimum 14°. The effect of the vertical angle of the chute on the action of the hydraulic jump could not be evaluated from the information available. However, this factor will be considered in Section 5 in connection with sloping apron design.

Column 39 designates the cross section of the basin. In all but three cases the basins were rectangular. The three cross sections that were trapezoidal had side slopes varying from 1/4:1 to 1/2:1. The generalized designs presented in this monograph are for stilling basins having rectangular cross sections. Where trapezoidal basins are contemplated a model study is strongly recommended.

Column 40, Table 2, indicates that in the majority of basins constructed for earth dam spillways the wing walls were normal to the training walls. Five basins were constructed without wing walls; instead a rock fill was used. The remaining basins utilized angling wing walls or warped transitions downstream from the basin. The latter are common on canal structures. The object, of course, is to build the cheapest wing wall that will afford the necessary protection. The type of wing wall is usually dictated by local conditions such as width of the channel downstream, depth to foundation rock, degree of protection needed, etc.; thus wing walls are not amenable to generalization.

Verification Tests

An inspection of the data shows that the structures listed in Table 2 do not cover the desired range of operating conditions. There is insufficient information to determine the length of basin for the larger values of the Froude number, there is little or no information on the tail water depth at which sweepout occurs, and the information available is of little value for generalizing the problem of determining water-surface profiles. Laboratory tests were therefore performed to extend the range and to supply the missing data. The experiments were made on 17 Type II basins, proportioned according to the above rules, and installed in Flumes B, C, D, and E (see Columns 1 and 2, Table 3). Each basin was judged at the discharge for which it was designed, the length was adjusted to the minimum that would produce satisfactory operation, and the absolute minimum tail water depth for acceptable operation was measured. The basin operation was also observed for flows less than the designed discharge and found to be satisfactory in each case.

Table 3 is quite similar to Table 2 with the exception that the length of Basin L_{II} (Col. 11) was determined by experiment, and the tail water depth at which the jump just began to sweep out of the basin was recorded (Col. 13).

Tail water depth. The solid line in Figure 11 was obtained from the hydraulic jump formula $\frac{D_2}{D_1}=1/2\ (\sqrt{1+8F^2}-1)$ and represents conjugate tail water depth. It is the same as the line shown in Figure 5. The dashed lines in Figure 11 are merely guides drawn for tail water depths other than conjugate depth. The points shown as dots were obtained from Column 13 of Table 2 and constitute the ratio of actual tail water depth to D_1 for each basin listed. It can be observed that the majority of the basins were designed for conjugate tail water depth or less. The minimum tail water depth for Basin II, obtained from Column 14 of Table 3, is shown in Figure 11. The curve labeled "Minimum TW Depth Basin II" indicates the point at which the front of the jump moves away from the chute blocks. In other words, any additional lowering of the tail water would cause the jump to leave the basin. Consulting Figure 11, it can be observed that the margin of safety for a Froude number of 2 is 0 percent; for a number of 6 it increases to 6 percent; for a number of 10 it diminishes to 4 percent; and for a number of 16 it is 2.5 percent. From a practical point of view this means that the jump will no longer operate properly when the tail water depth approaches $0.98D_2$ for a Froude number of 2, or $0.94D_2$ for a number of 6, or $0.96D_2$ for a number of 10, or $0.975D_2$ for a number of 16. The margin of safety is largest in the middle range. For the two extremes of the curve it is advisable to provide tail water greater than conjugate depth to be safe. For these reasons the Type II basin should never

TABLE 3.—*Verification tests on Type II basins*

Flume (1)	Test (2)	Q c.f.s. (3)	W = Width of basin ft. (4)	q per ft. of W c.f.s. (5)	TW depth (6)	V_1 ft./sec. (7)	D_1 ft. (8)	$\frac{D_2}{D_1}$ (9)	$F_1=\frac{V_1}{\sqrt{gD_1}}$ (10)	L_{II} ft. (11)	$\frac{L_{II}}{D_2}$ (12)	T_{so} TW at sweep out ft. (13)
B	1	2. 50	2. 00	1. 25	1. 120	17. 36	0. 072	15. 60	11. 39	4. 95	4. 42	1. 09
	2	4. 00		2. 00	1. 430	17. 54	. 114	12. 54	9. 16	6. 10	4. 27	1. 37
	3	6. 00		3. 00	1. 750	17. 65	. 170	10. 29	7. 54	7. 30	4. 17	1. 65
	4	8. 00		4. 00	2. 030	17. 86	. 224	9. 06	6. 64	8. 00	3. 94	1. 88
C	5	1. 60	1. 50	1. 07	1. 070	17. 49	. 061	17. 54	12. 48	4. 60	4. 30	1. 04
	6	2. 10		1. 40	1. 240	17. 94	. 078	15. 89	11. 32	5. 40	4. 35	1. 18
	7	2. 63		1. 75	1. 355	18. 26	. 096	14. 11	10. 39	5. 70	4. 21	1. 32
	8	2. 75		1. 83	1. 400	18. 33	. 100	14. 00	10. 21	6. 23	4. 45	1. 36
	9	4. 00		2. 67	1. 785	20. 36	. 131	13. 62	9. 91	7. 40	4. 15	1. 73
D	10	5. 00	3. 97	1. 26	1. 235	20. 30	. 062	19. 91	14. 38	5. 10	4. 13	1. 20
	11	6. 00		1. 51	1. 350	20. 41	. 074	18. 24	13. 21	5. 80	4. 30	1. 32
	12	9. 80		2. 47	1. 750	21. 84	. 113	15. 50	11. 45	7. 80	4. 46	1. 73
	13	11. 00		2. 77	1. 855	21. 15	. 131	14. 16	10. 29	8. 10	4. 37	1. 82
	14	13. 00		3. 27	2. 020	21. 39	. 153	13. 20	9. 64	8. 70	4. 31	1. 95
	15	20. 00		5. 04	2. 585	23. 00	. 319	11. 80	8. 66	10. 60	4. 10	2. 48
E	16	5. 00	3. 97	1. 26	0. 840	10. 49	. 120	7. 00	5. 33	3. 36	4. 00	0. 79
	17	10. 00		2. 52	1. 220	11. 09	. 227	5. 37	4. 10	4. 51	3. 70	1. 10

Flume	Test	$\frac{T_{so}}{D_1}$ (14)	$\frac{T_{so}}{D_2}$ (15)	h_1 Ht ft. (16)	$\frac{h_1}{D_1}$ (17)	$\frac{W_1}{h_1}$ (18)	$\frac{s_1}{h_1}$ (19)	h_2 Ht dentated sill ft. (20)	$\frac{h_2}{D_2}$ (21)	$\frac{W_2}{h_2}$ (22)	$\frac{s_2}{h_2}$ (23)	Slope water surface, ° (24)	Slope of chute (25)
B	1	15. 13	0. 97	0. 073	1. 01	1. 0	1. 0	0. 219	0. 196	0. 75	0. 75	10. 5	0. 7:1
	2	12. 02	. 96	. 114	1. 00	1. 0	1. 0	. 286	. 200	. 75	. 75	10. 0	
	3	9. 70	. 94	. 170	1. 00	1. 0	1. 0	. 352	. 201	. 75	. 75	9. 6	
	4	8. 39	. 93	. 229	1. 02	1. 0	1. 0	. 406	. 200	. 75	. 75	9. 0	
C	5	17. 04	. 97	. 062	1. 02	1. 0	1. 0	. 320	. 300	. 75	. 75	11. 3	2:1
	6	15. 12	. 95	. 078	1. 00	1. 0	1. 0	. 260	. 210	. 75	. 75	10. 8	
	7	13. 75	. 97	. 105	1. 09	1. 0	1. 0	. 250	. 185	. 75	. 75	10. 5	
	8	13. 60	. 97	. 100	1. 00	1. 0	1. 0	. 310	. 221	. 75	. 75	10. 0	
	9	13. 21	. 97	. 131	1. 00	1. 0	1. 0	. 446	. 250	. 75	. 75	10. 4	
D	10	19. 35	. 97	. 062	1. 00	1. 0	1. 0	. 250	. 203	1. 00	1. 00	12. 0	0. 6:1
	11	17. 83	. 98	. 074	1. 00	1. 0	1. 0	. 270	. 200	1. 00	1. 00	11. 2	
	12	15. 31	. 99	. 153	1. 35	1. 0	1. 0	. 400	. 229	1. 00	1. 00	10. 0	
	13	13. 89	. 98	. 131	1. 00	1. 0	1. 0	. 396	. 214	. 75	. 75	10. 2	
	14	12. 75	. 97	. 153	1. 00	1. 0	1. 0	. 400	. 198	. 75	. 75	8. 3	
	15	11. 32	. 96	. 219	1. 00	1. 0	1. 0	. 517	. 200	. 75	. 75	9. 5	
E	16	6. 58	. 94	. 122	1. 02	1. 0	1. 0	. 200	. 238	. 75	. 75	6. 5	Varied
	17	9. 02	. 90	. 235	1. 04	1. 0	1. 0	. 270	. 221	. 75	. 75	5. 3	

be designed for less than conjugate depth, and a minimum safety factor of 5 percent of D_2 is recommended.

Several precautions should be taken when determining tail water elevations. First, tail water curves are usually extrapolated for the discharges encountered in design, so they can be in error. Second, the actual tail water depth usually lags, in a temporal sense, that of the tail water curve for rising flow and leads the curve for a falling discharge. Extra tail water should therefore be provided if reasonable increasing increments of discharge limit the performance of the structure because of a lag in building up tail water depth. Third, a tail water curve may be such that the most adverse condition occurs at less than the maximum designed discharge; and fourth, temporary or permanent retrogression of

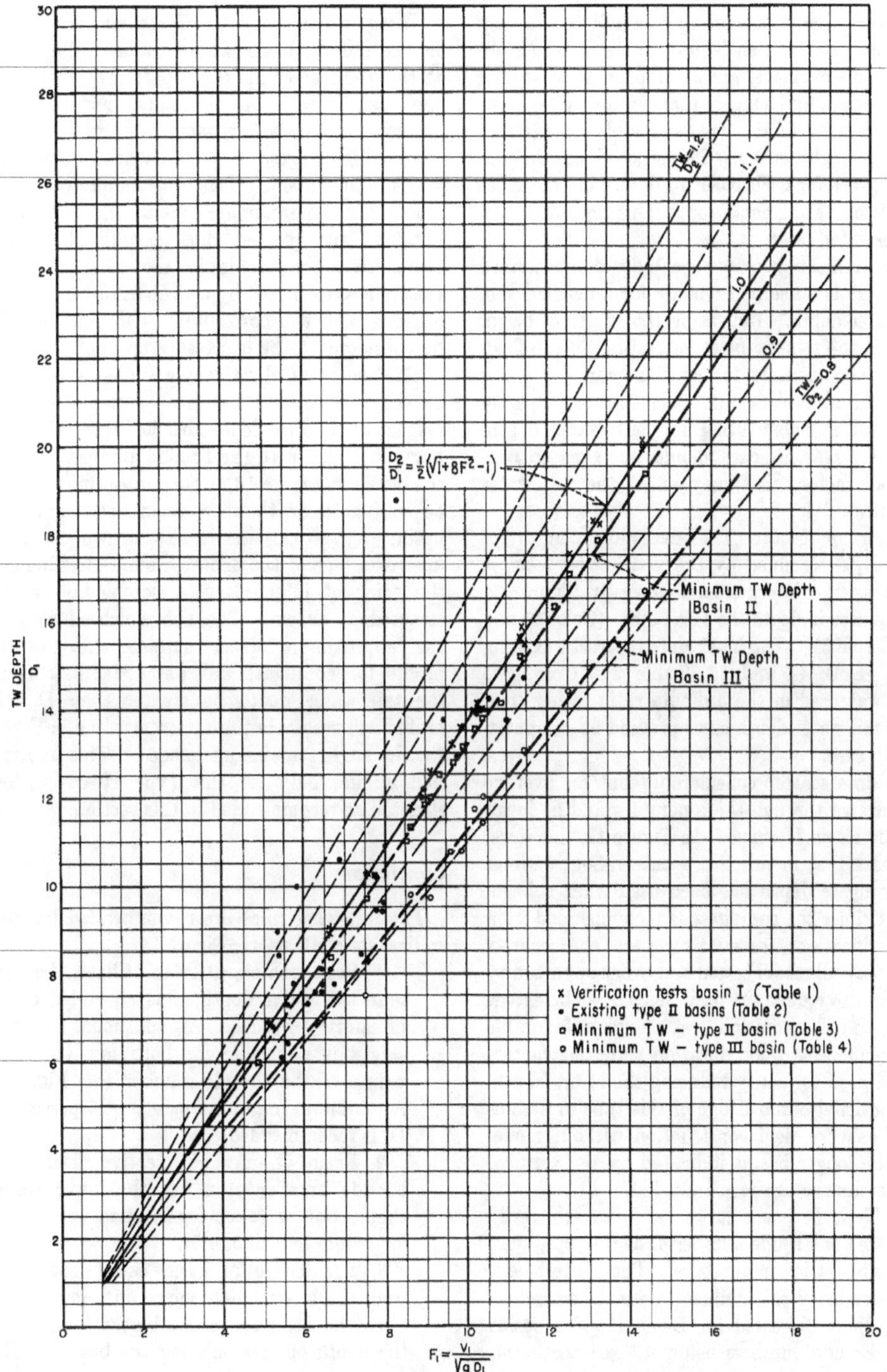

FIGURE 11.—*Minimum tail water depths (Basins I, II, and III).*

the riverbed downstream may be a factor needing consideration. These factors, some of which are difficult to evaluate, are all important in stilling basin design, and suggest that an adequate factor of safety is essential. It is advisable to construct a jump height curve, superimposed on the tail water curve for each basin to determine the most adverse operating condition. This procedure will be illustrated later.

The verification tests repeatedly demonstrated that there is no simple remedy for a deficiency in tail water depth. Increasing the length of basin, which is the remedy often attempted in the field, will not compensate for deficiency in tail water depth. Baffle piers and sills are only partly successful in substituting for tail water depth. For these reasons, care should be taken to consider all factors that may affect the tail water at a future date.

Length of basin. The necessary length of Basin II, determined by the verification tests, is shown as the intermediate curve in Figure 12. The squares indicate the test points (Cols. 10 and 12 of Table 3). The black dots represent existing basins (Cols. 11 and 17, Table 2). Conjugate depth was used in the ordinate ratio rather than actual tail water depth since it could be computed for each case.

The dots scatter considerably but an average curve drawn through these points would be lower than the Basin II curve. In Figure 12, therefore, it appears that in practice a basin about 3 times the conjugate depth has been used when a basin about 4 times the conjugate is recommended from the verification tests. However, the shorter basins were all model tested and every opportunity was taken to reduce the basin length. The extent and depth of bed erosion, wave heights, favorable flood frequencies, flood duration and other factors were all used to justify reducing the basin length. Lacking definite knowledge of this type in designing a basin for field construction without model tests, the longer basins indicated by the verification tests curve are recommended.

The Type II basin curve has been arbitrarily terminated at Froude number 4, as the jump may be unstable at lower numbers. The chute blocks have a tendency to stabilize the jump and reduce the 4.5 limit discussed for Basin I. For basins having Froude numbers below 4.5 see Section 4.

Water-surface profiles. Water-surface profiles in the stilling basin were measured during the tests to aid in computing uplift pressures under the basin apron. As the water surface in the stilling basin tests fluctuated rapidly, it was felt that a high degree of accuracy in measurement was not necessary. This was found to be true when the approximate water-surface profiles obtained were plotted, then generalized. It was found that the profile in the basin could be closely approximated by a straight line making an angle α with the horizontal. This line can also be considered to be a pressure profile.

The angle α (Col. 24, Table 3) observed in each of the verification tests has been plotted with respect to the Froude number in Figure 13. The angle increases with the Froude number. To use the curve in Figure 13, a horizontal line is drawn at conjugate depth on a scale drawing of the basin. A vertical line is also drawn from the upstream face of the dentated sill. Beginning at the point of intersection, a sloping line is constructed as shown. The above procedure gives the approximate water surface and pressure profile for conjugate tail water depth. Should the tail water depth be greater than D_2, the profile will resemble the uppermost line in Figure 13; the angle remains unchanged. This information applies only for the Type II basin, constructed as recommended in this section.

Conclusions

The following rules are recommended for generalization of Basin II, Figure 14:

1. Set apron elevation to utilize full conjugate tail water depth, plus an added factor of safety if needed. An additional factor of safety is advisable for both low and high values of the Froude number (see Fig. 11). A minimum margin of safety of 5 percent of D_2 is recommended.

2. Basin II may be effective down to a Froude number of 4 but the lower values should not be taken for granted (see Sec. 4 for values less than 4.5).

3. The length of basin can be obtained from the intermediate curve on Figure 12.

4. The height of chute blocks is equal to the depth of flow entering the basin, or D_1,

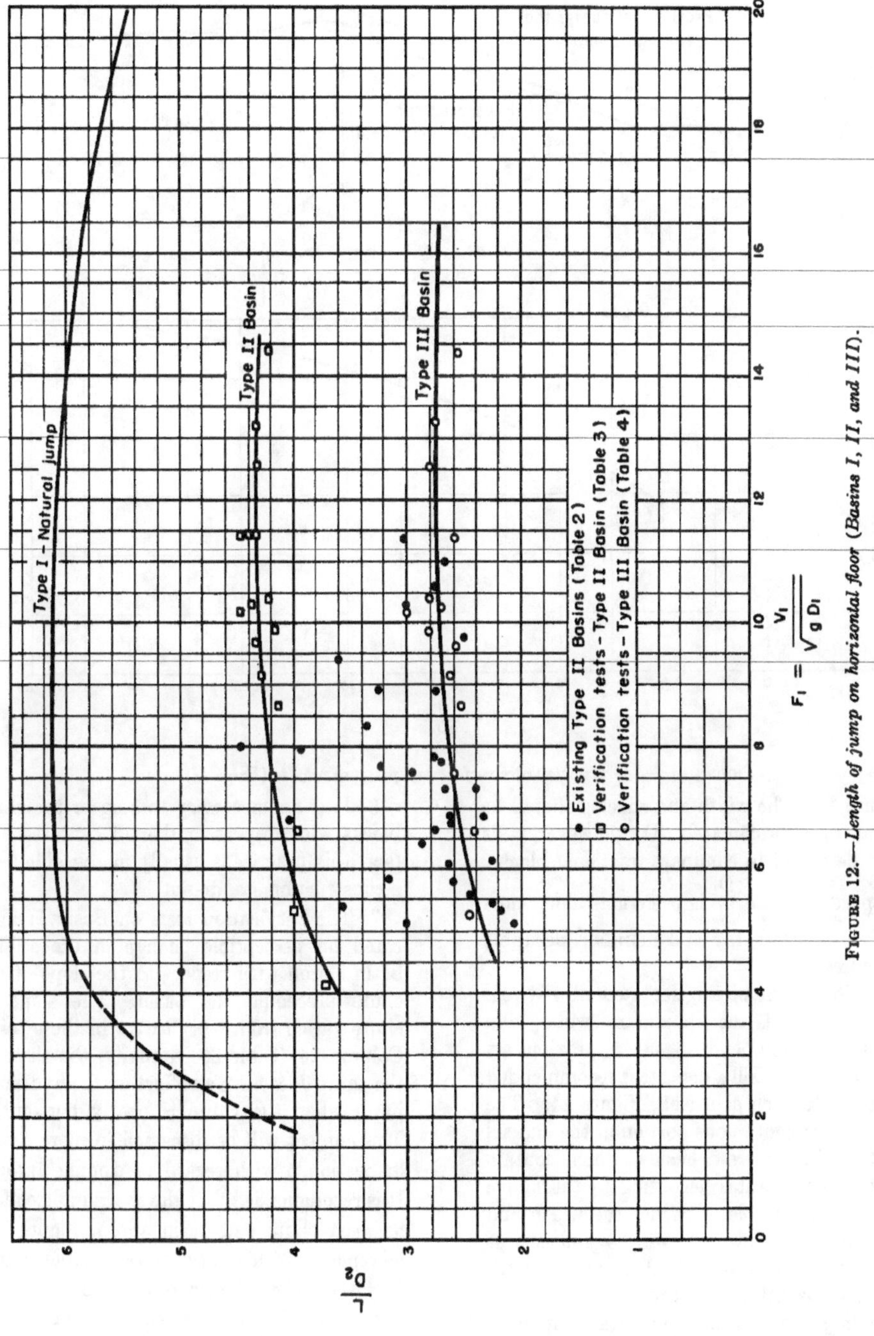

FIGURE 12.—*Length of jump on horizontal floor (Basins I, II, and III).*

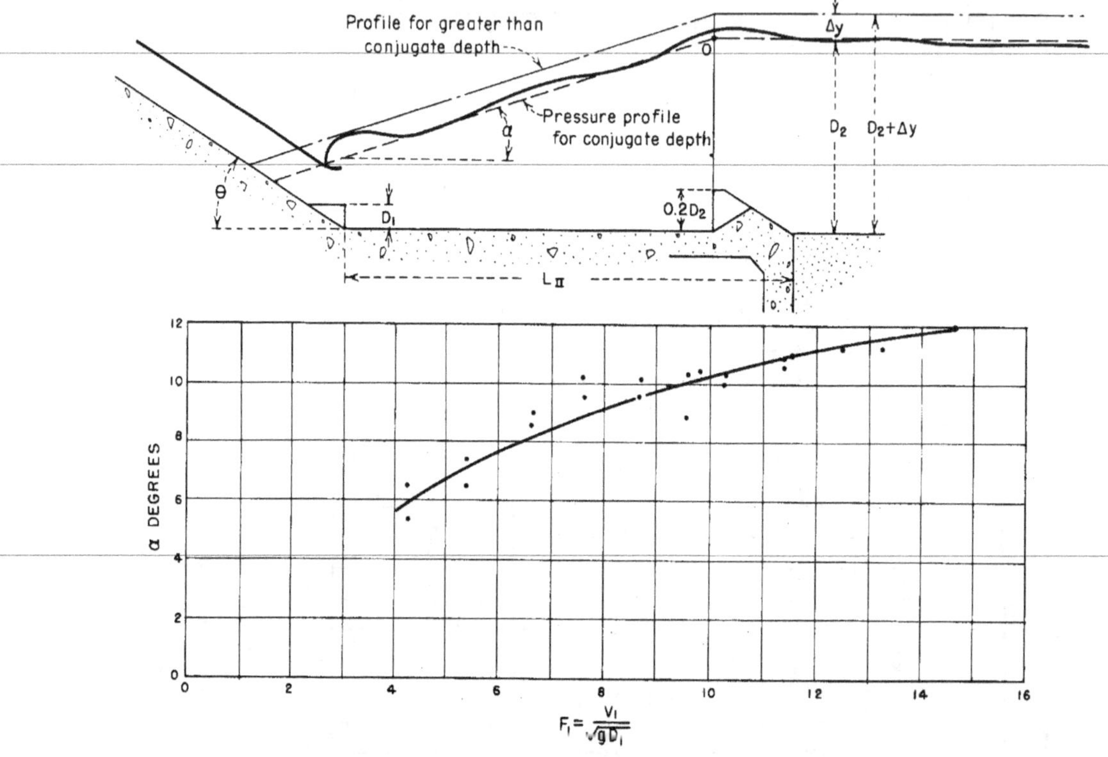

FIGURE 13.—*Approximate water surface and pressure profiles (Basin II).*

Figure 14. The width and spacing should be equal to approximately D_1; however, this may be varied to eliminate fractional blocks. A space equal to $\frac{D_1}{2}$ is preferable along each wall to reduce spray and maintain desirable pressures.

5. The height of the dentated sill is equal to $0.2D_2$, and the maximum width and spacing recommended is approximately $0.15D_2$. On the sill a dentate is recommended adjacent to each side wall, Figure 14. The slope of the continuous portion of the end sill is 2:1. For narrow basins, which contain only a few dentates according to the above rule, it is advisable to reduce the width and the spacing. However, widths and spaces should remain equal. Reducing the width and spacing actually improves the performance in narrow basins; thus, the minimum width and spacing of the dentates is governed only by structural considerations.

6. It is not necessary to stagger the chute blocks with respect to the sill dentates. In fact, this practice is usually inadvisable from a construction standpoint.

7. The verification tests on Basin II indicated no perceptible change in the stilling basin action with respect to the slope of the chute preceding the basin. The slope of chute varied from 0.6:1 to 2:1 in these tests, Column 25, Table 3. Actually, the slope of the chute does have an effect on the hydraulic jump when the chute is nearly horizontal. This subject will be discussed in more detail in Section 5 with regard to sloping aprons. It is recommended that the sharp intersection between chute and basin apron, Figure 14, be replaced with a curve of reasonable radius ($R \geqq 4D_1$) when the slope of the chute is 1:1 or greater. Chute blocks can be incorporated on the curved face as readily as on the plane surfaces.

Following the above rules will result in a safe, conservative stilling basin for spillways up to 200 feet high and for flows up to 500 c.f.s. per foot of basin width, provided the jet entering the basin is reasonably uniform both as to velocity and depth. For greater falls, larger unit discharges, or possible asymmetry, a model study of the specific design is recommended.

Aids in computation. Before presenting an example illustrating the method of proportioning Basin II, a chart will be presented which should be of special value for preliminary computations. The chart makes it possible to determine V_1 and D_1 with a fair degree of accuracy for chutes having slopes of 0.8:1 or steeper, where computation is a difficult and arduous procedure. The chart, Figure 15, represents a composite of experience, computation, and a limited amount of experimental information obtained from prototype tests on Shasta and Grand Coulee Dams. There is much to be desired in the way of experimental confirmation; however, the chart is sufficiently accurate for preliminary design.

The ordinate in Figure 15 is the fall from reservoir level to stilling basin floor, while the abscissa is the ratio of actual to theoretical velocity at the entrance to the stilling basin. The theoretical velocity $V_T=\sqrt{2g(Z-H/2)}$. The actual velocity is the term desired. The curves represent different heads, H, on the crest of the spillway. As is reasonable, the larger the head on the crest, the more nearly the actual velocity at the base of the spillway will approach the theoretical. For example, with H=40 feet and Z=230 feet, the actual velocity at the base of the dam would be 0.95 of the computed theoretical velocity; with a head of 10 feet on the crest the actual velocity would be 0.75 V_T. The value of D_1 may then be computed by dividing the unit discharge by the actual velocity obtained from Figure 15.

The chart is not applicable for chutes flatter than 0.6:1 as frictional resistance assumes added

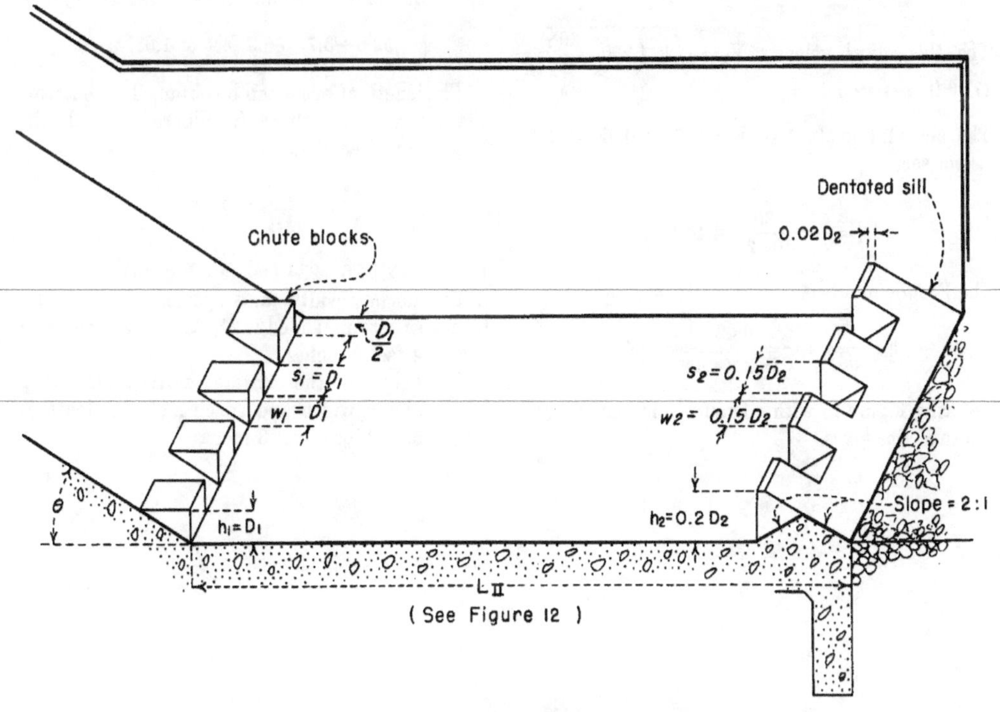

FIGURE 14.—*Recommended proportions (Basin II).*

importance in this range. Therefore, it will be necessary to compute the hydraulic losses starting at the gate section where critical depth is known.

Insufflation, produced by air from the atmosphere mixing with the sheet of water during the fall, need not be considered in the hydraulic jump computations. Insufflation is important principally in determining the height of chute and stilling basin walls. It is usually not possible to construct walls sufficiently high to confine all spray and splash; thus, wall heights are usually chosen commensurate with the material and terrain to be protected.

Application of results (*Example 2*). The crest of an overfall dam, having a downstream slope of 0.7:1, is 200 feet above the horizontal floor of the stilling basin. The head on the crest is 30 feet and the maximum discharge is 480 c.f.s. per foot of stilling basin width. Proportion a Type II stilling basin for these conditions.

Entering Figure 15 with a head of 30 feet over the crest and a total fall of 230 feet,

$$\frac{V_A}{V_T}=0.92$$

The theoretical velocity $V_T=\sqrt{2g\left(230-\frac{30}{2}\right)}=$ 117.6 ft. per sec.

The actual velocity $V_A=V_1=117.6\times0.92=108.2$ ft. per sec.

$$D_1=\frac{q}{V_1}=\frac{480}{108.2}=4.44 \text{ feet}$$

The Froude number

$$F_1=\frac{V_1}{\sqrt{gD_1}}=\frac{108.2}{\sqrt{32.2\times4.44}}=9.04$$

Entering Figure 11 with a Froude number of 9.04, the solid line gives

$$\frac{TW}{D_1}=12.3$$

As TW and D_2 are synonymous in this case, the conjugate tail water depth

$$D_2=12.3\times4.44=54.6 \text{ feet}$$

The minimum tail water line for the Type II basin on Figure 11 shows that a factor of safety of about 4 percent can be expected for the above Froude number.

Should it be desired to provide a margin of safety of 7 percent, the following procedure may be followed: Consulting the line for minimum TW depth for the Type II basin, Figure 11,

$\frac{TW}{D_1}=11.85$ for a Froude number of 9.04

The tail water depth at which sweepout is incipient:

$$T_{so}=11.85\times4.44=52.6 \text{ feet}$$

Adding 7 percent to this figure, the stilling basin apron should be positioned for a tail water depth of

$$52.6+3.7=56.3 \text{ feet or } 1.03D_2$$

The length of basin can be obtained by entering the intermediate curve in Figure 12 with the Froude number of 9.04

$$\frac{L_{II}}{D_2}=4.28$$

$L_{II}=4.28\times54.6=234$ feet (see Fig. 14).

The height, width, and spacing of the chute blocks as recommended is D_1; thus the dimension can be 4 feet 6 inches.

The height of the dentated sill is $0.2D_2$ or 11 feet, and the width and spacing of the dentates can be $0.15D_2$ or 8 feet 3 inches.

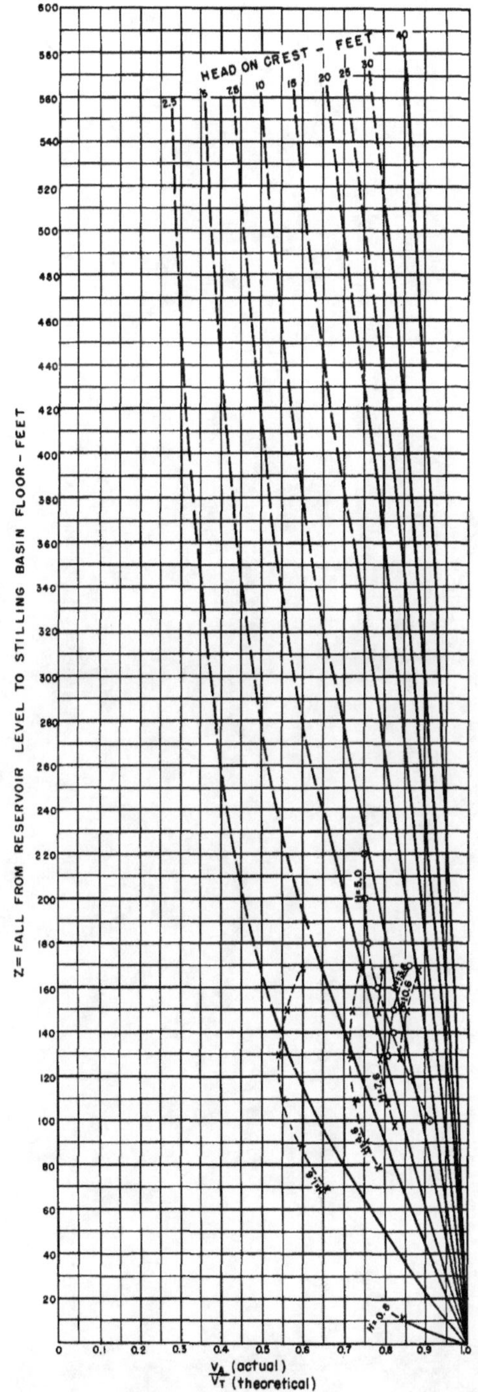

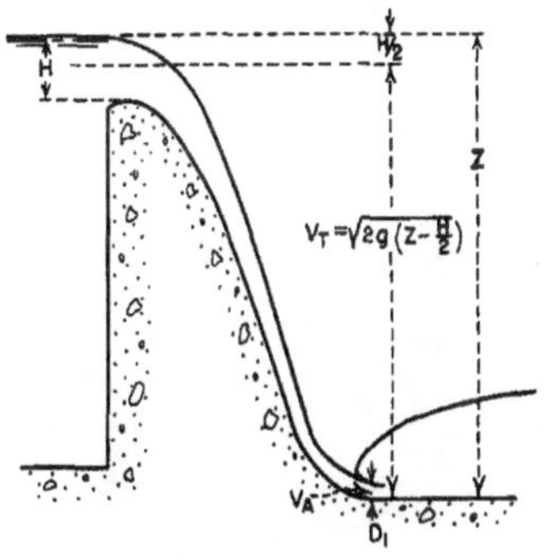

PROTOTYPE TESTS
X Shasta Dam
O Grand Coulee Dam

FIGURE 15.—*Curves for determination of velocity entering stilling basin for steep slopes 0.8:1 to 0.6:1.*

Section 3

Short stilling basin for canal structures, small outlet works, and small spillways (Basin III)

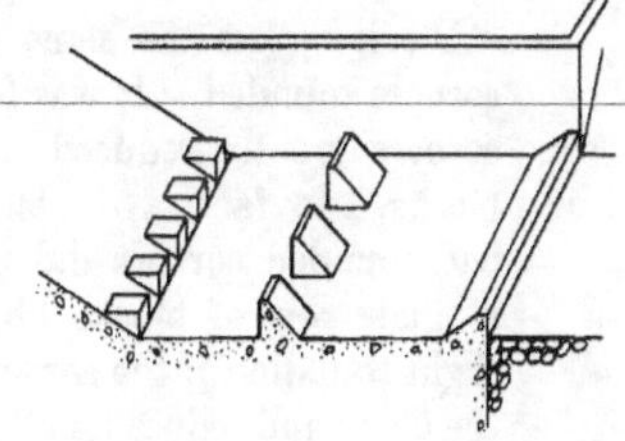

Introduction

BASIN II often is considered too conservative and consequently overcostly for structures carrying relatively small discharges at moderate velocities. A shorter basin having a simpler end sill may be used if baffle piers are placed downstream from the chute blocks. Because of the possibility of low pressures on the baffle piers and resulting cavitation, the incoming velocity and discharge per foot of width must be limited to reasonable values. In this section a minimum basin is developed for a class of smaller structures in which velocities at the entrance to the basin are moderate or low (up to 50–60 feet per second) and discharges per foot of width are less than 200 cubic feet per sec. Development tests and verification tests on 14 different basins are used to generalize the design and to determine the range over which Basin III will perform satisfactorily.

Development

The most effective way to shorten a stilling basin is to modify the jump by the addition of appurtenances in the basin. One restriction imposed on these appurtenances, however, is that they must be self-cleaning or nonclogging. This restriction thus limits the appurtenances to baffle piers or sills which can be incorporated on the stilling basin apron.

Numerous experiments were therefore performed using various types and arrangements of baffle piers and sills on the apron in an effort to obtain the best possible solution. Some of the arrangements tested are shown in Figure 16. The blocks were positioned in both single and double rows, the second row staggered with respect to the first. Arrangement "a" in Figure 16 consisted of a solid curved sill which was tried in several positions on the apron. This sill required an excessive tail water depth to be effective. The solid sill was then replaced with baffle piers. For certain heights, widths, and spacing, block "b" performed well, resulting in a water surface similar to that shown in Figure 19. Block "c" was ineffective for any height. The high-velocity jet passed over the block at about a 45° angle with little interference, and the water surface downstream was very turbulent with waves. Stepped block "d," both for single and double rows, was much the same as "c". The cube "e" was effective when the best height, width, spacing, and position on the apron were found. The front of the jump was almost vertical and the water surface downstream was quite flat and smooth, like the water surface shown in Figure 19. Block "f" performed identically with the cubical block "e." The important feature as to shape appeared to be the vertical upstream face. The foregoing blocks and others not mentioned here were all tested in single and double rows. The second row, sketch "h," Figure 16, in each case was of little value.

Block "g" is the same as block "f" with the corners rounded. It was found that rounding the corners greatly reduced the effectiveness of the blocks. In fact, a double row of blocks which had rounded corners did not perform as well as a single row of blocks "b," "e", or "f." Even slight rounding of the corners tended to streamline the block and reduce its effectiveness as an impact device. As block "f" is usually preferable from a construction standpoint, it was used throughout the remaining tests to determine a general design with respect to height, width, spacing, and position on the apron.

In addition to experimenting with the baffle piers, variations in the size and shape of the chute blocks and the end sill were also tested. It was found that the chute blocks should be kept small, no larger than D_1 if possible, to prevent the chute blocks from directing the flow over the baffle piers. The end sill had little or no effect on the jump proper when baffle piers are placed as recommended. Thus, there is no need for a dentated end sill and almost any type of solid end sill will suffice. The only purpose of the end sill in Basin III is to direct the remaining bottom currents upward and away from the river bed. The basin as finally developed is shown in Figure 17. This basin is principally an impact dissipation device whereby the baffle piers are called upon to do most of the work. The chute blocks aid in stabilizing the jump and the solid type end sill is for scour control.

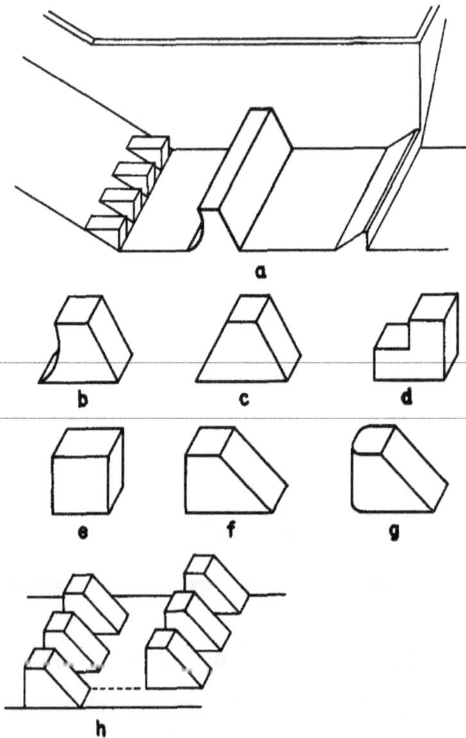

FIGURE 16.—*Record of appurtenances (Basin III).*

Verification Tests

At the conclusion of the development work, a set of verification tests was made to examine and record the performance of this basin, which will be designated as Basin III, over the entire range of operating conditions that may be met in practice. The tests were made on a total of 14 basins constructed in Flumes B, C, D, and E. The conditions under which the tests were run, the dimensions of the basin, and the results are recorded in Table 4. The headings are identical with those of Table 3 except for the dimensions of the baffle piers and end sills.

Stilling Basin Performance and Design

Stilling basin action was very stable for this design; in fact, more so than for either Basins I or II. The front of the jump was steep and there was less wave action to contend with downstream than in either of the former basins. In addition, Basin III has a large factor of safety against jump sweepout and operates equally well for all values of the Froude number above 4.0.

Basin III should not be used where baffle piers will be exposed to velocities above the 50 to 60 feet per second range without the full realization that cavitation and resulting damage may occur. For velocities above 50 feet per second, Basin II should be used or hydraulic model studies should be made.

Chute blocks. The recommended proportions for Basin III are shown in Figure 17. The height, width, and spacing of the chute blocks are equal to D_1, the same as for Basin II. Larger heights were tried, as can be observed from Column 18, Table 4, but are not recommended. The larger chute blocks tend to throw a portion of the high-velocity jet over the baffle piers. However, in some designs D_1 is less than 8 inches. The blocks may be made 8 inches high, which is considered by some designers to be the minimum size possible from a construction standpoint. The width and spacing of the blocks should be the same as the height. This may be varied but the aggregate width of spaces should equal, approximately, the total width of the blocks.

Baffle piers. The height of the baffle piers increases with the Froude number as can be observed from Columns 22 and 10, Table 4. The height, in terms of D_1, can be obtained from the upper line in Figure 18. The width and spacing may be varied but the total of the spaces should equal the total width of blocks. The most satisfactory width and spacing was found to be three-fourths of the height. It is not necessary to stagger the baffle piers with the chute blocks as it is often difficult to avoid construction joints and there is little to be gained from a hydraulic standpoint.

The most effective position of the baffle piers is $0.8D_2$ downstream from the chute blocks as shown

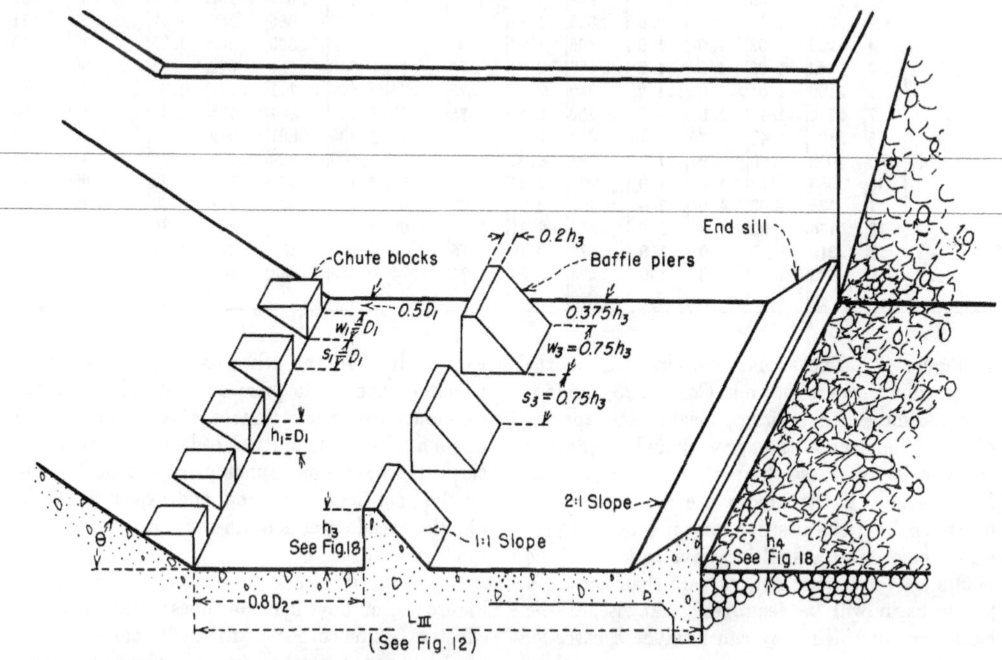

FIGURE 17.—*Recommended proportions (Basin III).*

TABLE 4.—*Verification tests on Type III basins*

Flume (1)	Test (2)	Q cfs (3)	W (4)	q per ft of W (5)	TW ft (6)	V_1 ft/sec (7)	D_1 ft (8)	$\frac{D_2}{D_1}$ (9)	$F_1=\frac{V_1}{\sqrt{gD_1}}$ (10)	L_{III} ft (11)	$\frac{L_{III}}{D_2}$ (12)	T_{so} TW at sweep out ft (13)	$\frac{T_{so}}{D_1}$ (14)	$\frac{T_{so}}{D_2}$ (15)	Slope of chute (16)
B	1	2.500	2.000	1.250	1.120	17.36	0.072	15.56	11.41	2.90	2.59	0.94	13.05	0.84	0.7:1
	2	4.000		2.000	1.430	17.54	.114	12.54	9.16	3.70	2.59	1.11	9.73	.78	
	3	6.000		3.000	1.750	17.65	.170	10.29	7.54	4.50	2.57	1.29	7.58	.74	
	4	8.000		4.000	2.030	17.86	.224	9.06	6.64	4.90	2.41	1.57	7.00	.77	
C	5	1.600	1.500	1.067	1.070	17.49	.061	17.54	12.48	3.00	2.80	.88	14.42	.82	2:1
	6	2.630		1.753	1.350	18.26	.096	14.06	10.39	3.80	2.81	1.16	12.08	.86	
	7	2.750		1.833	1.400	18.33	.100	14.00	10.21	4.20	3.00	1.17	11.70	.84	
	8	4.000		2.667	1.785	20.36	.131	13.62	9.91	5.00	2.80	1.42	10.84	.80	
D	9	5.000	3.970	1.259	1.250	20.30	.062	20.16	14.38	3.20	2.56	1.04	16.77	.83	0.6:1
	10	6.000		1.511	1.350	20.41	.074	18.24	13.21	3.70	2.74	1.12	15.13	.83	
	11	11.00		2.771	1.860	21.15	.131	14.20	10.29	5.00	2.69	1.50	11.45	.81	
	12	13.00		3.274	2.020	21.40	.153	13.20	9.64	5.20	2.57	1.65	10.78	.82	
	13	20.00		5.038	2.585	23.00	.219	11.80	8.66	6.46	2.50	2.15	9.82	.83	
E	14	5.000	3.970	1.259	0.840	10.49	.120	7.00	5.33	2.10	2.50	0.70	5.83	.83	Varied

Flume	Test	h_1 Ht of chute blocks ft (17)	$\frac{h_1}{D_1}$ (18)	$\frac{W_1}{h_1}$ (19)	$\frac{s_1}{h_1}$ (20)	h_3 Ht of baffle piers ft (21)	$\frac{h_3}{D_1}$ (22)	$\frac{W_3}{h_3}$ (23)	$\frac{s_3}{h_3}$ (24)	Distance to baffles ft (25)	$\frac{L_{III}}{3D_2}$ (26)	h_4 Ht of end sill ft (27)	$\frac{h_4}{D_1}$ (28)	Z Depth upstream from baffles ft (29)	$\frac{Z}{D_2}$ (30)
B	1	0.073	1.01	1.0	1.0	0.167	2.32	1.0	1.0	0.800	0.714	0.125	1.74	0.60	0.54
	2	.114	1.00	1.0	1.0	.218	1.91	1.0	1.0	0.920	.040	.187	1.64	.80	.56
	3	.333	1.96	.6	.6	.302	1.78	1.0	1.0	1.200	.686	.250	1.47	.95	.54
	4	.229	1.02	1.0	1.0	.396	1.77	1.0	1.0	1.340	.660	.302	1.35	1.20	.59
C	5	.062	1.02	1.0	1.0	.167	2.74	.75	.75	0.850	.794	.092	1.51	.60	.56
	6	.100	1.04	1.0	1.0	.240	2.50	.75	.75	1.000	.741	.146	1.52	.65	.48
	7	.146	1.46	1.0	1.0	.250	2.50	.75	.75	1.210	.864	.156	1.56	.70	.50
	8	.187	1.43	.75	.75	.312	2.38	.75	.75	1.430	.801	.219	1.67	.90	.50
D	9	.062	1.00	1.0	1.0	.188	3.03	1.0	1.0	1.000	.800	.125	2.02	.60	.48
	10	.083	1.12	1.0	1.0	.208	2.81	1.0	1.0	1.120	.830	.135	1.82	.65	.48
	11	.135	1.03	1.0	1.0	.302	2.31	1.0	1.0	1.250	.672	.208	1.59	.95	.51
	12	.156	1.02	1.0	1.0	.354	2.31	1.0	1.0	1.680	.832	.208	1.36	1.05	.52
	13	.219	1.00	1.0	1.0	.479	2.19	.75	.75	2.153	.833	.271	1.24	1.30	.50
E	14	.122	1.02	1.0	1.0	.215	1.79	.75	.75	0.672	.833	.150	1.25	.55	.65

in Figure 17. The actual positions used in the verification tests are shown in Column 25, Table 4. The recommended position, height, and spacing of the baffle piers on the apron should be adhered to carefully, as these dimensions are important. For example, if the blocks are set appreciably upstream from the position shown they will produce a cascade with resulting wave action. If the baffles are set farther downstream than shown, a longer basin will be required. Likewise, if the baffles are too high they can produce a cascade; if too low, jump sweepout or a rough water surface can result. On the other hand, the position or height of the baffle piers are not critical if the recommended proportions are followed. There exists a reasonable amount of leeway in all directions; however, one cannot place the baffle piers on the pool floor at random and expect anything like the excellent action otherwise associated with the Type III basin.

The baffle piers may be in the form shown in Figure 17, or they may be cubes; either shape is effective. The corners of the baffle blocks should not be rounded, as the edges are effective in pro-

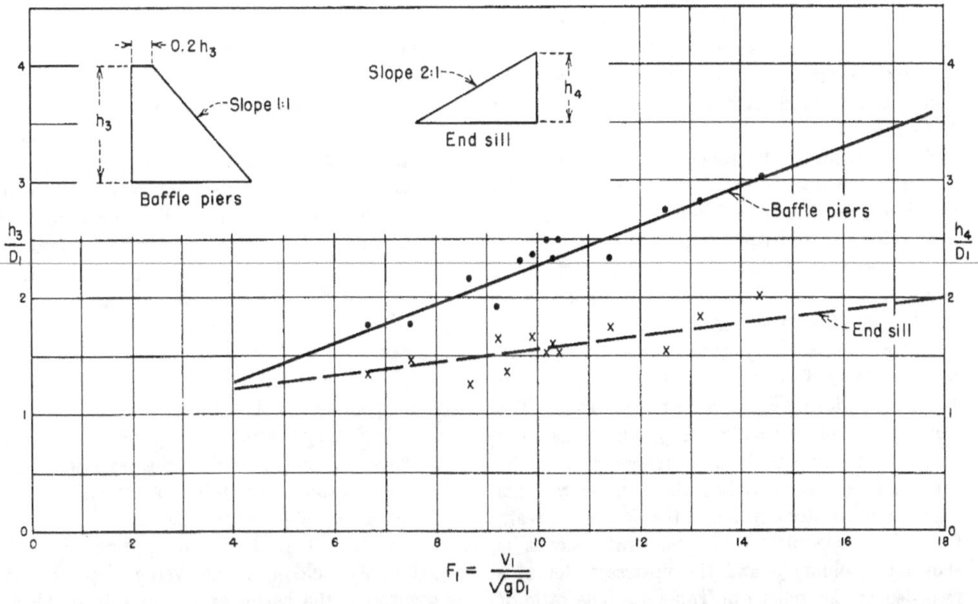

FIGURE 18.—*Height of baffle piers and end sill (Basin III).*

ducing eddies which in turn aid in the dissipation of energy. Small chamfers on the pier edges of the type used to obtain better forming of the concrete may be used.

End sill. The height of the solid end sill is also shown to vary with the Froude number, although there is nothing critical about this dimension. The heights of the sills used in the verification tests are shown in Columns 27 and 28 of Table 4. The height of the end sill in terms of D_1 is plotted with respect to the Froude number and shown as the lower line in Figure 18. A slope of 2:1 was used throughout the tests since previous sill experiments indicated that minimum wave heights and erosion could be expected with this slope.

Tail water depth. As in the case of Basin II, full conjugate depth, measured above the apron, is also recommended for Basin III. There are several reasons for this: First, the best operation for this stilling basin occurs at full conjugate tail water depth; second, if less than the conjugate depth is used, the surface velocities leaving the pool are high, the jump action is impaired, and there is greater chance for scour downstream; and third, if the baffle blocks erode with time, the additional tail water depth will serve to lengthen the interval between repairs. On the other hand, there is no hydraulic advantage in using greater than the conjugate depth, as the action in the pool will show little or no improvement. The same precautions should be considered when determining the tail water for Basin III that were discussed for Basin II.

The margin of safety for Basin III varies from 15 to 18 percent depending on the value of the Froude number, as can be observed by the dashed line labeled "Minimum Tail Water Depth—Basin III," in Figure 11. The points, from which the line was drawn, were obtained from the verification tests, Columns 10 and 14, Table 4. Again, this line does not represent complete jump sweep-out, but rather the tail water depth at which the front of the jump moves away from the chute blocks. In this position the jump is not fully developed and the stilling basin does not perform properly. In special cases it may be necessary to encroach on this wide margin of safety; however, it is not advisable as a general rule for the reasons stated above.

Length of basin. The length of Basin III, which is related to the Froude number, can be obtained by consulting the lower curve of Figure 12. The points, indicated by circles, were obtained from Columns 10 and 12, Table 4, and indicate the extent of the verification tests. The length is measured from the downstream end of the chute blocks to the downstream end of the end sill, Figure 17. Although this curve was determined conservatively, it will be found that the length of Basin III is less than one-half the length needed for a basin without appurtenances. Basin III, as was true of Basin II, may be effective for values of the Froude number as low as 4.5; thus the length curve was terminated at this value.

Water surface and pressure profiles. Approximate water-surface profiles were obtained for Basin III during the verification tests. The front of the jump was so steep, Figure 19, that only two measurements were necessary to define the water surface profile; these measurements were the tail water depth and the depth upstream from the baffle piers. The tail water depth is shown in Column 6 and the upstream depth is recorded in Column 29 of Table 4. The ratio of the upstream depth to conjugate depth is shown in Column 30. As can be observed, the ratio is much the same regardless of the value of the Froude number. The average of the ratios in Column 30 is 0.52. Thus it will be assumed that the depth upstream from the baffle blocks is one-half the tail water depth.

The profile represented by the crosshatched area, Figure 19, is for conjugate tail water depth. For a greater tail water depth, D_x, the upstream depth would be $\frac{D_x}{2}$. For a tail water depth less than conjugate, D_y, the upstream depth would be approximately $\frac{D_y}{2}$. There appears to be no particular significance in the fact that this ratio is one-half.

The information in Figure 19 applies only to Basin III, proportioned according to the rules set forth. It can be assumed that for all practical purposes the pressure and water-surface profiles are the same. There will be a localized increase in pressure on the apron immediately upstream from each baffle block, but this has been taken into account, more or less, by extending the diagram to full tail water depth beginning at the upstream face of the baffle blocks.

Recommendations

The following rules pertain to the design of the Type III basin, Figure 17:

1. The stilling basin operates best at full conjugate tail water depth, D_2. A reasonable factor of safety is inherent in the conjugate depth for all values of the Froude number (Fig. 11) and it is recommended that this margin of safety not be reduced.
2. The length of basin, which is less than one-half the length of the natural jump, can be obtained by consulting the curve for Basin III in Figure 12. As a reminder, an excess of tail water depth does not substitute for pool length or vice versa.
3. Stilling Basin III may be effective for values of the Froude number as low as 4.0, but this cannot be stated for certain (consult Sec. 4 for values under 4.5).
4. Height, width, and spacing of chute blocks should equal the average depth of flow entering the basin, or D_1. Width of blocks may be decreased, provided spacing is reduced a like amount. Should D_1 prove to be less than 8 inches, the blocks should be made 8 inches high.
5. The height of the baffle piers varies with the Froude number and is given in Figure 18. The blocks may be cubes or they may be constructed as shown in Figure 17; the upstream face should be vertical and in one plane. The vertical face is important. The width and spacing of baffle piers are also shown in Figure 17. In narrow structures where the specified width and spacing of blocks do not appear practical, block width and spacing may be reduced, provided both are reduced a like amount. A half space is recommended adjacent to the walls.
6. The upstream face of the baffle piers should be set at a distance of $0.8D_2$ from the downstream face of the chute blocks (Fig. 17). This dimension is also important.
7. The height of the solid sill at the end of the basin is given in Figure 18. The slope is 2:1 upward in the direction of flow.
8. It is undesirable to round or streamline the edges of the chute blocks, end sill, or baffle piers. Streamlining of baffle piers may

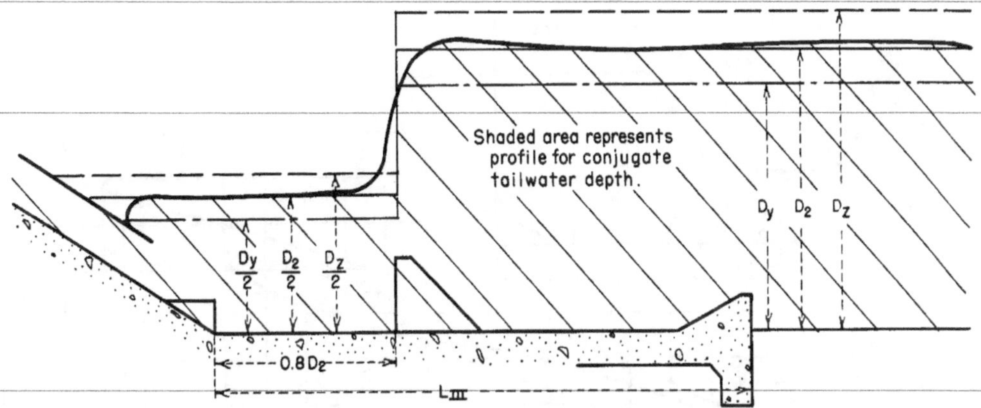

FIGURE 19.—*Approximate water surface and pressure profiles (Basin III).*

result in loss of half of their effectiveness. Small chamfers to prevent chipping of the edges may be used.

9. It is recommended that a radius of reasonable length ($R \geqq 4D_1$) be used at the intersection of the chute and basin apron for slopes of 45° or greater.

10. As a general rule, the slope of the chute has little effect on the jump unless long flat slopes are involved. This phase will be considered in Section 5 on sloping aprons.

Since Basin III is a short and compact structure, the above rules should be followed closely for its proportioning. If the proportioning is to be varied from that recommended, or if the limits given below are exceeded (as in the example below), a model study is advisable. Arbitrary limits for the Type III basin are set at 200 c.f.s. per foot of basin width and 50 to 60 feet per second entrance velocity until experience demonstrates otherwise.

Application of results (Example 3). Given the following computed values for a small overflow dam:

Q cfs	q cfs	V_1 ft/sec	D_1 ft
3, 900	78. 0	69	1. 130
3, 090	61. 8	66	. 936
2, 022	40. 45	63	. 642
662	13. 25	51	. 260

and the tail water curve for the river, identified by the solid line in Figure 20, proportion Basin III for the most adverse condition. The flow is symmetrical and the width of the basin is 50 feet. (The purpose of this example is to demonstrate the use of the jump elevation curve.)

The first step is to compute the jump elevation curve which in this case is D_2 plus the elevation of the basin floor. As V_1 and D_1 are given, the Froude number is computed and tabulated in Column 2, Table 5, below:

TABLE 5.—*Results of Example 3*

Q cfs	F_1	$\frac{D_2}{D_1}$	D_1 ft	D_2 ft	Jump elevation	
					Curve a	Curve a'
(1)	(2)	(3)	(4)	(4)	(6)	(7)
3, 900	11. 42	15. 75	1. 130	17. 80	617. 5	615. 0
3, 090	12. 02	16. 60	. 936	15. 54	615. 2	612. 7
2, 022	13. 85	19. 20	. 642	12. 33	612. 0	609. 5
662	17. 62	24. 5	. 260	6. 37	606. 1	603. 6

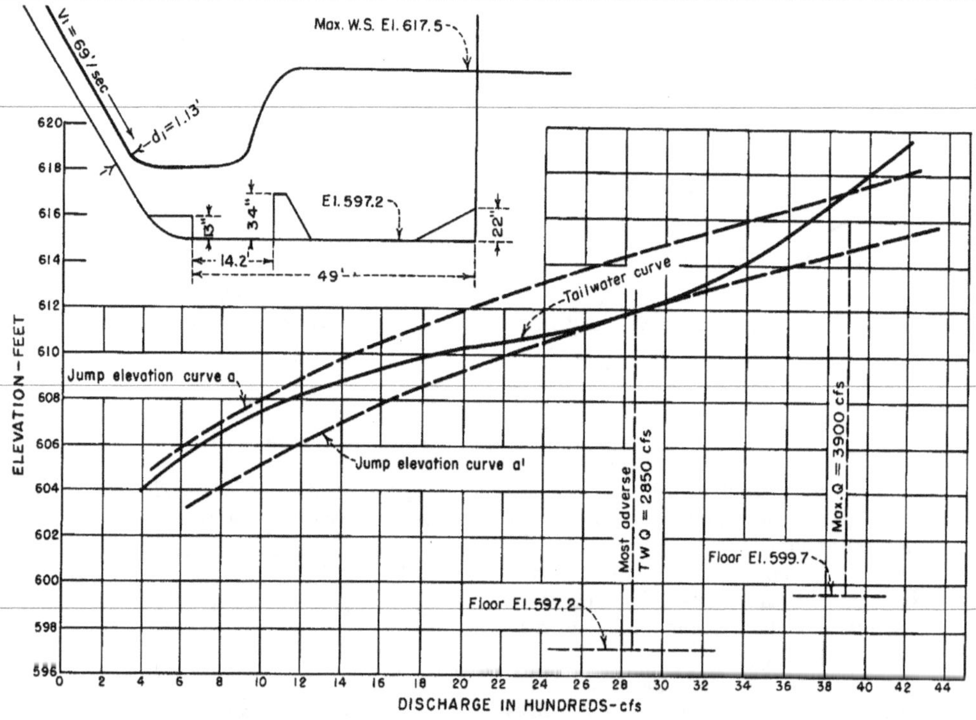

FIGURE 20.—*Tail water and jump elevation curve—Example 3 (Basin III).*

Entering Figure 11 with these values of the Froude number, values of $\frac{TW}{D_1}$ are obtained from the solid line. These values are also $\frac{D_2}{D_1}$ and are shown listed in Column 3 of Table 5. The conjugate depths for the various discharges, Column 5, were obtained by multiplying the values in Column 3 by those in Column 4.

If it is assumed that the most adverse operating condition occurs at the maximum discharge of 3,900 c.f.s., the stilling basin apron should be placed at elevation 617.5–17.8 or elevation 599.7.

With the apron at elevation 599.7, the tail water required for conjugate depth for each discharge would follow the elevations listed in Column 6. Plotting Columns 1 and 6 in Figure 20 results in Curve a, which shows that the tail water depth is inadequate for all but the maximum discharge.

The tail water curve is unusual in that the most adverse tail water condition occurs at a discharge of approximately 2,850 c.f.s. rather than maximum. As full conjugate depth is desired for the most adverse tail water condition, it is necessary to shift the jump elevation curve downward to match the tail water curve for a discharge of 2, 850 c.f.s. (see Curve a′, Fig. 20). The coordinates for Curve a′ are given in Columns 1 and 7, Table 5. This will place the basin floor 2.5 feet lower, or elevation 597.2 feet, as shown in the sketch in Figure 20.

Although the position of the basin floor was set for a discharge of 2,850 c.f.s., the remaining stilling basin details are proportioned for the maximum discharge 3,900 c.f.s.

Entering Figure 12 with a Froude number of 11.42,

$\frac{L_{III}}{D_2}=2.75$, and the length of basin required

$$L_{III}=2.75\times 17.80=48.95 \text{ feet.}$$

(Notice that conjugate depth was used, not tail water depth.)

The height, width, and spacing of chute blocks are equal to D_1 or 1.130 feet (use *13* or *14* inches).

The height of the baffle piers for a Froude number of 11.42 (Fig. 18) is $2.5D_1$.

$$h_3=2.5\times 1.130=2.825 \text{ feet}$$

(use *34* inches).

The width and spacing of the baffle piers are preferably three-fourths of the height or

$$0.75\times 34=25.5 \text{ inches.}$$

From Figure 17, the upstream face of the baffle piers should be $0.8D_2$ from the downstream face of the chute blocks, or

$$0.8\times 17.80=14.24 \text{ feet.}$$

The height of the solid end sill, Figure 18, is $1.60D_1$, or

$$h_4=1.60\times 1.130=1.81 \text{ feet}$$

(use *22* inches).

The final dimensions of the basin are shown in Figure 20.

Section 4

Stilling basin design and wave suppressors for canal structures, outlet works, and diversion dams (Basin IV)

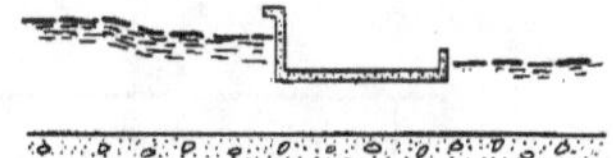

This section concerns the characteristics of the hydraulic jump for Froude numbers between 2.5 and 4.5 and the design of an adequate stilling basin, designated as Basin IV. The low Froude number range is encountered principally in the design of canal structures, but occasionally low dams and outlet works fall in this category. In the 2.5 to 4.5 Froude number range, Figure 9B, the jump is not fully developed and the previously discussed methods of design do not apply. The main problem concerns the waves created in the unstable hydraulic jump, making the design of a suitable wave suppressor a part of the stilling basin problem.

Four means of reducing wave heights are discussed. The first is an integral part of the stilling basin design and should be used only in the 2.5 to 4.5 Froude number range. The second may be considered to be an alternate design and may be used over a greater range of Froude numbers. These types are discussed as a part of the stilling basin design. The third and fourth devices are considered as appurtenances which may be included in an original design or added to an existing structure. Also, they may be used in any open channel flow-way without consideration of the Froude number. These latter devices are described under the heading Wave Suppressors.

Jump Characteristics—Froude Number 2.5 to 4.5

For low values of the Froude number, 2.5 to 4.5, the entering jet oscillates intermittently from bottom to surface, as indicated in Figure 9B, with no particular period. Each oscillation generates a wave which is difficult to dampen. In narrow structures, such as canals, waves may persist to some degree for miles. As they encounter obstructions in the canal, such as bridge piers, turnouts, checks, and transitions, reflected waves may

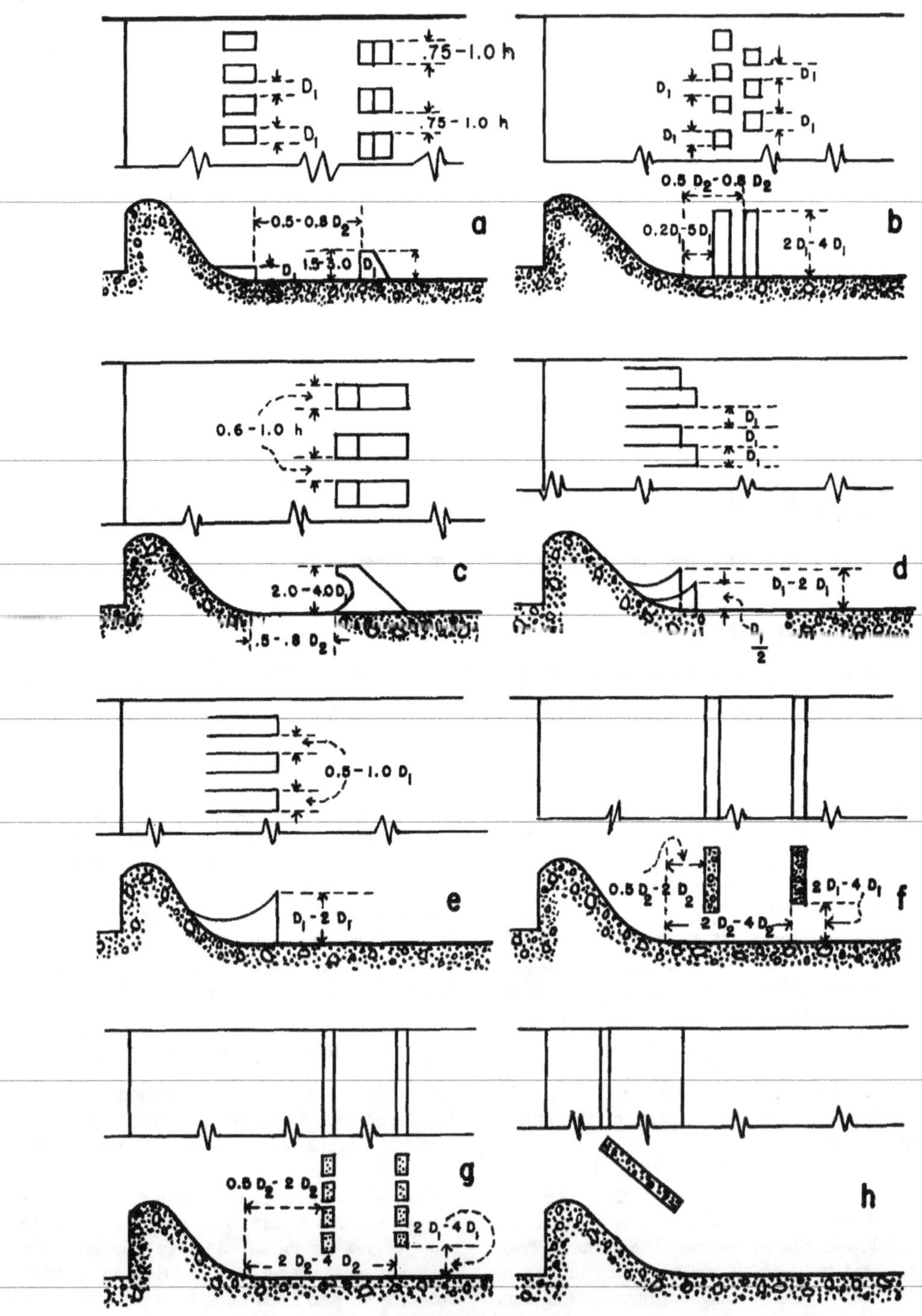

FIGURE 21.—*Record of appurtenances (Basin IV).*

be generated which tend to dampen, modify, or intensify the original wave. Waves are destructive to earth-lined canals and riprap and produce undesirable surges at gaging stations and in measuring devices. Structures in this range of Froude numbers are the ones which have been found to require the most maintenance.

On wide structures, such as diversion dams, wave action is not as pronounced when the waves can travel laterally as well as parallel to the direction of flow. The combined action produces some dampening effect but also results in a choppy water surface. These waves may or may not be dissipated in a short distance. Where outlet works operating under heads of 50 feet or greater fall within the range of Froude numbers between 2.5 and 4.5, a model study of the stilling basin is imperative. A model study is the only means of including preventive or corrective devices in the structure to assure proper performance.

Stilling Basin Design—Froude Number 2.5 to 4.5

Development tests. The best way to combat a wave problem is to eliminate the wave at its source by altering the condition which generates the wave. For the stilling basin preceded by an overfall or chute, two schemes were apparent for eliminating waves at their source. The first was to break up the entering jet by opposing it with directional jets deflected from baffle piers or sills. The second was to bolster or intensify the roller, shown in the upper portion of Figure 9B, by directional jets deflected from large chute blocks.

The first method was unsuccessful in that the number and size of appurtenances necessary to break up the jet occupied so much volume that the devices themselves posed an obstruction to the flow. This conclusion was based on tests in which various shaped baffle and guide blocks were systematically placed in a stilling basin in combination with numerous types of spreader teeth and deflectors in the chute. The program involved dozens of tests, and not until all possible ideas were tried was this approach abandoned. A few of the basic ideas tested are shown in Figure 21, a, b, c, f, g, and h.

Final Tests

Deflector blocks. The second approach, that of attempting to intensify the roller, yielded better results. Large blocks, similar to but larger than chute blocks, were placed on the chute; no changes were made in the stilling basin proper. The object was to direct a jet into the base of the roller in an attempt to strengthen it and thereby stabilize the jump. After a number of trials, using blocks with a curved top, the roller was actually intensified and the jump action was improved. Sketches d and e in Figure 21 indicate the only schemes that showed promise, although many variations were tried. Approximations of these curved top blocks were then tested to make the field construction as simple as possible. The dimensions and proportions of the adopted deflector blocks are shown in Figure 22.

The tests showed that it was desirable to place as few appurtenances as possible in the path of the flow, as volume occupied by appurtenances helps to create a backwater problem, thus requiring higher training walls. Also, random placement of blocks is apt to create a new wave problem in addition to the original problem. The number of deflector blocks shown in Figure 22 is a minimum requirement to accomplish the purpose set forth. The width of the blocks is shown equal to D_1 and this is the maximum width recommended. From a hydraulic standpoint it is desirable that the blocks be constructed narrower than indicated, preferably $0.75D_1$. The ratio of block width to spacing should be maintained as 1:2.5. The extreme

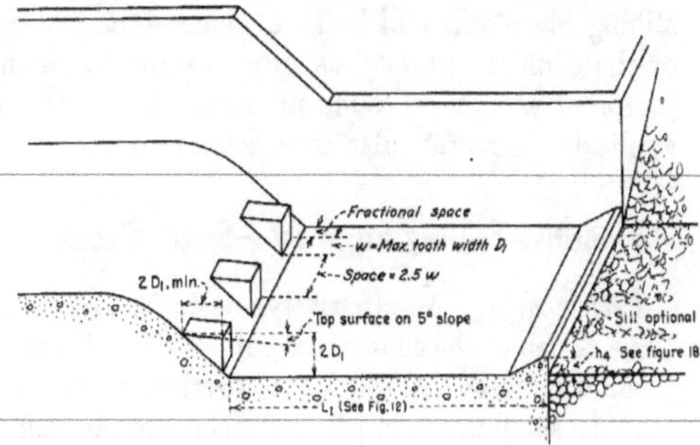

FIGURE 22.—*Proportions for Froude numbers 2.5 to 4.5 (Basin IV).*

tops of the blocks are $2D_1$ above the floor of the stilling basin. The blocks may appear to be rather high and, in some cases, extremely long, but this is essential as the jet leaving the top of the blocks must play at the base of the roller to be effective. To accommodate the various slopes of chutes and ogee shapes encountered, the horizontal top length of the blocks should be at least $2D_1$. The upper surface of each block is sloped at 5° in a downstream direction as it was found that this feature resulted in better operation, especially for discharges lower than the design flow.

Tail water depth. A tail water depth 5 to 10 percent greater than the conjugate depth is strongly recommended for Basin IV. Since the jump is very sensitive to tail water depth at these low values of the Froude number, a slight deficiency in tail water depth may allow the jump to sweep completely out of the basin. The jump performs much better and wave action is diminished if the tail water depth is increased to approximately $1.1D_2$.

Basin length and end sill. The length of Basin IV, which is relatively short, can be obtained from the upper curve in Figure 12. No baffle piers are needed in the basin, as these will prove a greater detriment than aid. The addition of a small triangular sill placed at the end of the apron for scour control is desirable. An end sill of the type used on Basin III is satisfactory, Figure 18.

Performance. If designed for the maximum discharge, Basin IV will perform satisfactorily for lesser flows. Waves downstream from the stilling basin will still be in evidence but will be of the ordinary variety usually encountered with jumps of a higher Froude number. Basin IV is applicable to rectangular cross sections only.

Alternative Stilling Basin IV—Small Drops

Performance. An alternative basin for reducing wave action at the source, for values of the Froude number between 2.5 and 4.5, is particularly applicable to small drops in canals. The Froude number in this case is computed for flow at the top of the drop rather than at the bottom and should be about 0.5. A series of steel rails, channel irons, or timbers in the form of a grizzly are installed at the drop, as shown in Figure 23. The overfalling jet is separated into

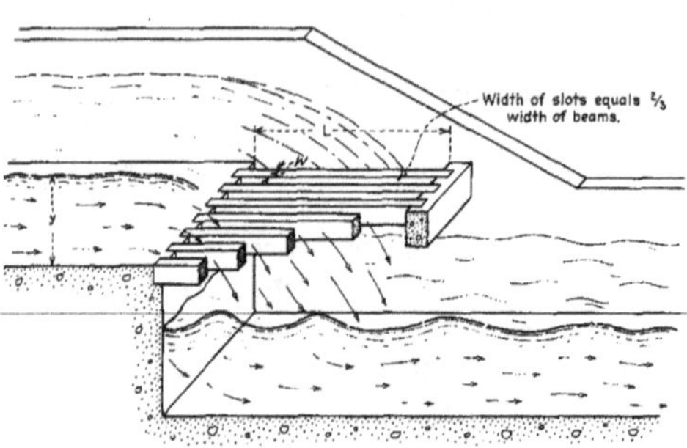

FIGURE 23.—*Drop-type energy dissipator for Froude numbers 2.5 to 4.5 (Alternative Basin IV).*

a number of long, thin sheets of water which fall nearly vertically into the canal below. Energy dissipation is excellent and the usual wave problem is avoided. If the rails are tilted downward at an angle of 3° or more, the grid is self-cleaning.

The use of this device is particularly justified when the Froude number is below 3.0. If use of a jump were possible the maximum energy loss would be less than 27 percent, as indicated in Figure 8. The suggested device accomplishes nearly as much energy loss and provides a smooth water surface in addition.

Design. Two spacing arrangements of the beams were tested in the laboratory: in the first, the spacing was equal to the width of the beams; in the second, the spacing was two-thirds of the beam width. The latter was the more effective. In the first, the length of beams required was about 2.9 times the depth of flow (y) in the canal upstream; in the second, it was necessary to increase the length to approximately 3.6y. The following expression can be used for computing the length of beams:

$$L=\frac{Q}{CSN\sqrt{2gy}} \tag{4}$$

where Q is the total discharge in c.f.s., C is an experimental coefficient, S is the width of a space in feet, N is the number of spaces, g is the acceleration of gravity, and y is the depth of flow in the canal upstream (see Fig. 23). The value of C for the two arrangements tested was 0.245.

Should it be desired to maintain a certain level in the canal upstream, the grid may be made adjustable and tilted upward to act as a check; however, this arrangement may introduce a cleaning problem.

Wave Suppressors

The two stilling basins described above may be considered to be wave suppressors, although the suppressor effect is obtained from the necessary features of the stilling basin. If greater wave reduction is required on a proposed structure, or if a wave suppressor is required to be added to an existing flow-way, the two types discussed below may prove useful. Both are applicable to most open channel flow-ways having rectangular, trapezoidal, or other cross-sectional shapes. The first or raft type may prove more economical than the second or underpass type, but rafts do not provide the degree of wave reduction obtainable with the underpass type. Both types may be used without regard to the Froude number.

Raft-type wave suppressor. In a structure of the type shown in Figure 24, there are no means for eliminating waves at their source. Tests showed that appurtenances in the stilling basin merely produced severe splashing and created a backwater effect, resulting in submerged flow at the gate for the larger flows. Submerged flow reduced the effective head on the structure, and in turn, the capacity. Tests on several suggested devices showed that rafts provided the best answer to the wave problem when additional submergence could not be tolerated. The general arrangement of the tested structure is shown in Figure 24. The Froude number varied from 3 to 7, depending on the head behind the gate and the gate opening. Velocities in the canal ranged from 5 to 10 feet per second. Waves were 1.5 feet high, measured from trough to crest.

During the course of the experiments a number of rafts were tested—thick rafts with longitudinal slots, thin rafts made of perforated steel plate,

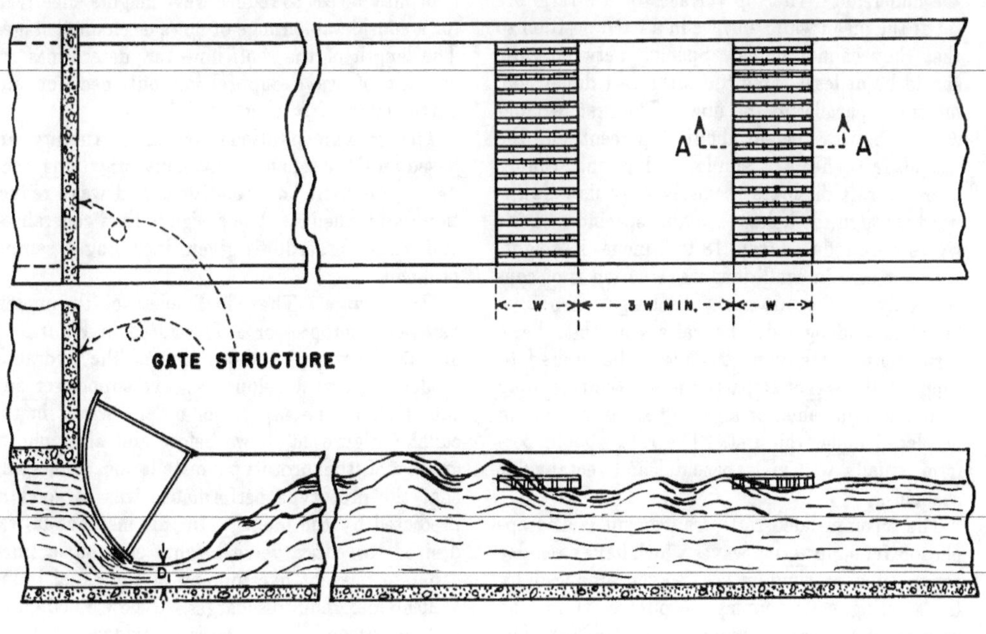

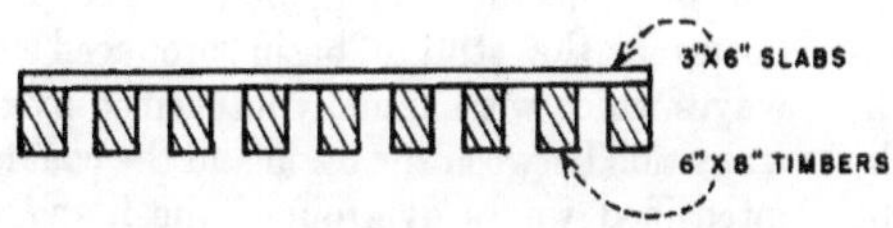

FIGURE 24.—*Raft wave suppressor (Type IV) for Froude numbers 2.5 to 4.5.*

and others, both floating and fixed. Rigid and articulated rafts were tested in various arrangements.

The most effective raft arrangement consisted of two rigid stationary rafts 20 feet long by 8 feet wide, made from 6- by 8-inch timbers, placed in the canal down-stream from the stilling basin, Figure 24. A space was left between timbers and lighter crosspieces were placed on the rafts parallel to the flow, giving the appearance of many rectangular holes. Several essential requirements for the raft were apparent: (1) that the rafts be perforated in a regular pattern; (2) that there be some depth to these holes; (3) that at least two rafts be used; and (4) that the rafts be rigid and held stationary.

It was found that the ratio of hole area to total area of raft could be from 1:6 to 1:8. The 8-foot width, W, in Figure 24, is a minimum dimension. The rafts must have sufficient thickness so that the troughs of the waves do not break free from the underside. The top surfaces of the rafts are set at the mean water surface in a fixed position so that they cannot move. Spacing between rafts should be at least three times the raft dimension, measured parallel to the flow. The first raft decreases the wave height about 50 percent, and the second raft effects a similar reduction. Surges over the raft dissipate themselves by flow downward through the holes. For this specific case the waves were reduced from 18 to 3 inches in height.

Under certain conditions wave action is of concern only at the maximum discharge when freeboard is endangered; the rafts can then be a permanent installation. Should it be desired to suppress the waves at partial flows, the rafts may be made adjustable, or a second set of rafts may be placed under the first. The rafts should perform equally well in trapezoidal and rectangular channels.

The recommended raft arrangement is also applicable for suppressing waves which have a regular period such as wind waves, waves produced by the starting and stopping of pumps, etc. The position of the down-stream raft is then very important. The second raft should be positioned downstream at some fraction of the wave length. Placing it at a full wave length could cause both rafts to be ineffective. Thus, for narrow canals it may be advisable to make the second raft portable. However, if it becomes necessary to make the rafts adjustable or portable, or if a moderate increase in depth in the stilling basin can be tolerated, consideration should be given to the type of wave suppressor discussed below.

Underpass-Type Wave Suppressor

General description. By far the most effective wave dissipator is the short-tube type of underpass suppressor. The name "short-tube" is used because the structure has many of the characteristics of the short-tube discussed in hydraulic textbooks. This wave suppressor may be added to an existing structure or included in the original construction. In either case it provides a sightly structure, which is economical to construct and effective in operation.

Essentially, the structure consists of a horizontal roof placed in the flow channel with a headwall sufficiently high to cause all flow to pass beneath the roof. The height of the roof above the channel floor may be set to reduce wave heights effectively for a considerable range of flows or channel stages. The length of the roof, however, determines the amount of wave suppression obtained for any particular roof setting.

The recommendations for this structure are based on three separate model investigations, each having different flow conditions and wave reduction requirements. The design is then generalized and design procedures given, including a sample problem.

Performance. The effectiveness of the underpass wave suppressor is illustrated in Figures 25 and 26. Figure 25 shows one of the hydraulic models used to develop the wave suppressor and the effect of the suppressor on the waves in the canal. Figure 26 shows before and after photographs of the prototype installation, indicating that the prototype performance was as good as predicted by the model. In this instance it was desired to reduce wave heights entering a lined canal to prevent overtopping of the canal lining at near maximum discharges. Below 3,000 cubic feet per second, waves were in evidence but did not overtop the lining. For larger discharges, however, the stilling basin produced moderate waves which were actually intensified by the short transition between the basin and the canal. These intensified waves overtopped the lining at 4,000 cubic feet per second and became a serious prob-

Without suppressor—waves overtop canal.

Suppressor in place—length 1.3 D_2, submerged 30 percent.

1:32 scale model.

Discharge 5,000 c.f.s.

FIGURE 25.—*Performance of underpass wave suppressor.*

Q=3,900 c.f.s.—before wave suppressor was installed.

Q=3,900 c.f.s.—after wave suppressor was installed.

FIGURE 26.—*Hydraulic performance of wave suppressor for Friant-Kern Canal.*

lem at 4,500 cubic feet per second. Tests were made with a suppressor 21 feet long using discharges from 2,000 to 5,000 c.f.s. The suppressor was located downstream from the stilling basin.

Figure 27, Test 1, shows the results of tests to determine the optimum opening between the roof and the channel floor using the maximum discharge, 5,000 c.f.s. With a 14-foot opening, waves were reduced from about 8 feet to about 3 feet. Waves were reduced to less than 2 feet with an opening of 11 feet. Smaller openings produced less wave height reduction because of the turbulence created at the underpass exit. Thus, it may be seen that an opening of from 10 to 12 feet produced optimum results.

With the opening set at 11 feet, the suppressor effect was then determined for other discharges. These results are shown in Figure 27, Test 2.

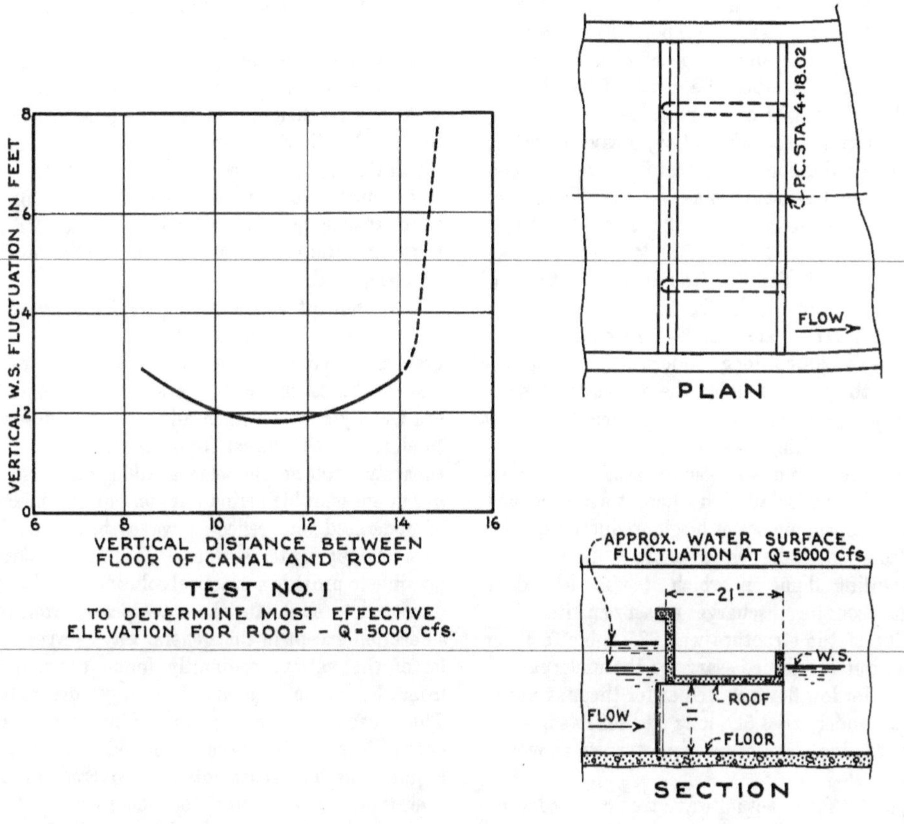

VERTICAL W.S. FLUCTUATION IN FEET

0 2 4 6 8

6 8 10 12 14 16

VERTICAL DISTANCE BETWEEN FLOOR OF CANAL AND ROOF

TEST NO. 1

TO DETERMINE MOST EFFECTIVE ELEVATION FOR ROOF - Q=5000 cfs.

P.C. STA. 4+18.02

FLOW

PLAN

APPROX. WATER SURFACE FLUCTUATION AT Q=5000 cfs

21'

W.S.

ROOF

FLOW

FLOOR

SECTION

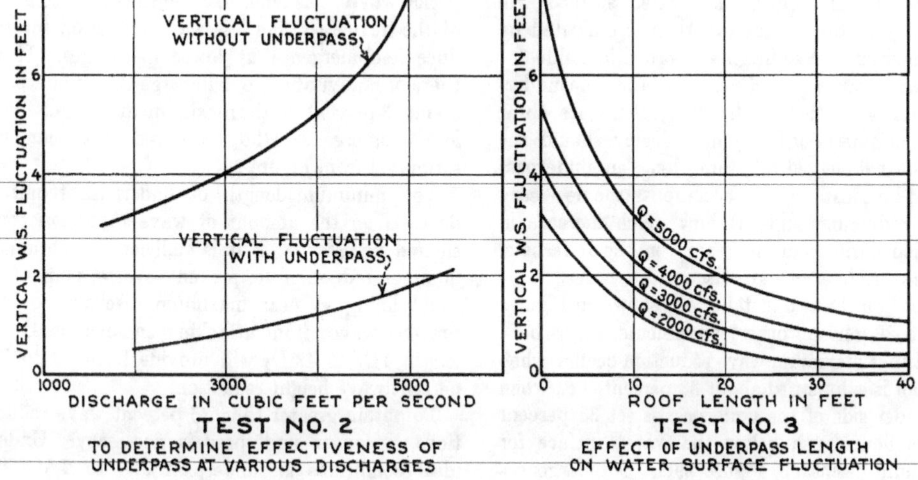

FIGURE 27.—*Wave suppressor for Friant-Kern Canal—results of hydraulic model tests.*

Wave height reduction was about 78 percent at 5,000 c.f.s., increasing to about 84 percent at 2,000 c.f.s. The device became ineffective at about 1,500 c.f.s. when the depth of flow became less than the height of the roof.

To determine the effect of suppressor length on the wave reduction, other factors were held constant while the length was varied. Tests were made on suppressors 10, 21, 30, and 40 feet long for discharges of 2,000, 3,000, 4,000, and 5,000 c.f.s., Figure 27, Test 3. Roof lengths in terms of the downstream depth, D_2, for 5,000 c.f.s. were $0.62D_2$, $1.31D_2$, and $2.5D_2$, respectively. In terms of a 20-foot-long underpass, halving the roof length almost doubled the downstream wave height and doubling the 20-foot length almost halved the resulting wave height.

The same type of wave suppressor was successfully used in an installation where it was necessary to obtain optimum wave height reductions, since flow from the underpass discharged directly into a measuring flume in which it was desired to obtain accurate discharge measurements. The capacity of the structure was 625 cubic feet per second, but it was necessary for the underpass to function for low flows as well as for the maximum. With an underpass $3.5D_2$ long and set as shown in Figure 28, the wave reductions were as shown in Table 6.

Figure 28 shows actual wave traces recorded by an oscillograph. Here it may be seen that the maximum wave height, measured from minimum trough to maximum crest, did not occur on successive waves. Thus, the water surface will appear smoother to the eye than is indicated by the maximum wave heights recorded in Table 6.

General design procedure. To design an underpass for a particular structure, there are three main considerations: (1) how deeply should the roof be submerged, (2) how long an underpass should be constructed to accomplish the necessary wave reduction, and (3) how much increase in flow depth will occur upstream from the underpass. These considerations are discussed in order.

Based on the two installations shown in Figures 27 and 28, and on other experiments, it has been found that maximum wave reduction occurs when the roof is submerged about 33 percent; i.e., when the under side of the underpass is set 33 percent of the flow depth below the water surface for maximum discharge, Figure 29C. Submergences greater than 33 percent produced undesirable turbulence at the underpass outlet resulting in less overall wave reduction. With the usual tail water curve, submergence and the percent reduction in wave height will become less, in general, for smaller-than-maximum discharges. This is illustrated by the upper curve in Figure 29C. The lower curve shows a near constant value for less submergence because the wave heights for less than maximum discharge were smaller and of shorter period.

It is known that the wave period greatly affects the performance of a given underpass. The greatest wave reduction occurs for short period waves. Since the wave periods to be expected are usually not known in advance, it is desirable to eliminate this factor from consideration. Fortunately, wave action below a stilling basin usually has no measurable period but consists of a mixture of generated and reflected waves best described as a choppy water surface. This fact makes it possible to provide a practical solution from limited data and to eliminate the wave period from consideration except in this general way: waves must be of the variety ordinarily found downstream from hydraulic jumps or energy dissipators. These usually have a period of not more than about 5 seconds. Longer period waves may require special treatment not covered in this discussion. Fortunately, too, there generally is a tendency for the wave period to become less with decreasing discharge. Since the suppressor provides a greater percentage reduction on shorter period waves, this tends to offset the characteristics of the device to give less wave reduction for reduced submergence at lower discharges. It is therefore advisable to submerge the underpass about 33 percent for the maximum discharge. For less submergence, the wave reduction can be estimated from Figure 29C.

The minimum length of underpass required depends on the amount of wave reduction considered necessary. If it is sufficient to obtain a nominal reduction to prevent overtopping of a canal lining at near maximum discharge or to prevent waves from attacking channel banks, a length $1D_2$ to $1.5D_2$ will provide from 60 to 75 percent wave height reduction.

To obtain greater than 75 percent wave reductions, a longer underpass is necessary. Under ideal conditions an underpass $2D_2$ to $2.5D_2$ in

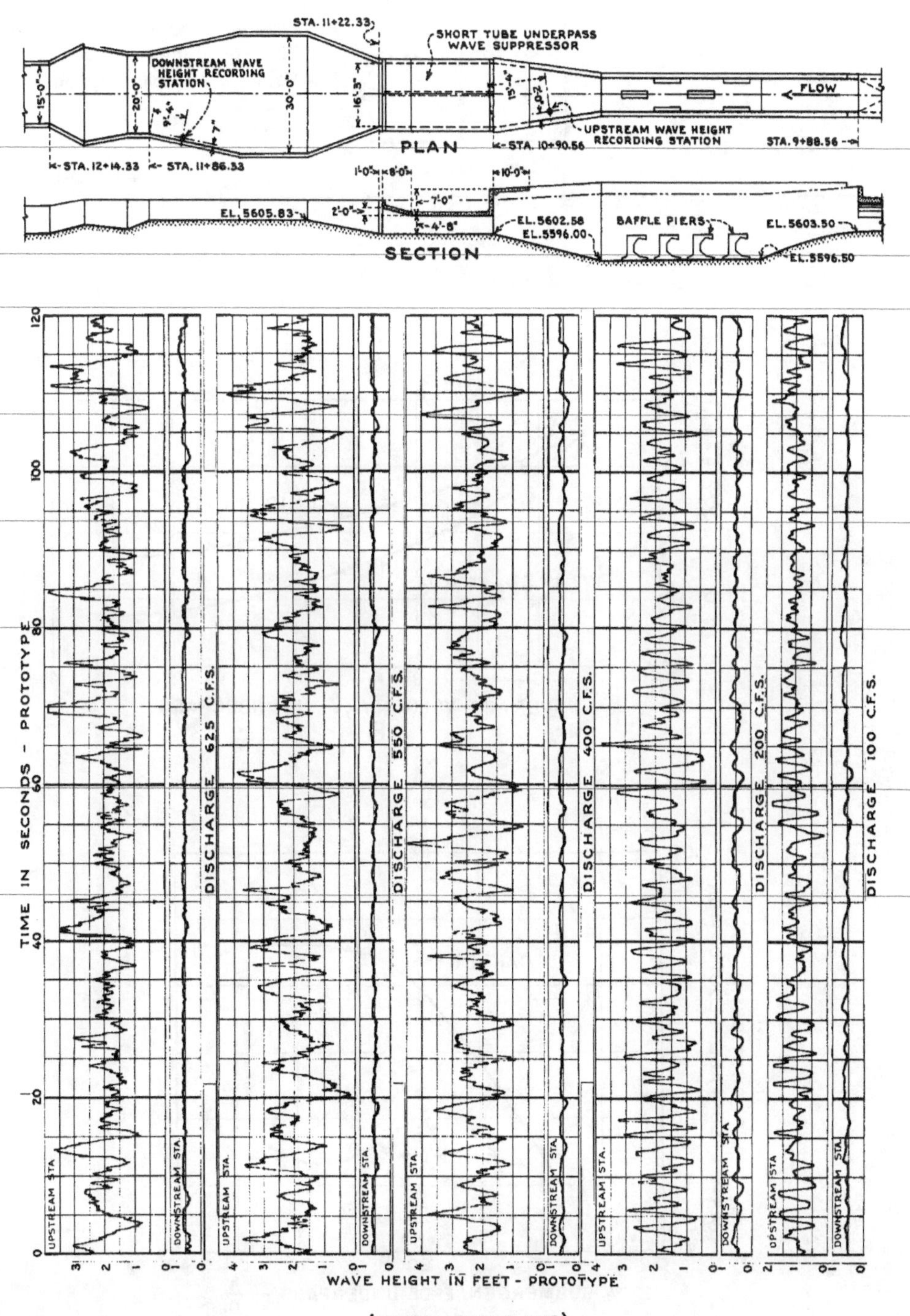

FIGURE 28.—*Wave height records for Carter Lake Dam No. 1 outlet works.*

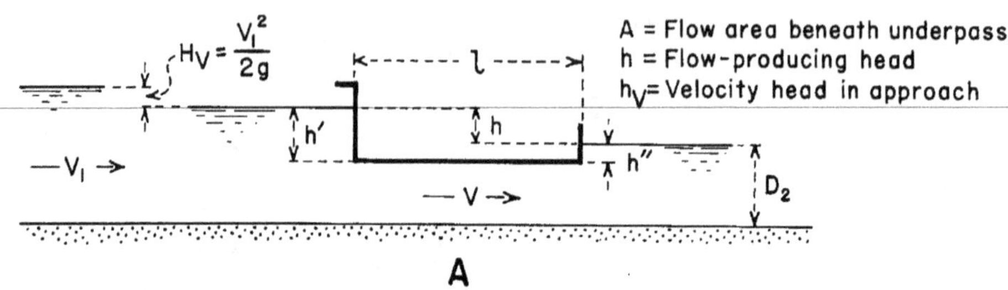

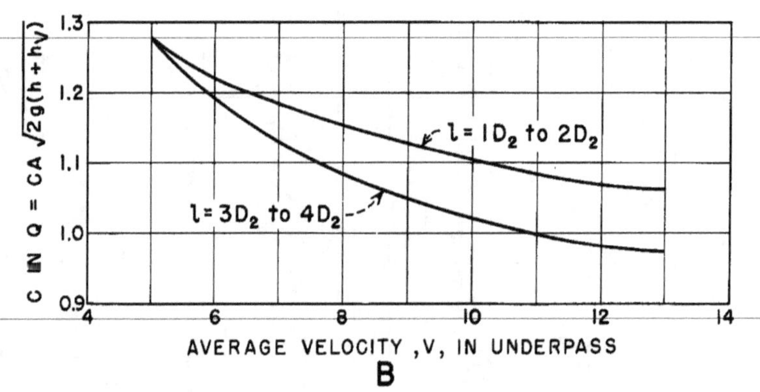

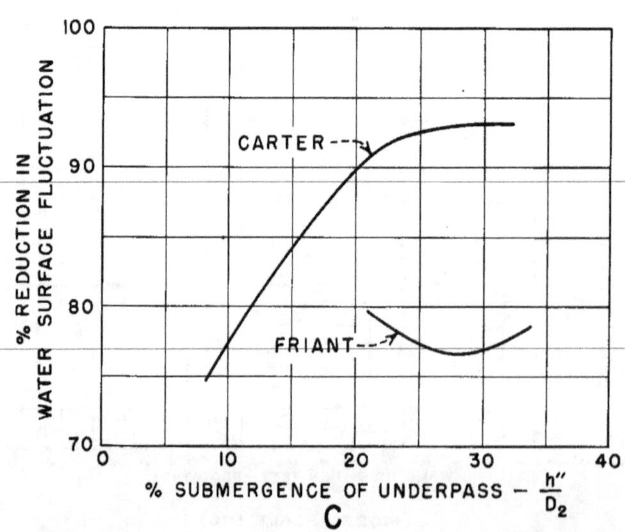

FIGURE 29.—*Hydraulic characteristics of underpass wave suppressor.*

TABLE 6.—*Wave heights in feet—prototype.*

Discharge in c.f.s.	625		550		400		200		100	
	Upstream [1]	Downstream [1]	U	D	U	D	U	D	U	D
Wave heights in feet	3.8 plus [2]	0. 3	4. 2	0. 3	4. 5	0. 4	3. 6	0. 4	1. 7	0. 3

[1] Upstream station is at end of stilling basin. Downstream station is in measuring flume.
[2] Recorder pen reached limit of travel in this test only.

length may provide up to 88 percent wave reduction for wave periods up to about 5 seconds. Ideal conditions include a velocity beneath the underpass of less than, say, 10 feet per second and a length of channel 3 to 4 times the length of the underpass downstream from the underpass which may be used as a quieting pool to still the turbulence created at the underpass exit.

Wave height reduction up to about 93 percent may be obtained by using an underpass 3.5 D_2 to $4D_2$ long. Included in this length is a 4:1 sloping roof extending from the underpass roof elevation to the tail water surface. The sloping portion should not exceed about one-quarter of the total length of underpass. Since slopes greater than 4:1 do not provide the desired draft tube action they should not be used. Slopes flatter than 4:1 provide better draft tube action and are therefore desirable.

Since the greatest wave reduction occurs in the first D_2 of underpass length, it may appear advantageous to construct two short underpasses rather than one long one. In the one case tested, two underpasses each $1D_2$ long, with a length $5D_2$ between them, gave an added 10-percent wave reduction advantage over one underpass $2D_2$ long. However, the extra cost of another headwall should be considered.

Table 7 summarizes the amount of wave reduction obtainable for various underpass lengths.

TABLE 7.—*Effect of underpass length on wave reduction*

[For underpass submergence 33 percent and maximum velocity less than 14 ft. per second]

Underpass length	Percent wave reduction [1]
$1D_2$ to $1.5D_2$	60 to 75.
$2D_2$ to $2.5D_2$	80 to 88.
3.5 to $4.0D_2$	90 to 93.[2]

[1] For wave periods up to 5 seconds.
[2] Upper limit only with draft tube type outlet.

To determine the backwater effect of placing the underpass in the channel, Figure 29B will prove helpful. Data from four different underpasses were used to obtain the two curves shown. Although the test points from which the curves were drawn showed minor inconsistencies, probably because factors other than those considered also affected the depth of water upstream from the underpass, the submitted curves are sufficiently accurate for design purposes. Figure 29B shows two curves of the discharge coefficient "C" versus average velocity beneath the underpass, one for underpass lengths of $1D_2$ to $2D_2$ and the other for lengths $3D_2$ to $4D_2$. Intermediate values may be interpolated although accuracy of this order is not usually required.

Pressures on the underpass were measured by means of piezometers to determine the direction and magnitude of the forces acting. Average pressures on the headwall were found to be distributed in a straightline variation from zero at the water surface to static pressure at the bottom. Pressures along the underside of the roof were found to be 1 to 2 feet below atmospheric; for design purposes they may be considered to be atmospheric. Pressures on the downstream vertical wall were equal to static pressures. In other words, there is only a slight tendency (except for the force of breaking waves which was not measured) to move the underpass downstream, and there is a slight resultant force tending to hold the underpass down.

Sample problem, Example 4. To illustrate the use of the preceding data in designing an underpass, a sample problem will be helpful.

A rectangular channel 30 feet wide and 14 feet deep flows 10 feet deep at maximum discharge, 2,400 c.f.s. It is estimated that waves will be 5 feet high and of the ordinary variety having a period less than 5 seconds. It is desired to reduce the height of the waves to approximately 1 foot at maximum discharge by installing an underpass-type wave suppressor without increasing the depth of water upstream from the underpass more than 15 inches.

To obtain maximum wave reduction at maximum discharge, the underpass should be submerged 33 percent. Therefore, the depth beneath the underpass is 6.67 feet with a corresponding velocity of 12 ft. per sec., $\left(V=\frac{2{,}400}{30\times 6.67}\right)$. To reduce the height of the waves from 5 feet to 1 foot, an 80-percent reduction in wave height is indicated, and, from Table 7, requires an underpass approximately $2D_2$ in length.

From Figure 29B, C=1.07 for $2D_2$ and a velocity of 12 ft. per sec.

From the equation given in Figure 29B:

$$\text{Total head, } h+h_v=\left(\frac{Q}{CA\sqrt{2g}}\right)^2$$

$$=\left(\frac{2{,}400}{8.02\times 1.07\times 200}\right)^2=1.95 \text{ feet}$$

$h+h_v$ is the total head required to pass the flow, and h represents the backwater effect of increase in depth of water upstream from the underpass. The determination of values for h and h_v is done by trial and error. As a first determination, assume that $h+h_v$ represents the increase in head.

$$\text{Then, channel approach velocity, } V_1=\frac{Q}{A}$$

$$=\frac{2{,}400}{(10+1.95)30}=6.7 \text{ ft. per sec.}$$

$$h_v=\frac{(V_1)^2}{2g}=\frac{(6.7)^2}{64.4}=0.70 \text{ foot}$$

and $h=1.95-0.70=1.25$ feet.

To refine the calculation, the above computation is repeated using the new head

$$V_1=\frac{2{,}400}{(10+1.25)30}=7.1 \text{ ft. per sec.}$$

$h_v=0.72$ foot and $h=1.17$ feet.

Further refinement is unnecessary.

Thus, the average water surface upstream from the underpass is 1.2 feet higher than the tail water which satisfies the assumed design requirement of a maximum backwater of 15 inches. The length of the underpass is $2D_2$ or 20 feet, and the waves are reduced 80 percent to a maximum height of approximately 1 foot.

If it is desired to reduce the wave heights still further, a longer underpass is required. Using Table 7 and Figure 29B as in the above problem, an underpass 3.5 to $4.0D_2$ or 35 to 40 feet in length reduces the waves 90 to 93 percent, making the downstream waves approximately 0.5 foot high and creating a backwater, h, of 1.61 feet.

In providing freeboard for the computed backwater, h, allowance should be made for waves and surges which, in effect, are above the computed water surface. One-half the wave height or more, measured from crest to trough, should be allowed above the computed surface. Full wave height would provide a more conservative design for the usual short period waves encountered in flow channels.

The headwall of the underpass should be extended to this same height and an overhang, Figure 29A, should be placed at the top to turn wave spray back into the basin. An alternative method would be to place a cover, say $2D_2$ long, upstream from the underpass headwall.

To insure obtaining the maximum wave reduction for a given length of underpass, a 4:1 sloping roof should be provided at the downstream end of the underpass, as indicated in Figure 28. This slope may be considered as part of the overall length. The sloping roof will help reduce the maximum wave height and will also reduce the frequency with which it occurs, providing in all respects a better appearing water surface. If the flow entering the underpass contains entrained air in the form of rising air bubbles, a few small vents in the underpass roof will relieve the possibility of air spurts and resulting surface turbulence at the underpass exit.

If the underpass is to be used downstream from a stilling basin the underpass must be placed sufficiently downstream to prevent turbulent flow, such as occurs at the end of a basin, from entering and passing through the wave suppressor. In highly turbulent flow the underpass is only partly effective.

A close inspection of the submitted data will reveal that slightly better results were obtained in the tests than are claimed in the example. This was done to illustrate the degree of conservatism required, since it should be understood that the problem of wave reduction can be very complex if unusual conditions prevail.

Section 5

Stilling basin with sloping apron (Basin V)

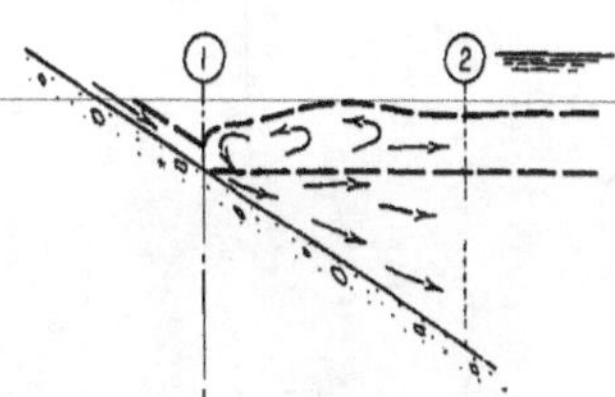

Much has been said concerning the advantages and disadvantages of stilling basins with sloping aprons. Previously there were not sufficient supporting data available from which to draw conclusions. In this study, therefore, the sloping apron was investigated sufficiently to answer many of the debatable questions and also to provide more definite design data.

Four flumes, A, B, D, and F, Figures 1, 2, and 3, were used to obtain the range of Froude numbers desired for the tests. In Flumes A, B, and D, floors were installed to the slope desired; Flume F could be tilted to obtain slopes from 0° to 12°. The slope in this discussion is the tangent of the angle between the floor and the horizontal and is designated as "tan ϕ." Five principal measurements were made in these tests, namely: the discharge, the average depth of flow entering the jump, the length of the jump, the tail water depth, and the slope of the apron. The tail water was adjusted so that the front of the jump formed either at the intersection of the spillway face and the sloping apron, or, in the case of the tilting flume, at a selected point.

The jump on a sloping apron takes many forms depending on the slope and arrangement of the apron, the value of the Froude number, and the concentration of flow (discharge per foot of width), but the dissipation is as effective as occurs in the true hydraulic jump on a horizontal apron.

Previous Experimental Work

Previous experimental work on the sloping apron has been carried on by several experimenters. In 1934, the late C. L. Yarnell of the U.S. Department of Agriculture supervised a series of experiments on the hydraulic jump on sloping aprons. Carl Kindsvater (*5*) later compiled these data and presented a rather complete picture, both experimentally and theoretically, for one slope, namely: 1:6 (tan ϕ=0.167). G. H. Hickox (*5*) presented data for a series of experiments on a slope of 1:3 (tan ϕ=0.333). Bakhmeteff (*1*) and Matzke (*6*) performed experiments on slopes of 0 to 0.07 in a flume 6 inches wide.

From an academic standpoint, the jump may occur in several ways on a sloping apron, as out-

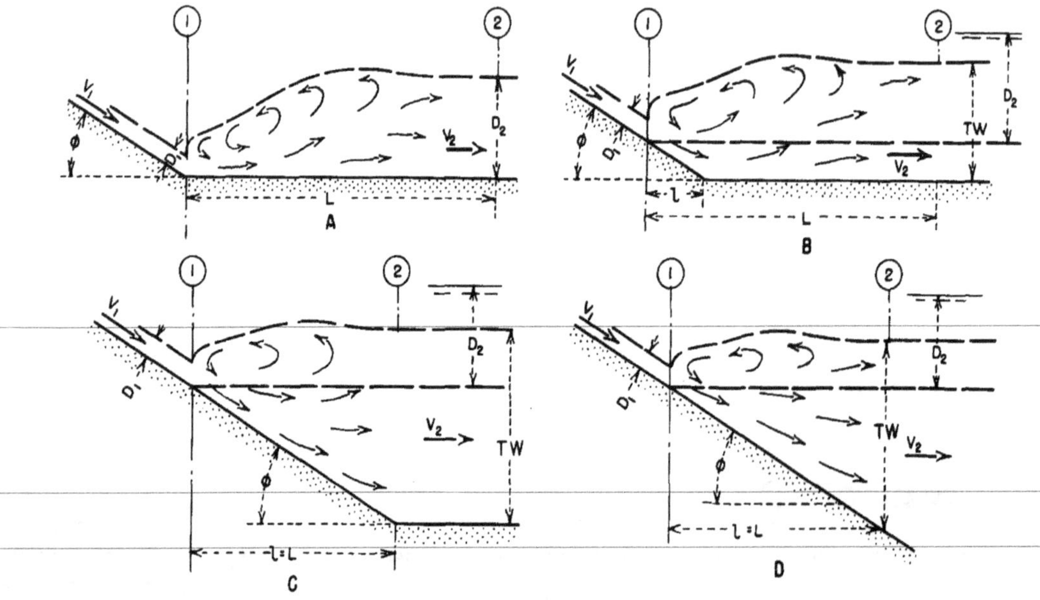

FIGURE 30.—*Sloping aprons (Basin V).*

lined by Kindsvater, presenting separate and distinct problems, Figure 30. Case A has the jump on a horizontal apron. In Case B, the toe of the jump forms on the slope, and the jump ends over the horizontal apron. In Case C, the toe of the jump is on the slope, and the end is at the junction of the slope and the horizontal apron; in Case D, the entire jump forms on the slope. With so many possibilities, it is easily understood why experimental data have been lacking on the sloping apron. Messrs. Yarnell, Kindsvater, Bakhmeteff, and Matzke limited their experiments to Case D. B. D. Rindlaub (7) of the University of California concentrated on the solution of Case B, but his experimental results are complete for only one slope, that of 12.33° ($\tan \phi = 0.217$).

Sloping Apron Tests

From a practical standpoint, the scope of the test program does not need to be as broad as outlined in Figure 30. For example, the action in Cases C and D is for all practical purposes the same, if it is assumed that a horizontal floor begins at the end of the jump for Case D. Sufficient tests were made on Case C to verify the above statement that Cases C and D can be considered as one.

The first experiments described in this section are for Case D. The second set of tests is for Case B. Case B is virtually Case A operating with excessive tail water depth. As the tail water depth is further increased, Case B approaches Case C. The results of Case A have already been discussed in the preceding chapters, and Cases D and B will be considered here in order.

Tail water depth (Case D). Data obtained from the four flumes used in the sloping apron tests (Case D experiments) are tabulated in Table 8. The headings are much the same as those in previous tables, but need some explanation. Column 2 lists the tangents of the angles of the slopes tested. The depth of flow entering the jump, D_1, Column 8, was measured at the beginning of the jump in each case, corresponding to Section 1, Figure 30. It represents the average of a generous number of point gage measurements. The velocity at this same point, V_1, Column 7, was computed by dividing the unit discharge, q (Col. 5), by D_1. The length of jump, Column 11, was measured in the flume, bearing in mind that the object of the test was to obtain practical

TABLE 8.—*Stilling basins with sloping aprons (Basin V, Case D)*

Test flume	Slope of apron tan φ	Total Q c.f.s.	W Width of basin ft.	q per ft. of W c.f.s.	TW ft.	V_1 ft. per sec	D_1 ft.	$\frac{TW}{D_1}$	$F_1=\frac{V_1}{\sqrt{gD_1}}$	L Length of jump ft.	$\frac{L}{TW}$	$\frac{D_2}{D_1}$	D_2 Conj. TW ft.	$\frac{TW}{D_2}$	$\frac{L}{D_2}$	K Shape factor
(1)	(2)	(3)	(4)	(5)	(6)	(7)	(8)	(9)	(10)	(11)	(12)	(13)	(14)	(15)	(16)	(17)
A	0. 067	2. 000	4. 880	0. 410	0. 520	7. 88	0. 052	10. 00	6. 09	2. 60	5. 00	8. 20	0. 426	1. 22	6. 11	2. 50
		2. 250		. 461	. 560	8. 09	. 057	9. 82	5. 97	2. 90	5. 18	7. 90	. 450	1. 24	6. 45	2. 50
		2. 500		. 512	. 589	8. 26	. 062	9. 50	5. 85	3. 10	5. 26	7. 85	. 486	1. 21	6. 38	2. 40
		2. 750		. 564	. 629	8. 42	. 067	9. 39	5. 73	3. 30	5. 25	7. 70	. 516	1. 22	6. 40	2. 45
		3. 000		. 615	. 660	8. 54	. 072	9. 17	5. 61	3. 40	5. 15	7. 55	. 544	1. 21	6. 25	2. 45
		3. 250		. 666	. 694	8. 65	. 077	9. 01	5. 49	3. 45	4. 97	7. 40	. 570	1. 22	6. 05	2. 50
		3. 500		. 717	. 744	8. 74	. 082	9. 07	5. 38	3. 60	4. 84	7. 20	. 590	1. 26	6. 10	2. 80
		1. 500	4. 350	. 345	. 474	7. 67	. 045	10. 53	6. 37	2. 40	5. 06	8. 60	. 387	1. 22	6. 20	2. 50
		2. 500		. 575	. 642	8. 46	. 068	9. 44	5. 72	3. 20	4. 98	7. 70	. 523	1. 23	6. 12	2. 50
		3. 500		. 805	. 792	8. 85	. 091	8. 70	5. 17	4. 00	5. 05	6. 90	. 628	1. 26	6. 37	2. 75
	0. 096	2. 000	4. 830	. 414	. 560	7. 96	. 052	10. 77	6. 15	2. 50	4. 47	8. 20	. 426	1. 31	5. 87	2. 04
		2. 500		. 518	. 652	7. 97	. 065	10. 03	5. 51	3. 60	5. 52	7. 45	. 484	1. 35	7. 44	2. 28
		3. 000		. 621	. 745	8. 28	. 075	9. 93	5. 33	3. 20	4. 30	7. 10	. 532	1. 40	6. 01	2. 40
		3. 500		. 725	. 835	8. 53	. 085	9. 82	5. 15	3. 60	4. 31	6. 90	. 586	1. 42	6. 15	2. 50
		4. 000		. 828	. 940	8. 63	. 096	9. 79	4. 90	4. 00	4. 26	6. 50	. 624	1. 51	6. 41	2. 75
	0. 135	2. 000	4. 810	. 416	. 620	6. 93	. 060	10. 33	4. 99	2. 50	4. 06	6. 60	. 396	1. 56	6. 32	2. 15
		2. 500		. 520	. 710	7. 54	. 069	10. 29	5. 06	3. 00	4. 23	6. 75	. 466	1. 52	6. 44	2. 07
		3. 000		. 624	. 895	7. 80	. 080	10. 06	4. 86	3. 20	3. 97	6. 40	. 512	1. 57	6. 25	2. 15
		3. 500		. 728	. 905	8. 09	. 090	10. 06	4. 75	3. 60	3. 98	6. 30	. 567	1. 60	6. 34	2. 22
		4. 000		. 832	. 985	8. 58	. 097	10. 15	4. 85	3. 90	3. 96	6. 40	. 621	1. 59	6. 28	2. 19
	0. 152	1. 500	4. 350	. 345	. 540	6. 27	. 055	9. 82	4. 71	2. 10	3. 89	6. 20	. 341	1. 58	6. 16	1. 94
		2. 000		. 460	. 663	6. 76	. 068	9. 75	4. 57	2. 55	3. 85	6. 10	. 415	1. 60	6. 15	2. 00
		2. 500		. 575	. 790	7. 57	. 076	10. 39	4. 84	3. 10	3. 92	6. 45	. 490	1. 61	6. 33	2. 00
		3. 000		. 690	. 900	7. 67	. 090	10. 00	4. 50	3. 40	3. 78	6. 00	. 540	1. 67	6. 30	2. 10
B	0. 102	5. 000	2. 000	2. 500	2. 300	16. 45	. 152	15. 13	7. 44	10. 00	4. 34	10. 10	1. 536	1. 50	6. 51	2. 75
		5. 500		2. 750	2. 450	16. 18	. 170	14. 41	6. 91	10. 60	4. 33	9. 35	1. 590	1. 54	6. 67	2. 85
	0. 164	2. 000		1. 000	1. 537	15. 38	. 065	23. 65	10. 64	6. 10	3. 97	14. 65	. 952	1. 61	6. 41	1. 88
		2. 500		1. 250	1. 737	14. 88	. 084	20. 68	9. 05	6. 90	3. 97	12. 40	1. 042	1. 67	6. 62	1. 95
		3. 000		1. 500	1. 940	14. 71	. 102	19. 02	8. 11	7. 50	3. 86	11. 05	1. 128	1. 72	6. 64	2. 02
		3. 500		1. 750	2. 120	14. 83	. 118	17. 97	7. 61	8. 20	3. 87	10. 30	1. 215	1. 74	6. 75	2. 03
		4. 000		2. 000	2. 270	15. 04	. 133	17. 07	7. 27	8. 70	3. 83	9. 85	1. 310	1. 73	6. 64	2. 01
		4. 500		2. 250	2. 420	14. 90	. 151	16. 03	6. 75	9. 20	3. 80	9. 10	1. 374	1. 76	6. 70	2. 08
		5. 000		2. 500	2. 590	14. 88	. 168	15. 42	6. 39	9. 70	3. 74	8. 65	1. 454	1. 78	6. 67	2. 08
		5. 500		2. 750	2. 750	14. 86	. 185	14. 86	6. 09	10. 20	3. 71	8. 20	1. 517	1. 81	6. 73	2. 10
	0. 213	2. 000		1. 000	1. 750	13. 33	. 075	23. 33	8. 60	6. 00	3. 43	11. 75	. 881	1. 99	6. 81	1. 71
		2. 500		1. 250	2. 000	13. 59	. 092	21. 74	7. 89	6. 60	3. 30	10. 70	. 984	2. 03	6. 71	1. 76
		3. 000		1. 500	2. 150	13. 51	. 111	19. 37	7. 15	7. 30	3. 40	9. 70	1. 077	2. 00	6. 78	1. 73
		3. 500		1. 750	2. 370	13. 57	. 129	18. 37	6. 65	8. 00	3. 38	9. 00	1. 161	2. 04	6. 89	1. 76
		4. 000		2. 000	2. 600	13. 51	. 148	17. 57	6. 19	8. 30	3. 19	8. 35	1. 236	2. 10	6. 71	1. 79

TABLE 8.—*Stilling basins with sloping apron (Basin V, Case D)*—Continued

Test flume	Slope of apron tan ϕ	Total Q c.f.s.	W Width of basin ft.	q per ft. of W c.f.s.	TW ft.	V_1 ft. per sec	D_1 ft.	$\frac{TW}{D_1}$	$F_1=\frac{V_1}{\sqrt{gD_1}}$	L Length of jump ft.	$\frac{L}{TW}$	$\frac{D_2}{D_1}$	D_2 Conj. TW ft.	$\frac{TW}{D_2}$	$\frac{L}{D_2}$	K Shape factor
(1)	(2)	(3)	(4)	(5)	(6)	(7)	(8)	(9)	(10)	(11)	(12)	(13)	(14)	(15)	(16)	(17)
B	0.213	4.500	2.000	2.250	2.720	13.55	.166	16.39	5.86	9.10	3.34	7.85	1.303	2.09	6.98	1.78
		5.000		2.500	2.890	13.59	.184	15.71	5.58	9.60	3.32	7.50	1.380	2.09	6.96	1.79
		5.500		2.750	3.100	13.55	.203	15.27	5.30	10.00	3.22	7.10	1.441	2.15	6.94	1.81
	0.263	2.000		1.000	1.900	11.63	.086	22.09	6.98	5.60	2.95	9.45	.813	2.34	6.89	1.55
		3.000		1.500	2.330	11.63	.129	18.06	5.70	6.90	2.96	7.65	.987	2.36	6.99	1.56
		4.000		2.000	2.820	12.35	.162	17.41	5.40	8.10	2.87	7.25	1.174	2.40	6.90	1.57
		5.000		2.500	3.270	12.38	.202	16.19	4.85	9.20	2.81	6.45	1.303	2.51	7.06	1.59
		6.000		3.000	3.602	12.35	.243	14.82	4.41	10.00	2.77	5.80	1.409	2.56	7.09	1.59
D	0.100	4.000	3.970	1.007	1.530	18.64	.054	28.33	14.14	6.60	4.31	19.50	1.053	1.45	6.27	2.65
		6.000		1.511	1.888	19.12	.079	23.90	11.99	8.20	4.34	16.50	1.303	1.45	6.29	2.65
		8.000		2.015	2.200	19.75	.102	21.57	10.90	9.70	4.41	14.95	1.525	1.44	6.36	2.65
		10.000		2.518	2.630	20.14	.125	21.04	10.04	11.50	4.37	13.75	1.719	1.53	6.69	2.85
		2.250		.567	1.200	18.90	.030	40.00	19.23	4.75	3.96	26.70	.801	1.50	5.93	2.75
A	0.185	1.500	4.350	.345	.600	6.05	.057	10.53	4.47	2.15	3.58	5.90	.336	1.78	6.40	1.83
		2.000		.460	.720	6.57	.070	10.29	4.38	2.60	3.61	5.80	.406	1.77	6.40	1.83
		2.500		.575	.840	7.01	.082	10.24	4.31	3.00	3.57	5.70	.467	1.80	6.42	1.85
	0.218	1.750	4.350	.402	.700	6.00	.067	10.45	4.08	2.30	3.29	5.45	.365	1.92	6.30	1.70
		2.250		.517	.862	6.63	.078	11.05	4.19	2.70	3.13	5.55	.433	1.99	6.24	1.73
	0.280	1.250	4.350	.287	.620	4.70	.061	10.16	3.35	1.60	2.58	4.25	.259	2.39	6.18	1.44
		1.500		.345	.675	4.79	.072	9.38	3.15	1.80	2.67	4.05	.292	2.31	6.17	1.44
		1.750		.402	.752	4.79	.084	8.95	2.91	1.95	2.59	3.70	.311	2.42	6.27	1.46
B	0.052	1.000	2.000	.500	.855	17.24	.029	29.48	17.85	4.10	4.79	24.75	.718	1.19	5.71	2.94
		1.500		.750	1.010	16.30	.046	21.96	13.40	5.10	5.05	18.45	.849	1.19	6.01	2.80
		2.000		1.000	1.160	16.39	.061	19.02	11.69	6.10	5.26	16.10	.982	1.18	6.21	2.78
		2.500		1.250	1.300	17.12	.073	17.81	11.16	6.50	5.00	15.35	1.121	1.16	5.80	2.45
		3.000		1.500	1.426	17.05	.088	16.20	10.13	7.50	5.26	13.85	1.218	1.17	6.15	2.70
		3.500		1.750	1.570	17.16	.102	15.39	9.46	8.00	5.10	12.95	1.321	1.19	6.06	2.80
		4.000		2.000	1.693	17.09	.117	14.47	8.80	8.90	5.26	12.10	1.416	1.20	6.28	2.92
		4.500		2.250	1.813	17.05	.132	13.73	8.27	9.60	5.29	11.30	1.492	1.22	6.44	3.10
		5.000		2.500	1.920	17.01	.147	13.06	7.82	9.80	5.10	10.60	1.558	1.23	6.29	3.20
		5.500		2.750	2.020	17.08	.161	12.55	7.50	10.50	5.20	10.20	1.642	1.23	6.40	3.20
		6.000		3.000	2.110	16.95	.177	11.92	7.10	11.00	5.21	9.65	1.708	1.24	6.44	3.30
	0.102	1.000		.500	.970	15.63	.032	30.31	15.40	4.20	4.33	21.25	.680	1.42	6.17	2.51
		1.500		.750	1.180	15.63	.048	24.58	12.57	5.20	4.41	17.30	.830	1.42	6.27	2.50
		2.000		1.000	1.354	15.87	.063	21.49	11.14	6.10	4.51	15.35	.967	1.40	6.31	2.44
		2.500		1.250	1.543	16.23	.077	20.04	10.30	6.80	4.40	14.15	1.088	1.42	6.24	2.50
		3.000		1.500	1.724	16.48	.091	18.95	9.63	7.60	4.41	13.20	1.200	1.44	6.34	2.56
		3.000		1.500	1.720	16.30	.092	18.70	9.47	7.50	4.36	12.95	1.191	1.44	6.30	2.58
		3.500		1.750	1.890	16.36	.107	17.66	8.81	8.20	4.34	12.10	1.293	1.46	6.34	2.75

		4.000		2.000	2.040	16.53	.121	16.86	8.37	8.80	4.31	11.40	1.379	1.48	6.38	2.72
		4.500		2.250	2.152	16.42	.137	15.71	7.82	9.40	4.37	10.60	1.452	1.48	6.47	2.70
D	0.100	4.500	3.970	1.134	1.710	18.29	.062	27.58	12.94	7.80	4.56	17.90	1.109	1.54	7.03	2.90
		6.750		1.700	2.100	19.54	.087	24.14	11.67	9.10	4.33	16.10	1.400	1.50	6.50	2.78
F	0.174	1.980	1.000	1.980	1.452	7.17	.276	5.26	2.41	4.30	2.96	3.00	.828	1.75	5.19	1.88
		2.800		2.800	1.663	7.69	.364	4.57	2.24	5.00	3.01	2.80	1.018	1.63	4.91	1.76
	0.200	2.980		2.980	2.035	8.32	.358	5.68	2.45	5.80	2.85	3.05	1.092	1.86	5.31	1.72
		3.850		3.850	2.460	8.48	.454	5.42	2.22	6.70	2.72	2.75	1.248	1.97	5.37	1.81
	0.150	3.850		3.850	2.095	7.97	.483	4.33	2.02	5.90	2.82	2.45	1.183	1.77	4.99	2.10
		1.780		1.780	1.260	6.93	.257	4.90	2.41	4.00	3.17	3.00	.771	1.63	5.19	2.00
	0.100	1.940		1.940	1.180	6.40	.303	3.89	2.05	3.70	3.14	2.50	.757	1.56	4.89	2.93
		3.870		3.870	1.648	7.38	.524	3.14	1.80	4.80	2.91	2.15	1.126	1.46	4.26	2.55
	0.050	3.620		3.620	1.357	7.62	.475	2.86	1.95	4.30	3.17	2.35	1.116	1.22	3.85	3.00
		1.820		1.820	1.306	12.38	.147	8.88	5.69	6.80	5.21	7.65	1.124	1.16	6.05	3.90
		3.910		3.910	1.291	6.66	.587	2.20	1.53	3.60	2.79	1.80	1.057	1.22	3.41	3.20
		2.300		2.300	.943	5.87	.392	2.41	1.65	2.80	2.97	1.95	.764	1.23	3.67	3.20

data for stilling basin design. The end of the jump was chosen as the point where the high velocity jet began to lift from the floor, or a point on the level tail water surface immediately downstream from the surface roller, whichever occurred farthest downstream. The length of the jump, as tabulated in Column 11, is the horizontal distance from Sections 1 to 2, Figure 30. The tail water depth, tabulated in Column 6, is the depth measured at the end of the jump, corresponding to the depth at Section 2 in Figure 30.

The ratio $\frac{TW}{D_1}$ (Col. 9, Table 8) is plotted with respect to the Froude number (Col. 10) for sloping aprons having tangents 0.05 to 0.30 in Figure 31. The plot for the horizontal apron ($\tan \emptyset=0$) is the same as shown in Figure 5. Superimposed on Figure 31 are data from Kindsvater (*5*), Hickox (*5*), Bakhmeteff (*1*), and Matzke (*6*). The agreement is within experimental error.

The small chart on Figure 31 was constructed using data from the larger chart, and shows, for a range of apron slopes, the ratio of tail water depth for a continuous sloping apron, to conjugate depth for a horizontal apron. D_2 and TW are identical for a horizontal apron. The conjugate depth, D_2, listed in Column 14, Table 6, is the depth necessary for a jump to form on an imaginary horizontal floor beginning at Section 1, Figure 31.

The small chart, therefore, shows the extra depth required for a jump of a given Froude number to form on a sloping apron rather than on a horizontal apron. For example, if the tangent of the slope is 0.10, a tail water depth equal to 1.4 times the conjugate depth (D_2 for a horizontal apron) will occur at the end of the jump; if the slope is 0.30, the tail water depth at the end of the jump will be 2.8 times the conjugate depth, D_2. The conjugate depth, D_2, used in connection with a sloping apron is merely a convenient reference figure which has no other meaning. It will be used throughout this discussion on sloping aprons.

Length of jump (*Case D*). The length of jump for the Case D experiments has been presented in two ways. First, the ratio length of jump to tail water depth, Column 12, was plotted with respect to the Froude number in Figure 32 for sloping aprons having tangents from 0 to 0.25. Second, the ratio length of jump to the conjugate tail water depth, Column 16, Table 8, has been plotted with respect to the Froude number for the same range of slopes in Figure 33. Although not evident in Figure 32, it can be seen in Figure 33 that the length of jump on a sloping apron is longer than the same jump which occurs on a horizontal floor. For example, for a Froude number of 8, the ratio $\frac{L}{D_2}$ varies from 6.1, for a horizontal apron, to 7.0, for an apron having a slope of 0.25. Length determinations from Kindsvater (*5*) for a slope of 0.167 are also plotted in Figure 32. The points show a wide spread.

Expression for jump on sloping apron (*Case D*). Several mathematicians and experimenters have developed expressions for the hydraulic jump on sloping aprons (*2, 5, 6, 13*) so there is no need to repeat any of these derivations here. An expression presented by Kindsvater (*5*) is the more common and perhaps the more practical to use:

$$\frac{D_2}{D_1}=\frac{1}{2\cos\emptyset}\left[\sqrt{\frac{8F_1^2\cos^3\emptyset}{1-2K\tan\emptyset}+1}-1\right] \qquad (5)$$

All symbols have been referred to previously, except for the coefficient K, a dimensionless parameter called the shape factor, which varies with the Froude number and the slope of the apron. Kindsvater and Hickox evaluated this coefficient from the profile of the jump and the measured floor pressures. Surface profiles and pressures were not measured in the current tests, but, as a matter of interest, K was computed from Equation 5 by substituting experimental values and solving for K. The resulting values of K are listed in Column 17 of Table 8, and are shown plotted with respect to the Froude number for the various slopes in Figure 34A. Superimposed in Figure 34A are data from Kindsvater for a slope of 0.167, and data from Hickox on a slope of 0.333. The agreement is not particularly striking nor do the points plot well, but it should be remembered that the value K is dependent on the method used for determining the length of jump. The current experiments indicate that the Froude number has little effect on the value of K. Assuming this to be true, values of individual points for each slope were averaged and K is shown plotted with respect to $\tan \emptyset$ in Figure 34B. The

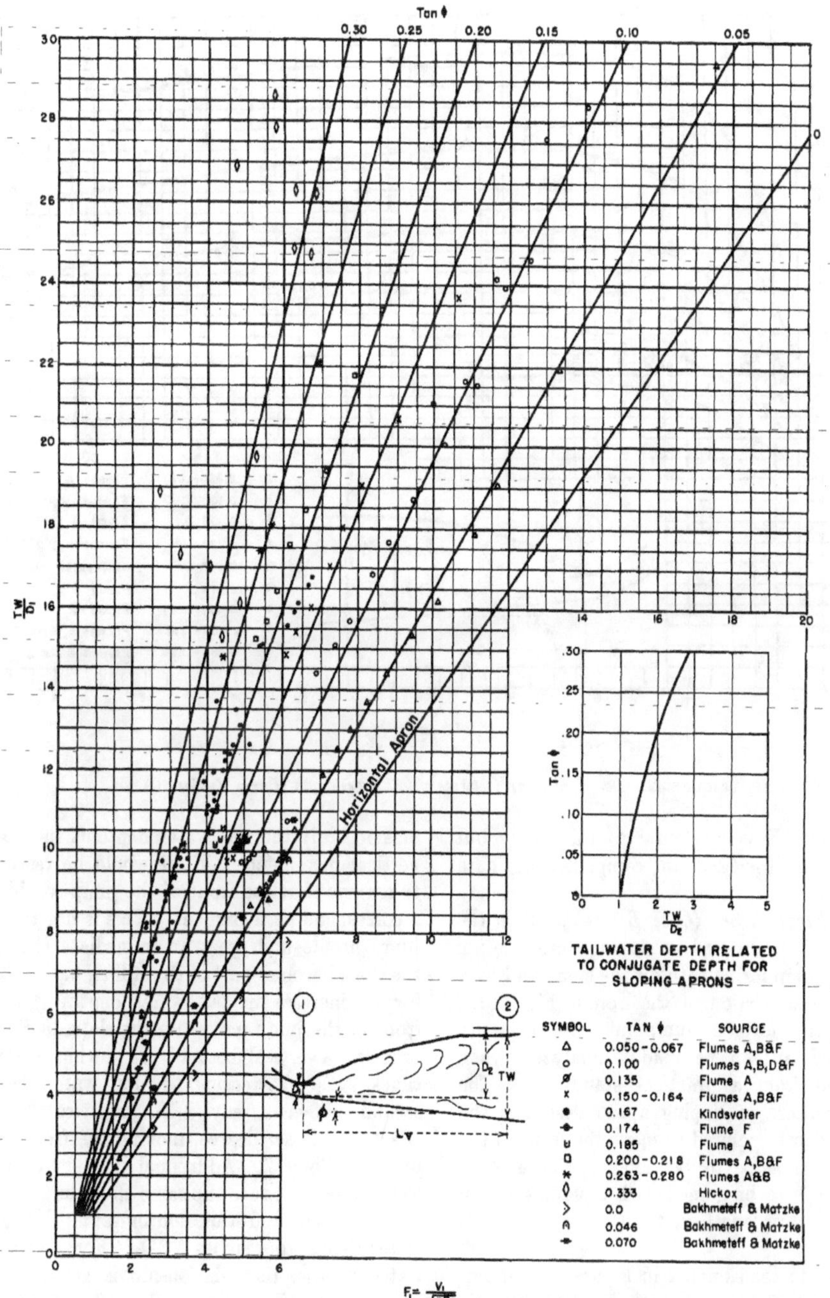

FIGURE 31.—*Ratio of tail water depth to D_1 (Basin V, Case D).*

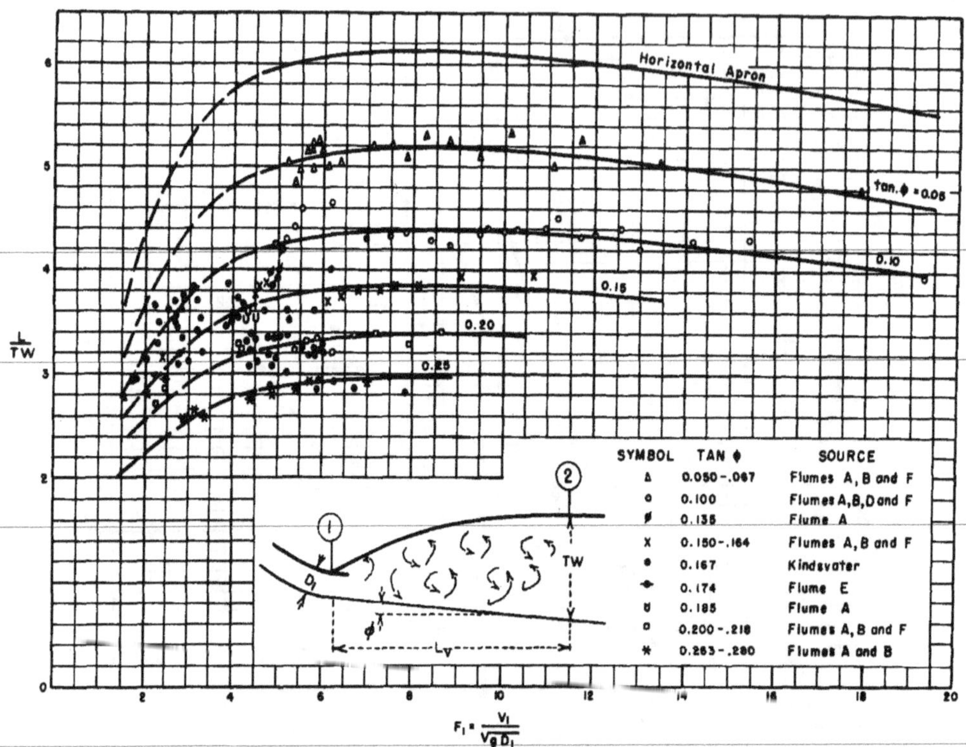

FIGURE 32.—*Length of jump in terms of tail water depth (Basin V, Case D).*

evaluation of K is incidental to this study but it has been discussed to complete the data analysis.

Jump characteristics (Case B). Case B is the one usually encountered in sloping apron design where the jump forms both on the slope and over the horizontal portion of the apron (Fig. 30B). Although this form of jump may appear quite complicated, it can be readily analyzed when approached from a practical standpoint. The primary concern in sloping apron design is the tail water depth required to move the front of the jump up the slope to Section 1, Figure 30B. There is little to be gained with a sloping apron unless the entire length of the sloping portion is utilized.

Referring to the sketches in Figure 35A, it can be observed that for tail water equal to the conjugate depth, D_2, the front of the jump will occur at a point 0, a short distance up the slope. This distance is noted as l_0 and varies with the degree of slope. If the tail water depth is increased a vertical increment, ΔY_1, it would be reasonable to assume that the front of the jump would raise a corresponding increment. This is not true; the jump profile undergoes an immediate change as the slope becomes part of the stilling basin. Thus, for an increase in tail water depth, ΔY_1, the front of the jump moves up the slope to Point 1, or moves a vertical distance $\Delta Y'_1$, which is several times ΔY_1. Increasing the tail water depth a second increment, say ΔY_2, produces the same effect to a lesser degree, moving the front of the jump to Point 2. Additional increments of tail water depth produce the same effect but to a still lesser degree, and this continues until the tail water depth approaches $1.3D_2$. For greater tail water depths, the relationship is geometric; an increase in tail water depth, ΔY_4, moves the front of the jump up the slope an equal vertical distance $\Delta Y'_4$, from Point 3 to 4. Should the slope be very flat, as in Figure 35B, the horizontal movement of the front of the jump is even more pronounced.

The following studies were made to tabulate the characteristics described above for conditions encountered in design since it has been necessary in the past to check practically all sloping apron designs by model studies to be certain that the entire sloping portion of the apron was utilized.

Experimental results (*Case B*). The experiments for determining the magnitude of the profile characteristics were carried out on a large scale in Flume D, and the results are recorded in Table 9. A sloping floor was placed in the flume as in Figure 30B. A discharge was established (Col. 3, Table 9) and the depth of flow, D_1 (Col. 6), was measured immediately upstream from the front of the jump in each instance. The velocity entering the jump, V_1 (Col. 7), and the Froude number (Col. 8) were computed. Entering Figure 31 with the computed values of F_1, the ratio $\frac{D_2}{D_1}$ (Col. 9) was obtained from the line labeled "Horizontal apron." Multiplying this ratio by D_1 gives the conjugate depth for a horizontal apron which is listed in Column 10 of Table 9. The tail water was then set at conjugate depth (Point 0, Figure 35) and the distance, l_0, measured and tabulated.

The distance, l_0, gives the position of the front of the jump on the slope, measured from the break in slope, for conjugate depth. The tail water was then increased, moving the front of the jump up to Point 1, Figure 35. Both the distance, l_1, and the tail water depth were measured, and these are recorded in Columns 11 and 12, respectively, of Table 9. The tail water was then raised, moving the front of the jump to Point 2 while the length, l_2, and the tail water depth were recorded. The same procedure was repeated until the entire apron was utilized by the jump. In each case, D_1 was measured immediately upstream from the front of the jump, thus compensating for frictional resistance on the slope. The velocity, V_1, and the Froude number were computed at the same location. The tests were made for slopes with tangents varying from 0.05 to 0.30, and in some cases, several lengths of floor were used for each slope, as indicated in Column 15 of Table 9.

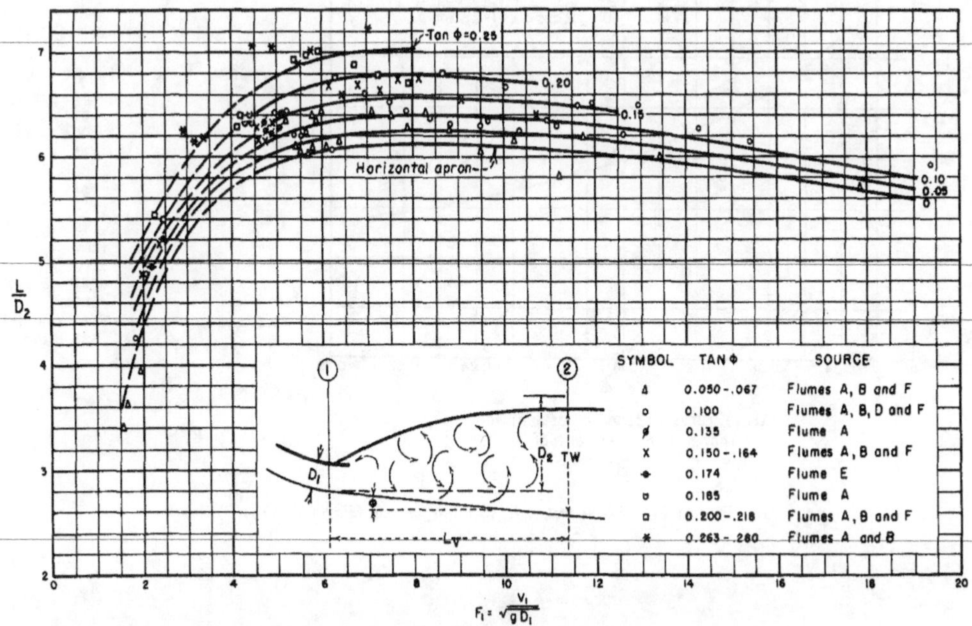

FIGURE 33.—*Length of jump in terms of conjugate depth, D_2 (Basin V, Case D).*

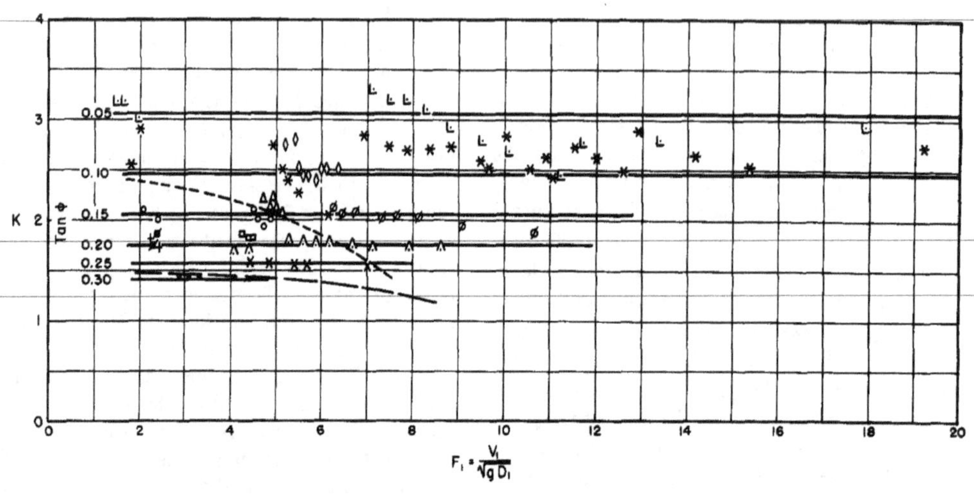

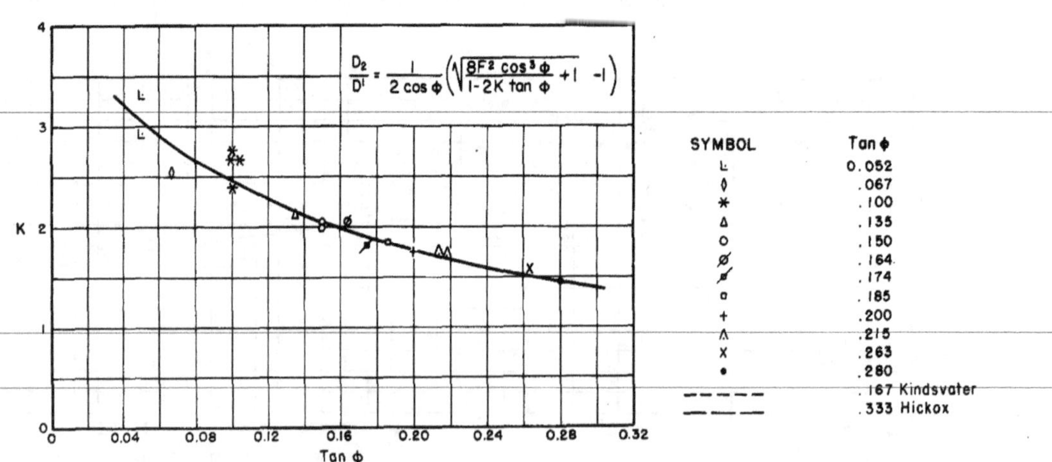

FIGURE 34.—*Shape factor, K, in jump formula (Basin V, Case D).*

TABLE 9.—*Stilling basins with sloping aprons (Basin V, Case B)*

Test flume	Slope of apron tan ø	Total Q c.f.s.	W Width of basin ft.	q per ft. of W c.f.s.	D_1 ft.	V_1 ft. per sec.	$F_1=\frac{V_1}{\sqrt{gD_1}}$	$\frac{D_2}{D_1}$	D_2 Conj. TW ft.	l Length of jump on slope ft.	TW ft.	$\frac{l}{D_2}$	$\frac{TW}{D_2}$	L_s Length of sloping floor ft.
(1)	(2)	(3)	(4)	(5)	(6)	(7)	(8)	(9)	(10)	(11)	(12)	(13)	(14)	(15)
D	0.05	5.050	3.970	1.272	0.063	20.19	14.18	19.51	1.229	6.00	1.390	4.88	1.13	4.0
		8.070		2.033	.101	20.13	11.16	15.30	1.545	6.00	1.745	3.88	1.13	
		11.555		2.910	.139	20.94	9.90	13.60	1.890	6.00	2.040	3.17	1.08	
	.10	5.255		1.324	.067	19.76	13.46	18.60	1.246	4.80	1.440	3.85	1.16	4.0
		8.090		2.038	.103	19.79	10.87	15.00	1.545	4.80	1.750	3.11	1.13	
		11.560		2.911	.140	20.79	9.80	13.40	1.876	4.80	2.080	2.56	1.11	
		5.000		1.259	.064	19.67	13.70	18.90	1.210	8.10	1.830	6.69	1.51	8.0
					.065	19.37	13.38	18.40	1.196	6.30	1.660	5.26	1.39	
					.067	18.79	12.80	17.65	1.183	4.70	1.510	3.97	1.28	
					.068	18.51	12.50	17.20	1.169	4.00	1.410	3.42	1.21	
					.070	17.98	11.99	16.50	1.155	3.20	1.340	2.77	1.16	
		7.850		1.977	.101	19.57	10.86	15.00	1.515	7.80	2.070	5.15	1.37	
					.102	19.38	10.70	14.70	1.499	6.00	1.940	4.00	1.29	
					.103	19.19	10.54	14.50	1.494	5.30	1.880	3.55	1.26	
					.104	19.01	10.39	14.25	1.482	4.40	1.770	2.96	1.19	
		11.218		2.825	.139	20.32	9.61	13.15	1.828	8.30	2.410	4.54	1.32	
					.141	20.04	9.41	12.88	1.816	6.20	2.260	3.41	1.24	
					.142	19.89	9.30	12.80	1.818	4.80	2.180	2.64	1.20	
		6.000		1.511	.076	19.88	12.70	17.50	1.330	2.20	1.375	1.65	1.03	2.0
								17.50	1.330	1.70	1.340	1.28	1.01	
					.077	19.62	12.46	17.15	1.321	.80	1.305	.61	.99	
					.078	19.37	12.23	16.80	1.310	0	1.280	0	.98	
		8.057		2.029	.098	20.70	11.66	16.00	1.568	2.40	1.625	1.53	1.04	
					.099	20.49	11.48	15.80	1.564	1.90	1.600	1.22	1.02	
					.100	20.29	11.31	15.60	1.560	1.60	1.585	1.02	1.02	
					.101	20.09	11.14	15.40	1.555	.60	1.550	.39	1.00	
					.102	19.89	10.98	15.15	1.545	0	1.530	0	.99	
	.15	6.000		1.511	.075	20.15	12.96	17.85	1.339	.50	1.335	.37	1.00	1.2
								17.85	1.339	1.10	1.365	.82	1.02	
		8.057		2.029	.099	20.49	11.48	15.80	1.564	.60	1.564	.38	1.00	
								15.80	1.564	1.20	1.600	.77	1.02	
		11.535		2.905	.136	21.36	10.21	14.00	1.904	.50	1.905	.26	1.00	
								14.00	1.904	1.50	1.970	.79	1.03	
		5.295		1.333	.069	19.32	12.97	17.85	1.232	4.00	1.530	3.25	1.24	
		8.080		2.035	.104	19.57	10.70	14.70	1.529	4.20	1.810	2.75	1.18	
		11.553		2.910	.141	20.64	9.69	13.25	1.868	4.20	2.150	2.25	1.15	
		4.976		1.253	.064	19.58	13.64	18.75	1.200	5.30	1.660	4.42	1.38	5.3
					.065	19.27	13.32	18.35	1.193	5.10	1.590	4.27	1.33	
					.066	18.98	13.02	18.00	1.188	4.00	1.505	3.37	1.27	

TABLE 9.—*Stilling basins with sloping aprons (Basin V, Case B)*—Continued

Test flume	Slope of apron tan ø	Total Q c.f.s.	W Width of basin ft.	q per ft. of W c.f.s.	D_1 ft.	V_1 ft. per sec.	$F_1=\frac{V_1}{\sqrt{gD_1}}$	$\frac{D_2}{D_1}$	D_2 Conj. TW ft.	I Length of jump on slope ft.	TW ft.	$\frac{1}{D_2}$	$\frac{TW}{D_2}$	L_s Length of sloping floor ft.
(1)	(2)	(3)	(4)	(5)	(6)	(7)	(8)	(9)	(10)	(11)	(12)	(13)	(14)	(15)
D	.15	4.976	3.970	1.253	.067	18.70	12.74	17.55	1.176	3.10	1.420	2.64	1.21	5.3
								17.55	1.176	2.60	1.375	2.21	1.17	
					.068	18.42	12.45	17.15	1.166	2.20	1.305	1.89	1.12	
								17.15	1.166	1.80	1.230	1.54	1.05	
		8.025		2.021	.103	19.62	10.77	14.85	1.530	5.30	1.940	3.46	1.27	
								14.85	1.530	4.30	1.875	2.81	1.23	
								14.85	1.530	3.80	1.800	2.48	1.18	
					.104	19.43	10.62	14.60	1.518	2.80	1.705	1.84	1.12	
								14.60	1.518	2.20	1.640	1.45	1.08	
					.105	19.25	10.47	14.40	1.512	1.20	1.580	.79	1.05	
		11.530		2.904	.142	20.45	9.57	13.10	1.860	5.30	2.260	2.85	1.22	
								13.10	1.860	4.30	2.190	2.31	1.18	
								13.10	1.860	3.60	2.120	1.94	1.14	
	.20	5.393		1.358	.071	19.13	12.65	17.35	1.232	4.60	1.790	3.73	1.45	5.3
								17.35	1.232	4.40	1.720	3.58	1.40	
								17.35	1.232	4.00	1.680	3.25	1.36	
					.072	18.86	12.40	17.05	1.228	3.60	1.605	2.93	1.31	
								17.05	1.228	3.00	1.550	2.44	1.26	
								17.05	1.228	2.60	1.490	2.12	1.21	
					.073	18.60	12.13	16.65	1.212	2.30	1.420	1.90	1.17	
								16.60	1.215	1.50	1.350	1.24	1.11	
								16.60	1.215	1.20	1.280	.99	1.05	
		8.080		2.035	.105	19.38	10.54	14.50	1.523	4.60	2.010	3.02	1.32	
								14.50	1.523	4.00	1.955	2.63	1.28	
								14.50	1.523	3.30	1.890	2.17	1.24	
					.104	19.57	10.70	14.70	1.529	3.10	1.830	2.03	1.20	
								14.70	1.529	2.50	1.730	1.64	1.13	
								14.70	1.529	1.80	1.670	1.18	1.09	
		11.573		2.915	.145	20.10	9.30	12.80	1.856	4.40	2.310	2.37	1.24	
								12.80	1.856	3.70	2.230	1.99	1.20	
								12.80	1.856	3.30	2.175	1.78	1.17	
		4.820		1.214	.063	19.27	13.53	18.70	1.178	3.70	1.605	3.14	1.36	4.0
		8.089		2.037	.105	19.40	10.55	14.50	1.523	3.90	1.900	2.56	1.25	
		11.565		2.913	.143	20.37	9.50	13.05	1.866	3.90	2.180	2.09	1.17	

.25	5.344		1.346	.071	18.96	12.54	17.25	1.225	4.20	1.825	3.43	1.49	5.3
							17.25	1.225	4.10	1.755	3.35	1.43	
							17.25	1.225	3.50	1.680	2.86	1.37	
				.070	19.22	12.81	17.75	1.242	3.00	1.600	2.42	1.29	
							17.75	1.242	2.60	1.525	2.09	1.23	
							17.75	1.242	2.20	1.445	1.77	1.16	
							17.75	1.242	1.60	1.375	1.29	1.11	
							17.75	1.242	.90	1.290	.72	1.04	
	8.080		2.035	.107	19.02	10.25	14.05	1.503	4.30	2.100	2.86	1.40	
				.106	19.20	10.40	14.30	1.516	3.40	1.960	2.24	1.29	
							14.30	1.516	2.90	1.860	1.91	1.23	
							14.30	1.516	2.10	1.740	1.38	1.15	
				.105	19.38	10.54	14.50	1.523	1.30	1.650	.85	1.08	
							14.50	1.523	.50	1.550	.33	1.02	
	11.553		2.910	.147	19.80	9.10	12.45	1.830	4.30	2.410	2.35	1.32	
				.146	19.93	9.20	12.60	1.840	4.10	2.320	2.22	1.26	
								1.840	3.20	2.230	1.74	1.21	
				.145	20.07	9.29	12.75	1.849	2.30	2.140	1.24	1.16	
								1.849	1.60	2.090	.87	1.13	
				.144	20.21	9.39	12.85	1.850	1.20	2.010	.65	1.09	
								1.850	.80	1.950	.43	1.05	
	6.005		1.512	.079	19.14	12.00	16.50	1.306	3.60	1.760	2.76	1.35	4.0
	8.057		2.029	.105	19.32	10.51	14.45	1.517	3.60	1.925	2.37	1.27	
	11.535		2.905	.144	20.17	9.37	12.85	1.850	3.60	2.210	1.95	1.19	
.30	8.105		2.041	.110	18.55	9.86	13.50	1.485	4.50	2.300	3.03	1.55	5.3
	5.410		1.362	.074	18.40	11.92	16.40	1.214	4.50	2.070	3.71	1.71	
	11.553		2.910	.150	19.40	8.83	12.05	1.808	4.50	2.570	2.49	1.42	
	5.980		1.506	.079	19.06	11.95	16.45	1.300	3.40	1.840	2.62	1.42	4.0
	8.050		2.028	.106	19.13	10.36	14.25	1.510	3.40	2.025	2.25	1.34	
	11.538		2.906	.146	19.90	9.18	12.55	1.832	3.40	2.300	1.85	1.26	

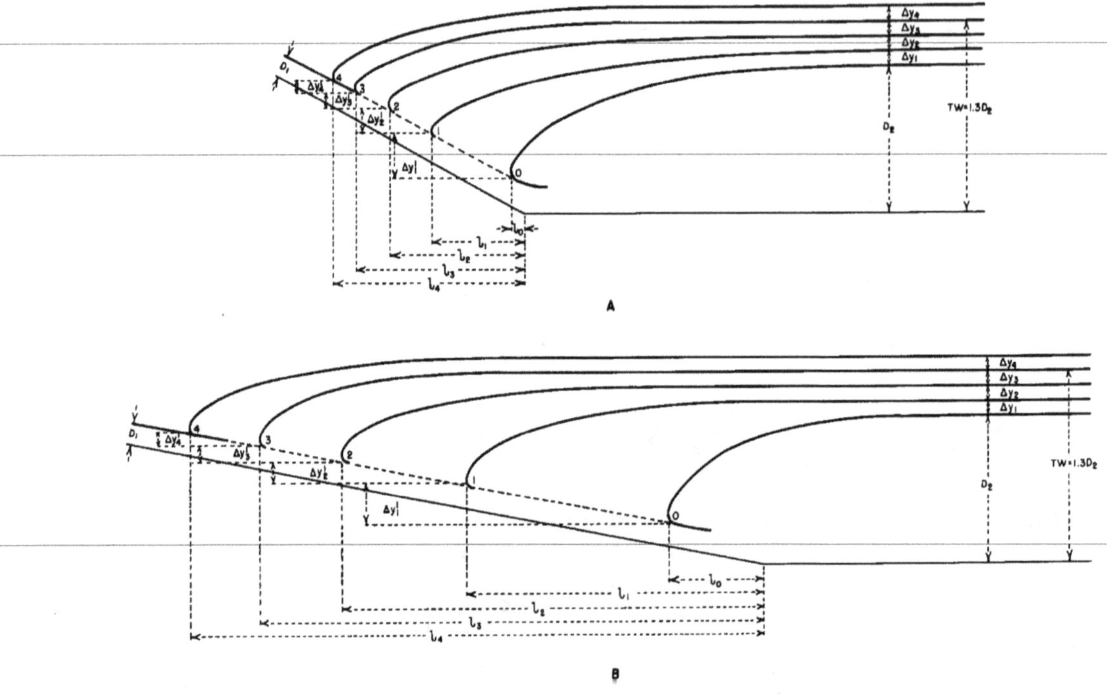

FIGURE 35.—*Profile characteristics (Basin V, Case B).*

The resulting lengths and tail water depths, divided by the conjugate depth, are shown in Columns 13 and 14 of Table 9, and these values have been plotted in Figure 36. The horizontal length has been used rather than the vertical distance, ΔY, as the former dimension is more convenient to use. Figure 36 shows that the straight lines for the geometric portion of the graph tend to intersect at a common point, $\frac{l}{D_2}=1$ and $\frac{TW}{D_2}=0.92$, indicated by the circle on the graph. The change in the profile of the jump as it moves from a horizontal floor to the slope is evidenced by the curved portion of the lines.

Case C, Figure 30, is the upper extreme of Case B; and as there is practically no difference in the performance for Cases D and C, data for Case D (Table 8) can again be utilized. By assuming that a horizontal floor begins at the end of the jump in Case D, Columns 15 and 16 of Table 8 can be plotted in Figure 36. In addition, data from experiments by B. D. Rindlaub of the University of California, for a slope of 0.217, have been plotted in Figure 36. The agreement of the information from the three sources is very satisfactory.

Length of jump (*Case B*). It is suggested that the length of jump for Case B be obtained from Figure 33. Actually, Figure 33 is for continuous sloping aprons, but these lengths can be applied to Case B with but negligible error. In some cases the length of jump is not of particular concern because it may not be economically possible to design the basin to confine the entire jump. This is especially true when sloping aprons are used in conjunction with medium or high overfall spillways where the rock in the riverbed is in fairly good condition. When sloping aprons are designed shorter than the length indicated in Figure 33, the riverbed downstream must act as part of the stilling basin. On the other hand, when the quality of foundation material is questionable, it is advisable to make the apron sufficiently long to confine the entire jump, Figure 33.

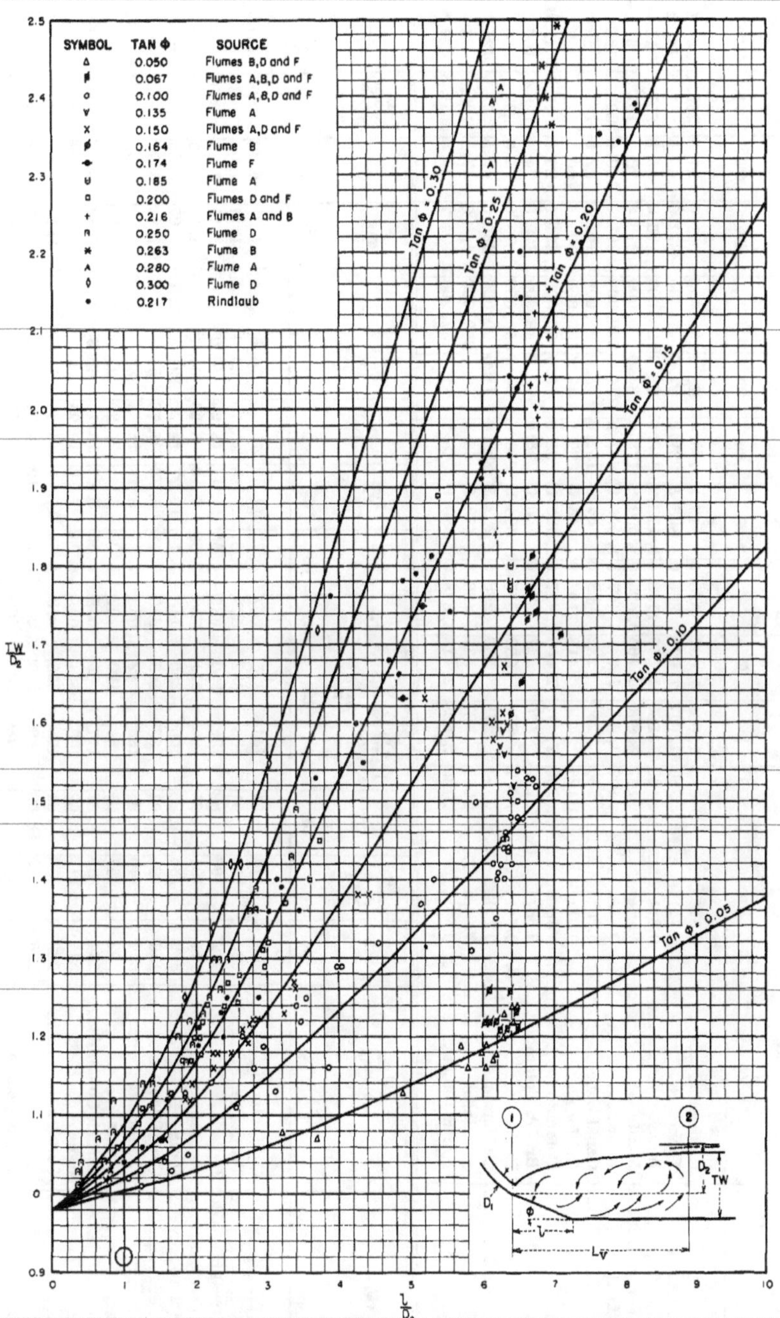

FIGURE 36.—*Tail water requirement for sloping aprons (Basin V, Case B).*

TABLE 10.—*Existing stilling basins with sloping aprons.*

Dam	Location	Slope of dam face	Slope of apron tan φ	Res el ft.	Crest el ft	El up end of apron ft.	El down end of apron ft.	Fall HW to up end of apron ft.	Head on crest ft.	Q Max c.f.s.	Max TW el ft.	TW depth ft.	l Length of sloping apron ft.	h Length of horizontal apron ft.
(1)	(2)	(3)	(4)	(5)	(6)	(7)	(8)	(9)	(10)	(11)	(12)	(13)	(14)	(15)
Shasta	California	0.8:1	0. 083	1, 065	1, 037	570. 6	549. 5	494. 4	28. 0	250, 000	631. 0	81. 5	256. 7	51. 9
Norris	Tennessee	0.7:1	. 250	1, 047	1, 020	826. 0	805. 5	221. 0	27. 0	197, 600	872. 0	66. 5	81. 5	142. 5
Bhakra (prelim)	India	0.8:1	. 100	1, 580	1, 552	1, 139. 4	1, 112. 2	440. 6	28. 0	189, 600	1, 196. 0	83. 8	257	15
Canyon Ferry	Montana	Varies	. 167	3, 800	3, 766	3, 625. 0	3, 600. 0	175. 0	34. 0	200, 000	3, 670. 0	70. 0	137	57
Bhakra (final)	India	0.8:1	. 100	1, 685	1, 645	1, 117. 5	1, 095. 0	567. 5	40. 0	290, 000	1, 205. 0	110. 0	224. 5	165
Madden	Canal Zone	0.75:1	. 250	250	232	98. 0	64. 5	152. 0	18. 0	280, 000	141. 5	77. 0	142	8
Folsom	California	0.67:1	. 125	466	418	137. 0	115. 0	329. 0	48. 0	250, 000	205. 5	90. 5	177	147
Olympus	Colorado	Varies	. 250	7, 475	7, 460	7, 417. 0	7, 405. 0	58. 0	15. 0	20, 000	7, 431. 0	26. 0	48. 5	43. 6
Capilano	British Columbia	0.65:1	. 222	570	547	274. 0	246. 0	296. 0	23. 0	43, 000	320. 0	74. 0	128	106
Rihand	India	0.7:1	. 077	888	852	647. 6	604. 0	240. 4	36. 0	455, 000	679. 0	75. 0	325	10
Friant	California	0.7:1	. 143	578	560	296. 0	282. 5	282. 0	18. 0	90, 000	330. 0	47. 5	97	125
Keswick	do	Varies	. 062	587	537	488. 6	483. 8	98. 4	50. 0	250, 000	541. 0	57. 2	105	23
Dickinson	North Dakota	0.5:1	. 125	2, 428	2, 416	2, 388. 0	2, 380. 0	40. 9	12. 4	33, 200	2, 404. 7	23. 0	61. 5	9. 5

Dam	Location	l+h ft.	V_{th} Theoretical vel entering basin ft./sec.	$\frac{V_A}{V_{th}}$	$V_A=V_1$ Act vel entering basin ft./sec.	W Width of basin ft.	q per ft. of W c.f.s.	D_1 ft.	$F_1=\frac{V_1}{\sqrt{gD_1}}$	$\frac{D_2}{D_1}$	D_2 Conj TW depth ft.	$\frac{l}{D_2}$	$\frac{TW}{D_2}$	$\frac{L}{D_2}$	L Actual length of jump ft.	$\frac{l+h}{L}$
		(16)	(17)	(18)	(19)	(20)	(21)	(22)	(23)	(24)	(25)	(26)	(27)	(28)	(29)	(30)
Shasta	California	308. 6	176	0. 81	141	375	667	4. 73	11. 42	15. 75	74. 5	3. 44	1. 09	6. 30	469	0. 66
Norris	Tennessee	224	116	. 91	106	332	595	5. 61	7. 88	10. 70	60. 0	1. 36	1. 11	7. 03	422	. 53
Bhakra (prelim)	India	272	166	. 82	136	300	632	4. 65	11. 11	15. 25	70. 9	3. 63	1. 18	6. 35	450	. 60
Canyon Ferry	Montana	194	101	. 95	96	271	738	7. 69	6. 10	8. 20	63. 1	2. 17	1. 11	6. 55	413	. 47
Bhakra (final)	India	389. 5	188	. 85	156	260	1, 115	7. 15	10. 27	14. 20	101. 5	2. 21	1. 08	6. 36	646	. 60
Madden	Canal Zone	150	96	. 91	87	448	625	7. 18	5. 72	7. 70	55. 3	2. 57	1. 39	6. 90	382	. 39
Folsom	California	324	140	. 95	133	242	1, 033	7. 76	8. 41	11. 45	88. 9	1. 99	1. 02	6. 50	578	. 56
Olympus	Colorado	92. 1	57	. 95	54	120	167	3. 09	5. 41	7. 30	22. 6	2. 15	1. 15	6. 86	155	. 59
Capilano	British Columbia	234	135	. 87	117	80	538	4. 60	9. 62	13. 10	60. 3	2. 12	1. 23	6. 85	384	. 61
Rihand	India	335	117	. 92	108	664	685	6. 34	7. 56	10. 25	65. 0	5. 00	1. 15	6. 30	413	. 82
Friant	California	222	133	. 81	108	330	273	2. 53	11. 97	16. 45	41. 6	2. 33	1. 14	6. 40	266	. 83
Keswick	do	128	69	1. 00	69	240	1, 042	15. 15	3. 12	4. 00	60. 6	1. 73	0. 94	5. 45	330	. 39
Dickinson	North Dakota	71	51	-----	48	200	166	3. 50	4. 44	5. 70	20. 0	3. 08	1. 19	6. 00	120	. 59

Average 0.60

Applications

Existing structures. To determine the value of the methods given for the design of sloping aprons, existing basins employing sloping aprons were, in effect, redesigned using the current experimental information. Pertinent data for 13 existing spillways are tabulated in Table 10. The slope of the spillway face is listed in Column 3; the tangent of the sloping stilling basin apron is listed in Column 4; the elevation of the upstream end of the apron, or front of the jump, is listed in Column 7; the elevation of the end of the apron is listed in Column 8; the fall from headwater to upstream end of the apron is tabulated in Column 9; and the total discharge is shown in Column 11. Where outlets discharge into the spillway stilling basin, that discharge has also been included in the total. The length of the sloping portion of the apron is given in Column 14; the length of the horizontal portion of the apron is given in Column 15; and the overall length is given in Column 16. Columns 17 through 27 show computed values similar to those in the previous table.

The lower portions of the curves of Figure 36 have been reproduced to a larger scale in Figure 37. The coordinates from Columns 26 and 27 of Table 10 have been plotted in Figure 37 for each of the 13 spillways. Longitudinal sections through the basins are shown in Figures 38 and 39.

Each point in Figure 37 has been connected with an arrow to the tan ∅ curve corresponding to the apron slope. Points which lie to the right and below the corresponding tan ∅ curve indicate that if the tail water depth is correct the sloping portion of the apron is excessively long; if the length of the slope is correct the tail water is insufficient to move the jump upstream to

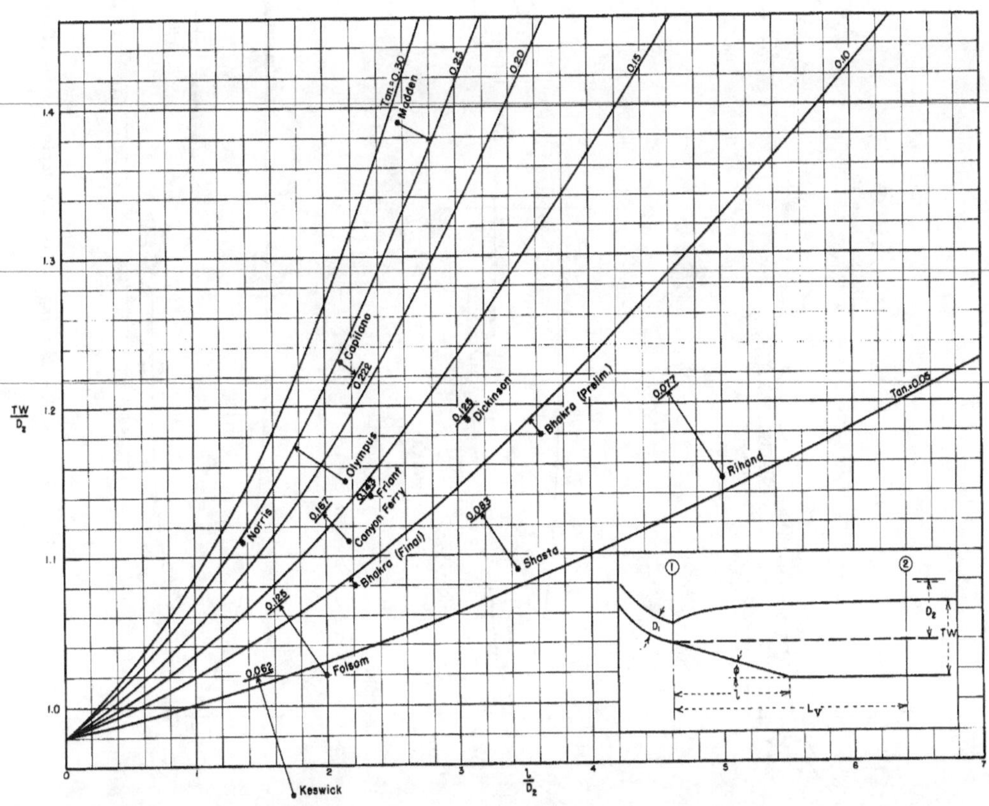

FIGURE 37.—*Comparison of existing sloping apron designs with experimental results (Basin V, Case B).*

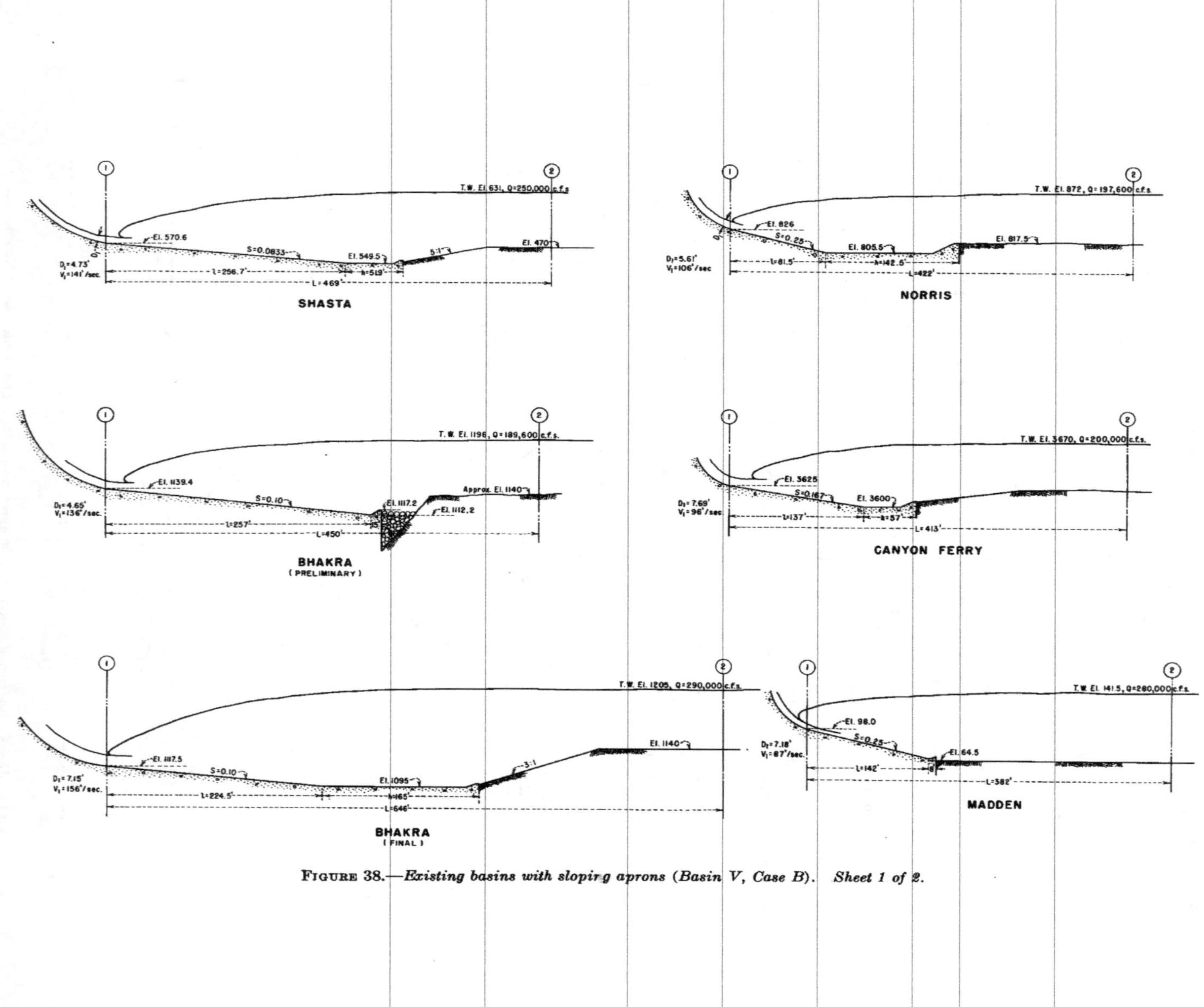

FIGURE 38.—*Existing basins with sloping aprons (Basin V, Case B). Sheet 1 of 2.*

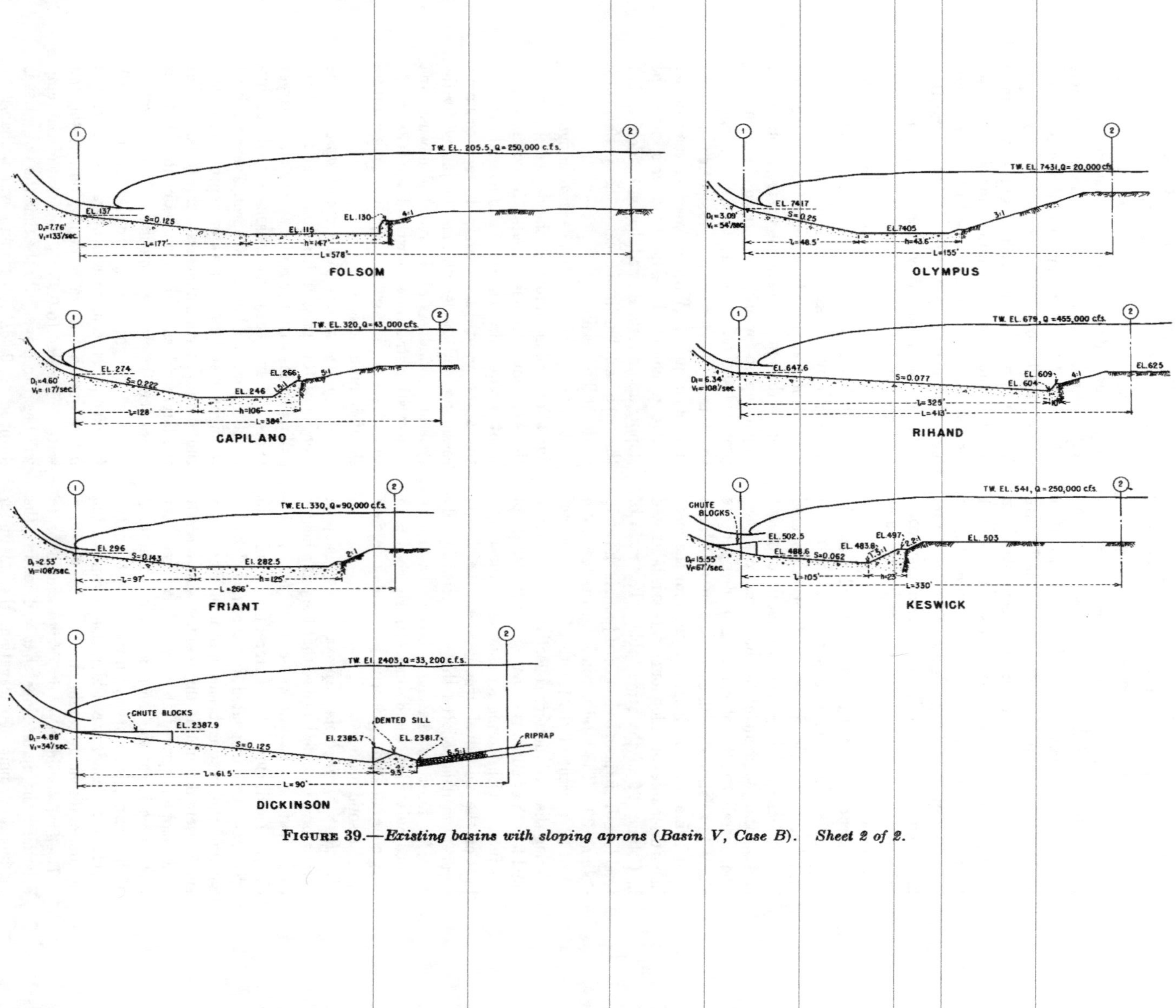

FIGURE 39.—*Existing basins with sloping aprons (Basin V, Case B). Sheet 2 of 2.*

Section 1 on the slope. Only the points for Capilano and Madden Dams show an excess of tail water depth for the length of slope used. On both these aprons the jump will occur upstream from Section 1 as shown in Figures 38 and 39. Friant and Dickinson Dams show almost perfect agreement with the derived curves while Bhakra (final) and Norris Dams show agreement within practical limits. All other points indicate that the tail water depth is insufficient to move the toe of the jump upstream to Section 1. The rather large chute blocks on Keswick Dam may compensate for the discrepancy indicated by the point in the margin of Figure 37.

All structures listed in Table 10 and shown in Figures 38 and 39 were designed with the aid of model studies. The degree of conservatism used in each case was dependent on local conditions and on the judgment of the individual designer. The overall lengths of aprons provided for the above 13 existing structures are shown in Column 16 of Table 10. The length of jump for the maximum discharge condition for each case is tabulated in Column 29 of the same table. The ratio of total length of apron to length of jump is shown in Column 30. The total apron length ranges from 39 to 83 percent of the length of jump; or considering the 13 structures collectively, the average total length of apron is 60 percent of the length of the jump. Considering all aspects of the model tests on the individual structures and the sloping apron tests it is believed that 60 percent is sufficient for most installations. Longer basins are needed only when the downstream riverbed is in very poor condition. Shorter basins may be used where a solid bed exists.

Evaluation of sloping aprons. Many sloping aprons have been designed so that the jump height curve matches the tail water curve for all discharge conditions. This procedure results in what has been designated a "tailormade" basin. Some of the existing basins shown in Figures 38 and 39 were designed in this manner. As a result of the sloping apron tests it was discovered that this course is not the most desirable. Matching of the jump height curve with the tail water curve should be a secondary consideration, except for the maximum discharge condition.

The first consideration in design should be to determine the apron slope that will require the minimum amount of excavation, the minimum amount of concrete, or both, for the maximum discharge and tail water condition. This is the prime consideration. Only then is the jump height checked to determine whether the tail water depth is adequate for the intermediate discharges. It will be found that the tail water depth usually exceeds the required jump height for the intermediate discharges resulting in a slightly submerged condition for intermediate discharges, but performance will be very acceptable. The extra depth will provide a smoother water surface in and downstream from the basin and greater stability at the toe of the jump. Should the tail water depth be insufficient for intermediate flows, it will be necessary to increase the depth by increasing the slope, or reverting to a horizontal apron. It is not necessary that the front of the jump form at the upstream end of the sloping apron for low or intermediate discharges, provided the tail water depth and the length of basin available for energy dissipation are considered adequate. With this method, the designer is free to choose the slope he desires, since the sloping apron tests showed, beyond a doubt, that the slope itself has little effect on the performance of the stilling basin.

It is not possible to standardize design procedures for sloping aprons to the degree shown for the horizontal aprons; greater individual judgment is required. The slope and overall shape of the apron must be determined from economic reasoning, and the length must be judged by the type and soundness of the riverbed downstream. The existing structures shown in Figures 38 and 39 should serve as a guide in proportioning future sloping apron designs.

Sloping apron versus horizontal apron. The Bureau of Reclamation has constructed very few stilling basins with horizontal aprons for its larger dams. It has been the consensus that the hydraulic jump on a horizontal apron is very sensitive to slight changes in tail water depth. The horizontal apron tests demonstrate this to be true for the larger values of the Froude number, but this characteristic can be remedied. If a horizontal apron is designed for a Froude number of 10, for example, the basin will operate satisfactorily for conjugate tail water depth, but as the tail water is lowered to 0.98 D_2 the front of the jump will begin to move. By the time the tail water is dropped to $0.96D_2$, the jump will probably be completely out of the basin. Thus, to design a stilling basin in this range the tail water depth must be known with certainty or a factor of safety provided in the design.

To guard against deficiency in tail water depth, the same procedure used for Basins I and II is suggested here. Referring to the minimum tail water curve for Basins I and II in Figure 11, the margin of safety can be observed for any value of the Froude number. It is recommended that the tail water depth for maximum discharge be at least 5 percent larger than the minimum shown in Figure 11. For values of the Froude number greater than 9, a 10-percent factor of safety may be advisable as this will not only stabilize the jump but will improve the performance of the basin. With the additional tail water depth, the horizontal apron will perform on a par with the sloping apron. Thus, the primary consideration in design need not be hydraulic but structural. The basin, with either horizontal or sloping apron, which can be constructed at the least cost is the most desirable.

Effect of slope of chute. A factor which occasionally affects stilling basin operation is the slope of the chute upstream from the basin. The foregoing experimentation was sufficiently extensive to shed some light on this factor. The tests showed that the slope of chute upstream from the stilling basin was unimportant, as far as jump performance was concerned, provided the velocity distribution in the jet entering the jump was reasonably uniform. For steep chutes or short flat chutes, the velocity distribution can be considered normal. Difficulty is experienced, however, with long flat chutes where frictional resistance on the bottom and side walls is sufficient to produce a center velocity greatly exceeding that on the bottom or sides. When this occurs, greater activity results in the center of the stilling basin than at the sides, producing an asymmetrical jump with strong side eddies. This same effect is also witnessed when the angle of divergence of a chute is too great for the water to follow properly. In either case the surface of the jump is unusually rough and choppy and the position of the front of the jump is not always predictable.

When long chutes precede a stilling basin the practice has been to make the upstream portion unusually flat, then increase the slope to 2:1, or that corresponding to the natural trajectory of the jet, immediately preceding the stilling basin. Figure 1A, which shows the model spillway for Trenton Dam, illustrates this practice. Bringing an asymmetrical jet into the stilling basin at a steep angle usually helps to redistribute the flow to stabilize the jump. This is not effective, however, where very long flat slopes have caused the velocity distribution to be completely out of balance.

The most adverse condition has been observed where long canal chutes terminate in stilling basins. A typical example is the chute and basin at Station 25+19 on the South Canal, Uncompahgre project, Colorado, Figure 40. The operation of this stilling basin is not particularly objectionable, but it will serve as an illustration. The above chute is approximately 700 feet long and has a slope of 0.0392. The stilling basin at the end is also shown in Figure 40. A photograph of the prototype basin operating at normal capacity is shown in Figure 41. The action is of the surging type; the jump is unusually rough, and has a great amount of splash and spray. Two factors contribute to the rough operation: the unbalanced velocity distribution in the entering jet, and excessive divergence of the chute in the steepest portion.

A definite improvement can be accomplished in future designs where long flat chutes are involved by utilizing the Type III basin described in Section 3. The baffle piers on the floor tend to alter the asymmetrical jet, resulting in an overall improvement in operation.

Recommendations. The following rules have been devised for the design of the sloping aprons developed from the foregoing experiments:

1. Determine an apron arrangement which will give the greatest economy for the maximum discharge condition. This is a governing factor and the only justification for using a sloping apron.

2. Position the apron so that the front of the jump will form at the upstream end of the slope for the maximum discharge and tail water condition by means of the information in Figure 37. Several trials will usually be required before the slope and location of the apron are compatible with the hydraulic requirement. It may be necessary to raise or lower the apron, or change the original slope entirely.

3. The length of the jump for maximum or partial flows can be obtained from Figure 33.

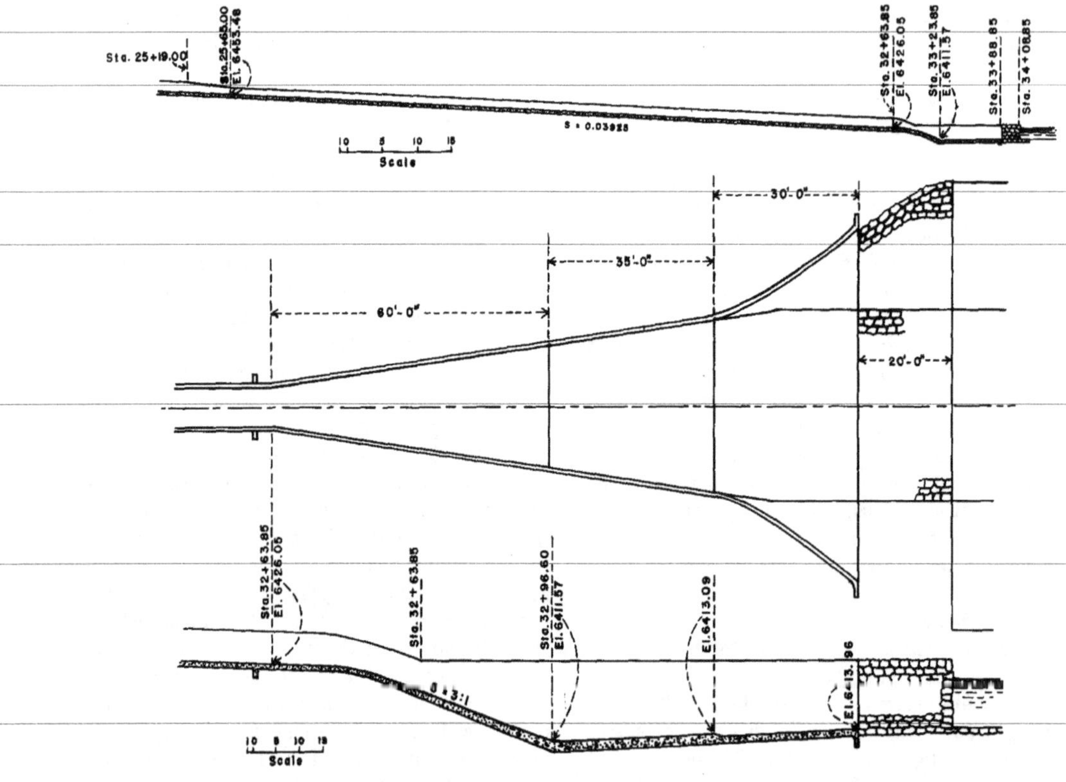

FIGURE 40.—*South Canal chute, Sta. 25+19, Uncompahgre project, Colorado.*

The portion of the jump to be confined on the stilling basin apron is a decision for the designer. In making this decision, Figures 38 and 39 may be helpful. The average over-all apron in Figures 38 and 39 averages 60 percent of the length of jump for the maximum discharge condition. The apron may be lengthened or shortened, depending upon the quality of the rock in the riverbed and other local conditions. If the apron is set on loose material and the downstream channel is in poor condition, it may be advisable to make the total length of apron the same as the length of jump.

4. With the apron designed properly for the maximum discharge condition, it should then be determined that the tail water depth and length of basin available for energy dissipation are sufficient for, say, ¼, ½, and ¾ capacity. If the tail water depth is sufficient or in excess of the jump height for the intermediate discharges, the design is acceptable. If the tail water depth is deficient, it may then be necessary to try a different slope or reposition the sloping portion of the apron. It is not necessary that the front of the jump form at the upstream end of the sloping apron for partial flows. In other words, the front of the jump may remain at Section 1 (Fig. 30B), move upstream from Section 1, or move down the slope for partial flows, provided the tail water depth and length of apron are considered sufficient for these flows.

5. Horizontal and sloping aprons will

perform equally well for high values of the Froude number if the proper tail water depth is provided.

6. The slope of the chute upstream from a stilling basin has little effect on the hydraulic jump when the velocity distribution and depth of flow are reasonably uniform on entering the jump.

7. A small solid triangular sill, placed at the end of the apron, is the only appurtenance needed in conjunction with the sloping apron. It serves to lift the flow as it leaves the apron and thus acts to control scour. Its dimensions are not critical; the most effective height is between $0.05D_2$ and $0.10D_2$ and a slope of 3:1 to 2:1 (see Figs. 38 and 39).

8. The spillway should be designed to operate with as nearly symmetrical flow in the stilling basin as possible. (This applies to all stilling basins.) Asymmetry produces large horizontal eddies that can carry riverbed material on to the apron. This material, circulated by the eddies, can abrade the apron and appurtenances in the basin at a very surprising rate. Eddies can also undermine wing walls and riprap. Asymmetrical operation is expensive operation, and operating personnel should be continually reminded of this fact.

9. Where the discharge over high spillways exceeds 500 c.f.s. per foot of apron width, where there is any form of asymmetry involved, and for the higher values of the Froude number where stilling basins become increasingly costly and the performance relatively less acceptable, a model study is advisable.

FIGURE 41 *Chute stilling basin on South Canal, Uncompahgre project, Colorado.*

Section 6

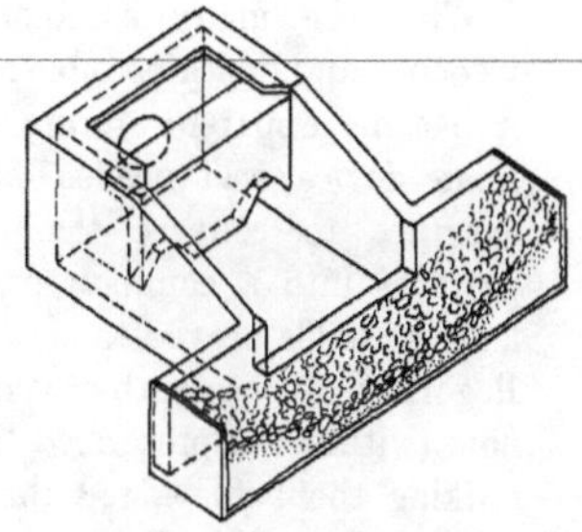

Stilling basin for pipe or open channel outlets (Basin VI)

The stilling basin developed in these tests is an impact-type energy dissipator, contained in a relatively small boxlike structure, which requires no tail water for successful performance. Although the emphasis in this discussion is placed on use with pipe outlets, the entrance structure may be modified for use with an open channel entrance.

Generalized design rules and procedures are presented to allow determining the proper basin size and all critical dimensions for a range of discharges up to 339 cubic feet per second and velocities up to about 30 feet per second. Greater discharges may be handled by constructing multiple units side by side. The efficiency of the basin in accomplishing energy losses is greater than a hydraulic jump of the same Froude number.

The development of this short impact-type basin was initiated by the need for some 50 or more stilling structures on a single irrigation project. The need was for relatively small basins providing energy dissipation independent of a tail water curve or tail water of any kind.

Since individual model studies on 50 small stilling structures were too costly a procedure, tests were made on a single setup which was modified as necessary to generalize the design for the range of expected operations.

Test Procedure

Hydraulic models. Hydraulic models were used to develop the stilling basin, determine the discharge limitations, and obtain dimensions for the various parts of the basin. Basins 1.6 to 2.0 feet wide were used in the tests. The inlet pipe was 6⅝ inches, inside diameter, and was equipped with a slide gate well upstream from the basin entrance so that the desired relations between head, depth, and velocity could be obtained. The pipe was transparent so that backwater effects in the pipe could be studied. Discharges of over 3 cubic feet per second and velocities up to 15 feet per second could be obtained during the tests. Hydraulic model-prototype relations were used to scale up the results to predict performance for discharges up to 339 second-feet and velocities up to 30 feet per second.

The basin was tested in a tail box containing gravel formed into a trapezoidal channel. The size of the gravel was changed several times during the tests. The outlet channel bottom was slightly wider than the basin and had 1:1 side slopes. A tail gate was provided at the downstream end to evaluate the effects of tail water.

Development of basin. The shape of the basin evolved from the development tests was the result of extensive investigations on many different arrangements. These tests are discussed briefly to show the need for the various parts of the adopted design.

With the many combinations of discharge, velocity, and depth possible for the incoming flow, it became apparent during the early tests that some device was needed at the stilling basin entrance to convert the many possible flow patterns into a common pattern. The vertical hanging baffle proved to be this device, Figure 42. Regardless of the depth or velocity of the incoming flow (within the prescribed limits) the flow after striking the baffle acted the same as any other combination of depth and velocity. Thus, some of the variables were eliminated from the problem.

The effect of velocity alone was then investigated, and it was found that for velocities 30 feet per second and below the performance of the structure was primarily dependent on the discharge. Actually, the velocity of the incoming flow does affect the performance of the basin, but from a practical point of view it could be eliminated from consideration. Had this not been done, an excessive amount of testing would have been required to evaluate and express the effect of velocity.

For velocities of 30 feet per second or less the basin width W was found to be a function of the discharge, Figure 42. Other basin dimensions are related to the width. To determine the necessary width, erosion test results, judgment, and operating experiences were all used, and the advice of laboratory and design personnel was used to obtain the finally determined limits. Since no definite line of demarcation between a "too wide" or "too narrow" basin exists, it was necessary to work between two more definite lines, shown in Figure 42 as the upper and lower limits. These lines required far less judgment to determine than a single intermediate line.

Various basin sizes, discharges, and velocities were tested taking note of the erosion, wave heights, energy losses, and general performance. When the upper and lower limit lines had been established, a line about midway between the two was used to establish the proper width of basin for various discharges. The exact line is not shown because strict adherence to a single curve would result in difficult-to-use fractional dimensions. Accuracy of this degree is not justifiable. Figure 43 shows typical performance of the recommended stilling basin for the three limits discussed. It is evident that the center photograph (B) represents a compromise between the upper limit operation which is very mild and the lower limit operation which is approaching the unsafe range.

Using the middle range of basin widths, other basin dimensions were determined, modified, and made minimum by means of trial and error tests on the several models. Dimensions for nine different basins are shown in Table 11. These should not be arbitrarily reduced since in the interests of economy the dimensions have been reduced as much as is safely possible.

Performance of basin. Energy dissipation is initiated by flow striking the vertical hanging baffle and being turned upstream by the horizontal portion of the baffle and by the floor, in vertical eddies. The structure, therefore, requires no tail water for energy dissipation as is necessary for a hydraulic jump basin. Tail water as high as $d+\frac{g}{2}$, Figure 42, however, will improve the performance by reducing outlet velocities, providing a smoother water surface, and reducing tendencies toward erosion. Excessive tail water, on the other hand, will cause some flow to pass over the top of the baffle. This should be avoided if possible.

The effectiveness of the basin is best illustrated by comparing the energy losses within the structure to those which occur in a hydraulic jump. Based on depth and velocity measurements made in the approach pipe and in the downstream channel (no tail water), the change in momentum was computed as explained in Section 1 for the hydraulic jump. The Froude number of the in-

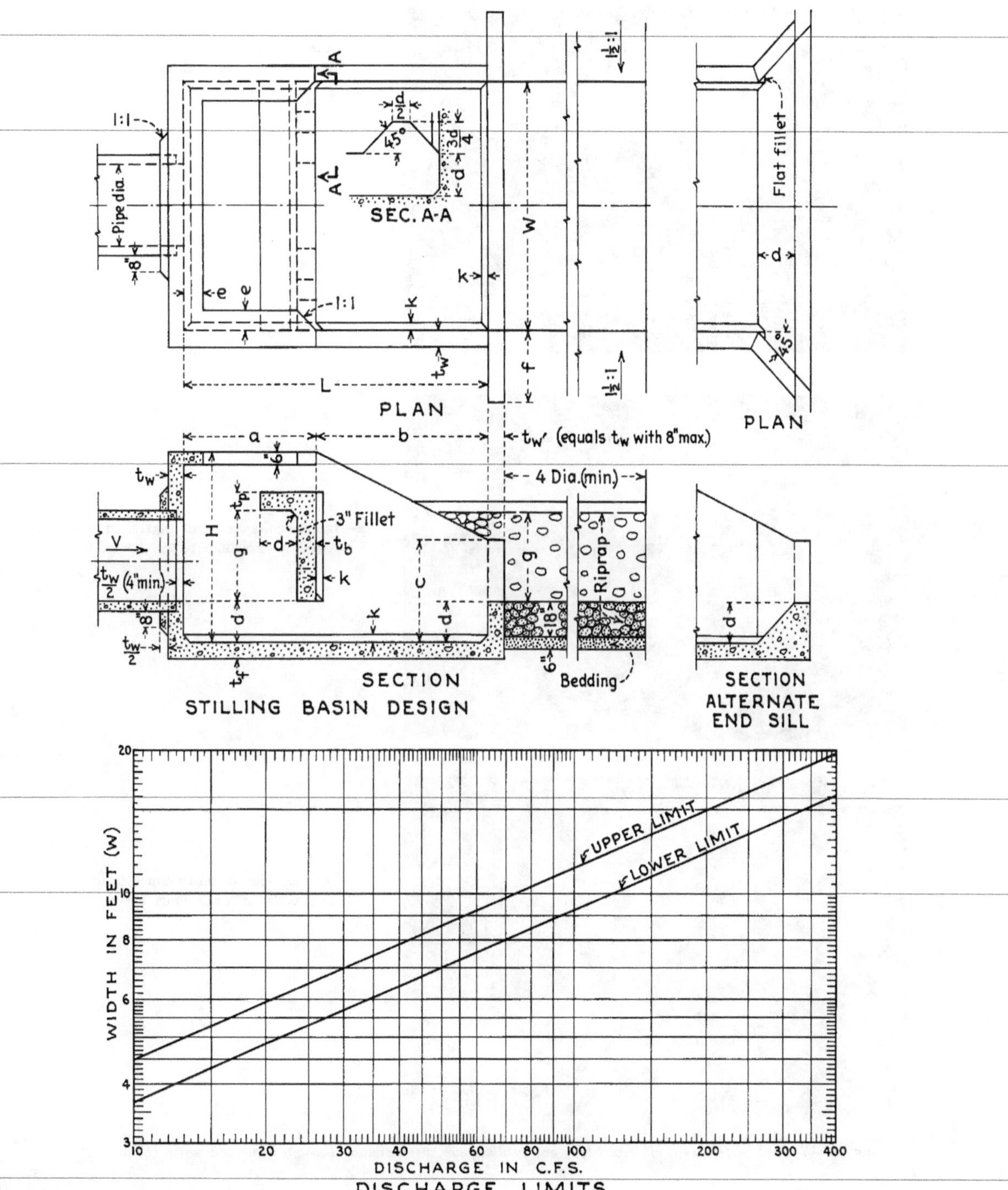

FIGURE 42.—*Impact-type energy dissipator (Basin VI).*

A—Lowest value of maximum discharge. Corresponds to upper limit curve.

B—Intermediate value of maximum discharge. Corresponds to tabular values.

C—Largest value of maximum discharge. Corresponds to lower limit curve.

FIGURE 43.—*Typical performance of impact-type energy dissipator at maximum discharges—no tail water (Basin VI).*

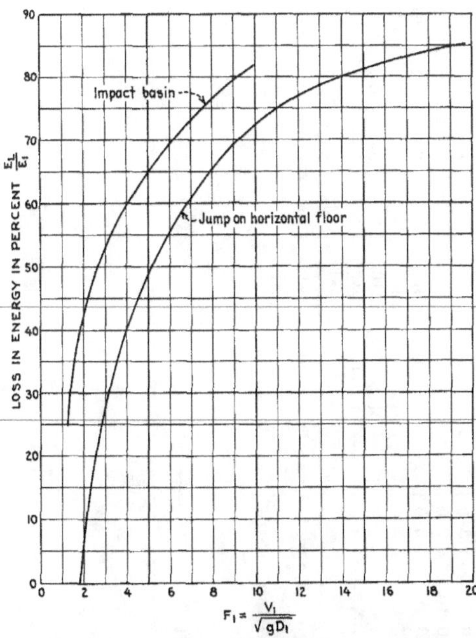

FIGURE 44.—*Comparison of energy losses—impact basin and hydraulic jump.*

coming flow was computed using D_1, obtained by converting the flow area in the partly full pipe into an equivalent rectangle as wide as the pipe diameter. Compared to the losses in the hydraulic jump, Figure 44, the impact basin shows greater efficiency in performance. Inasmuch as the basin would have performed just as efficiently had the flow been introduced in a rectangular cross section, the above conclusion is valid.

Basin Design

Table 11 and the key drawing, Figure 42, may be used to obtain dimensions for the usual structure operating within usual ranges. However, a further understanding of the design limitations may help the designer to modify these dimensions when necessary for special operating conditions.

The basin dimensions, Columns 4 to 13, are a function of the maximum discharge to be expected, Column 3. Velocity at the stilling basin entrance need not be considered, except that it should not greatly exceed 30 feet per second.

Columns 1 and 2 give the pipe sizes which have been used in field installations. However, these may be changed as necessary. The suggested sizes were obtained by assuming the velocity of flow to be 12 feet per second. The pipes shown would then flow full at maximum discharge or they would flow half full at 24 feet per second. The basin operates as well whether a small pipe flowing full or a larger pipe flowing partially full is used. The pipe size may therefore be modified to fit existing conditions, but the relation between structure size and discharge should be maintained as given in the table. In fact, a pipe need not be used at all; an open channel having a width less than the basin width will perform equally as well.

The invert of the entrance pipe, or open channel, should be held at the elevation shown on the drawing of Figure 42, in line with the bottom of the baffle and the top of the end sill, regardless of the size of the pipe selected. The entrance pipe may be tilted downward somewhat without affecting performance adversely. A limit of 15° is a suggested maximum although the loss in efficiency at 20° may not cause excessive erosion. For greater slopes use a horizontal or sloping pipe (up to 15°) two or more diameters long just upstream from the stilling basin.

For submerged conditions a hydraulic jump may be expected to form in the downstream end of the pipe sealing the exit end. If the upper end of the pipe is also sealed by incoming flow, a vent may be necessary to prevent pressure fluctuation in the system. A vent to the atmosphere, say one-sixth the pipe diameter, should be installed upstream from the jump.

The notches shown in the baffle are provided to aid in cleaning out the basin after prolonged nonuse of the structure. When the basin has silted level full of sediment before the start of the spill, the notches provide concentrated jets of water to clean the basin. If cleaning action is not considered necessary the notches need not be constructed. However, the basin is designed to carry the full discharge, shown in Table 11, over the top of the baffle if for any reason the space beneath the baffle becomes clogged, Figure 45C. Although performance is obviously not as good, it is acceptable.

TABLE 11.—*Stilling basin dimensions (Basin VI). Impact-type energy dissipator.*

Suggested pipe size [1]		Max discharge Q	Feet and inches										Inches					
Dia in.	Area (sq ft)		W	H	L	a	b	c	d	e	f	g	t_w	t_f	t_b	t_p	K	Suggested riprap size
(1)	(2)	(3)	(4)	(5)	(6)	(7)	(8)	(9)	(10)	(11)	(12)	(13)	(14)	(15)	(16)	(17)	(18)	(19) [3]
18	1. 77	[2] 21	5–6	4–3	7–4	3–3	4–1	2–4	0–11	0–6	1–6	2–1	6	6½	6	6	3	4. 0
24	3. 14	38	6–9	5–3	9–0	3–11	5–1	2–10	1–2	0–6	2–0	2–6	6	6½	6	6	3	7. 0
30	4. 91	59	8–0	6–3	10–8	4–7	6–1	3–4	1–4	0–8	2–6	3–0	6	6½	7	7	3	8. 5
36	7. 07	85	9–3	7–3	12–4	5–3	7–1	3–10	1–7	0–8	3–0	3–6	7	7½	8	8	3	9. 0
42	9. 62	115	10–6	8–0	14–0	6–0	8–0	4–5	1–9	0–10	3–0	3–11	8	8½	9	8	4	9. 5
48	12. 57	151	11–9	9–0	15–8	6–9	8–11	4–11	2–0	0–10	3–0	4–5	9	9½	10	8	4	10. 5
54	15. 90	191	13–0	9–9	17–4	7–4	10–0	5–5	2–2	1–0	3–0	4–11	10	10½	10	8	4	12. 0
60	19. 63	236	14–3	10–9	19–0	8–0	11–0	5–11	2–5	1–0	3–0	5–4	11	11½	11	8	6	13. 0
72	28. 27	339	16–6	12–3	22–0	9–3	12–9	6–11	2–9	1–3	3–0	6–2	12	12½	12	8	6	14. 0

[1] Suggested pipe will run full when velocity is 12 feet per second or half full when velocity is 24 feet per second. Size may be modified for other velocities by Q=AV, but relation between Q and basin dimensions shown must be maintained.

[2] For discharges less than 21 second-feet, obtain basin width from curve of Fig. 42. Other dimensions proportional to W; $H=\frac{3W}{4}$, $L=\frac{4W}{3}$, $d=\frac{W}{6}$, etc.

[3] Determination of riprap size explained in Sec. 10.

A—*Erosion of channel bed—standard wall and end sill.*

B—*Less erosion occurs with alternative end sill and wall design.*

C—*Flow appearance when entire maximum discharge passes over top of baffle during emergency operation.*

FIGURE 45. *Channel erosion and emergency operation for maximum tabular discharge—impact type energy dissipator—no tail water (Basin VI).*

With the basin operating normally, the notches provide some concentration of flow passing over the end sill, resulting in some tendency to scour, Figure 45A.. Riprap as shown on the drawing will provide ample protection in the usual installation, but if the best possible performance is desired, it is recommended that the alternate end sill and 45° end walls be used, Figures 45B and 42. The extra sill length reduces flow concentration, scour tendencies, and the height of waves in the downstream channel.

Figure 46 shows the performance of a prototype structure designed from Table 11. The basin, designed for a maximum discharge of 165 second-feet, is shown discharging 130 second-feet at a higher than recommended entrance velocity of about 39 feet per second. Performance is entirely satisfactory.

Conclusions and Recommendations

The following procedures and rules pertain to the design of Basin VI:

1. Use of Basin VI is limited to installations where the velocity at the entrance to the stilling basin does not greatly exceed 30 feet per second.

2. From the maximum expected discharge, determine the stilling basin dimensions, using Table 11, Columns 3 to 13. The use of multiple units side by side may prove economical in some cases.

3. Compute the necessary pipe area from the velocity and discharge. The values in Table 11, Columns 1 and 2, are suggested sizes based on a velocity of 12 feet per second and the desire that the pipe run full at the

discharge given in Column 3. Regardless of the pipe size chosen, maintain the relation between discharge and basin size given in the table. An open channel entrance may be used in place of a pipe. The approach channel should be narrower than the basin with invert elevation the same as the pipe.

4. Although tail water is not necessary for successful operation, a moderate depth of tail water will improve the performance. For best performance set the basin so that maximum tail water does not exceed $d+\frac{g}{2}$, Figure 42.

Discharge 130 c.f.s. (80 percent of maximum)

FIGURE 46.—*Prototype performance of Basin VI.*

5. Suggested thicknesses of various parts of the basin are given in Columns 14 to 18, Table 11.

6. The suggested sizes for the riprap protective blanket, given in Column 19 of Table 11, show the minimum size of individual stones which will resist movement when critical velocity occurs over the end sill. Since little is known regarding the effect of interlocking rock pieces, most of the riprap should consist of the sizes given or larger. An equation (*34*), (*35*) for determining minimum stone sizes, which appears from a limited number of experiments and observations to be accurate, is given below

$$V_b=2.6\sqrt{d}$$

where

V_b=bottom velocity in feet per second

d=diameter of rock in inches

The rock is assumed to have a specific gravity of about 2.65. The accuracy of the equation is not known for velocities above 16 feet per second.

7. The entrance pipe or channel may be tilted downward about 15° without affecting performance adversely. For greater slopes use a horizontal or sloping pipe (up to 15°) two or more diameters long just upstream from the stilling basin. Maintain proper elevation of invert at entrance as shown on the drawing.

8. If a hydraulic jump is expected to form in the downstream end of the pipe and the pipe entrance is sealed by incoming flow, install a vent about one-sixth the pipe diameter at any convenient location upstream from the jump.

9. For best possible operation of basin use, an alternative end sill and 45° wall design are shown in Figure 42. Erosion tendencies will be reduced as shown in Figure 45.

Section 7

Slotted and solid buckets for high, medium, and low dam spillways (Basin VII)

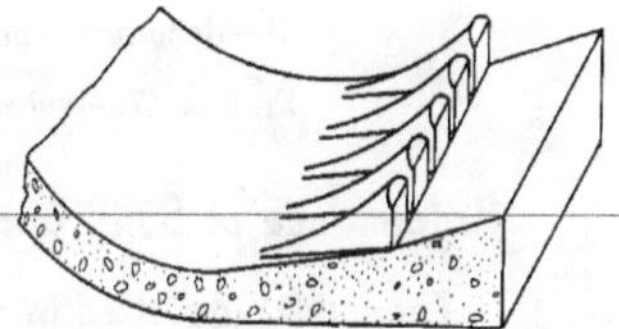

THE development of submerged buckets has been in progress for many years. Several types have been proposed, tested, and rejected for one reason or another. In 1933, with the aid of hydraulic models, the Bureau of Reclamation developed a solid bucket of the type shown in Figure 47A for use at Grand Coulee Dam.[1]

In 1945, a submerged slotted bucket of the type shown in Figure 47B was developed by the Bureau for use at Angostura Dam.[2] In 1953 and 1954, extensive hydraulic model tests, covering a complete range of bucket sizes and tail water elevations, were conducted to verify the bucket dimensions and details obtained in 1945 and to establish general relations between bucket size, discharge capacity, height of fall, and the maximum and minimum tail water depth limits. The 1945 and 1953–54 studies are the subject of this section.

Using the 1953–54 data, dimensionless curves were plotted which may be used in the hydraulic design of slotted buckets for most combinations of spillway height and discharge capacity without the need for individual hydraulic model tests. Strict adherence to the charts and rules presented will provide the designer with the smallest possible structure consistent with good performance and a moderate factor of safety. It is suggested, however, that confirming hydraulic model tests be performed whenever: (a) sustained operation near the limiting conditions is expected, (b) discharges per foot of width exceed 500 to 600 c.f.s., (c) velocities entering the bucket are over 75 feet per second, (d) eddies appear to be possible at the ends of the spillway, and (e) waves in the downstream channel would be a problem.

[1] Grand Coulee Dam, on the Columbia River in northeastern Washington, is a major feature of the Columbia Basin project. It is a concrete gravity-type dam having an overfall spillway 1,650 feet wide by 390 feet high from the bucket invert to crest elevation. The spillway is designed for 1 million cubic feet per second.

[2] Angostura Dam is a principal structure of the Angostura Unit of the Missouri River Basin project. It is on the Cheyenne River in southwestern South Dakota, and is an earthfill structure having a concrete overfall spillway 274 feet wide by 117.2 feet high from the bucket invert to crest elevation. The spillway is designed for 247,000 cubic feet per second.

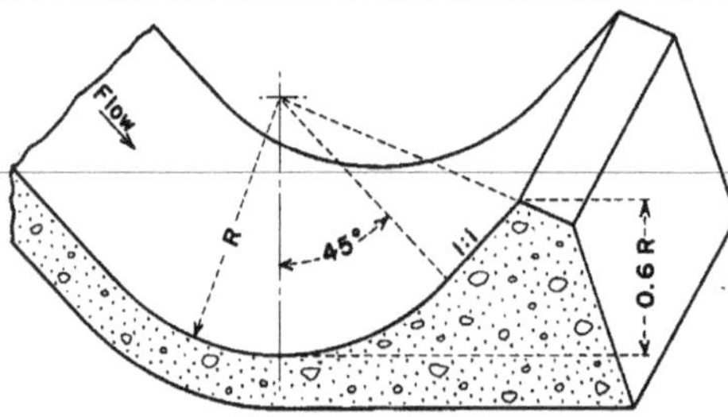

A—*Grand Coulee type solid bucket*

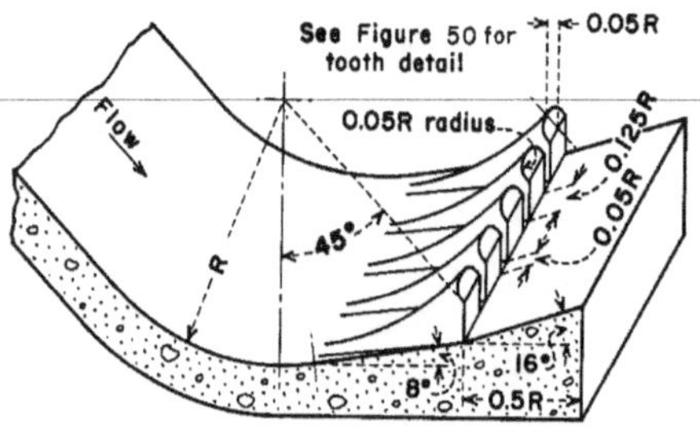

B—*Angostura type slotted bucket*

FIGURE 47.—*Submerged buckets.*

Performance of Solid and Slotted Buckets

The solid and slotted buckets are shown operating in Figure 48.[3] The hydraulic action and the resulting performance of the two buckets are quite different. Both types require more tail water depth than a hydraulic jump basin. In the solid bucket, all of the flow is directed upward by the bucket lip to create a boil on the water surface and a violent ground roller on the riverbed. The severity of the high boil and the ground roller depends upon tail water depth. Low tail water produces the most violent boils and ground rollers. The upstream current in the ground roller moves bed material from downstream and deposits it at the bucket lip. Here, it is picked up, carried away, and dropped again. The constant motion of the loose material against the concrete lip and the fact that unsymmetrical spillway operation can cause eddies to sweep the piled-up material into the bucket make this bucket undesirable in some installations. Trapped material can cause abrasion damage in the bucket itself. With the slotted bucket, part of the flow passes through the slots, spreads laterally, and is lifted away from the channel bottom by the apron. Thus, the flow is dispersed and distributed over a greater area, providing less violent flow concentrations than occur with a solid bucket. Bed material is neither deposited nor carried away from the bucket lip. Debris that might get into the bucket is immediately washed out.

[3] Fig. 48 and other drawings showing flow currents have been traced from one or more photographs.

With the slotted bucket, sweepout occurs at a slightly higher tail water elevation than with the solid bucket, and if the tail water is extremely high, the flow may dive from the apron lip to scour the channel bed, as shown in Figure 49. With the solid bucket, diving does not occur. In general, however, the slotted bucket is an improvement over the solid type, particularly for lower ranges of tail water depths.

Slotted Bucket Development Tests

General. The basic concept of the slotted bucket was the result of tests made to adapt the solid bucket for use at Angostura Dam. These tests, made on a 1:42 scale sectional model, are summarized in the following paragraphs.

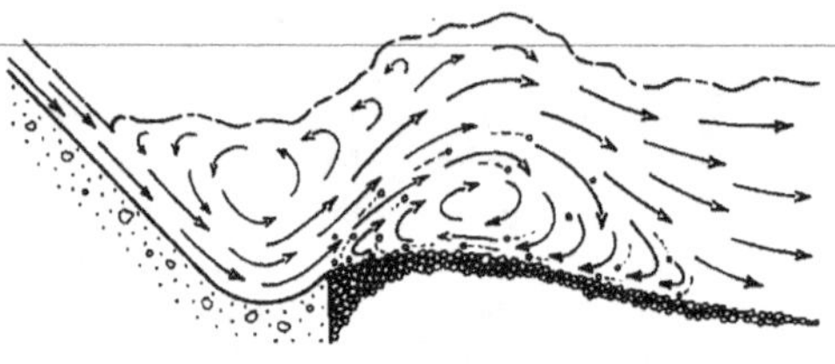

A—*Solid type bucket*

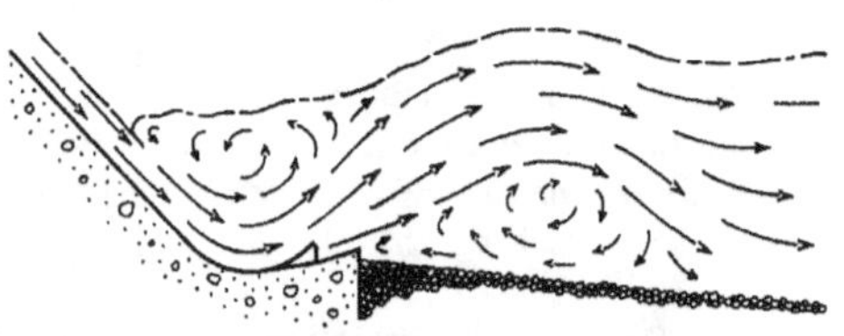

B—*Angostura type slotted bucket*

Bucket radius=12″, Discharge (q)=3 c.f.s.
Tailwater depth=2.3′
Crest elevation to bucket invert=5.0′

FIGURE 48.—*Performance of solid and slotted buckets.*

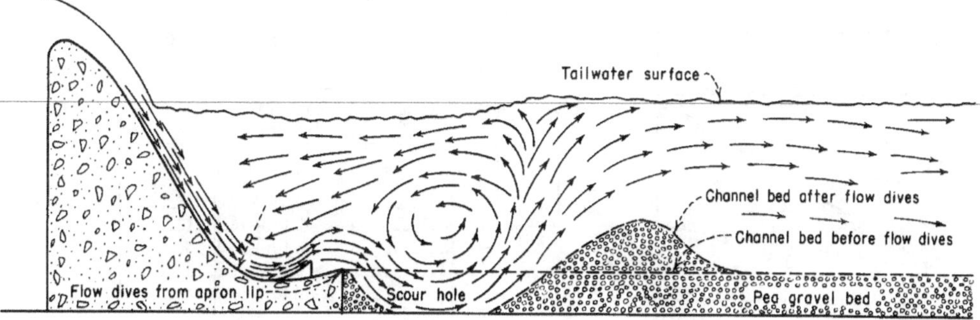

Note: The diving flow condition occurs with the slotted bucket only when the tailwater depth becomes too great.

FIGURE 49.—*Diving flow condition—slotted bucket.*

Development from solid bucket. The first tests were undertaken to determine the minimum radius of bucket required for the maximum flow and to determine the required elevation of the bucket invert for the existing tail water conditions. Solid type buckets were used in the model to determine these approximate values, since the slotted bucket had not yet been anticipated. The 42-foot-radius bucket was found to be the smallest bucket which would provide satisfactory performance for 1,010 c.f.s. per foot of width and a velocity of 75 feet per second.

Best performance occurred when the bucket invert was 77 feet below tail water elevation. For all invert elevations tested, however, a ground roller, Figure 48A, moved bed material from downstream and deposited it against the bucket lip.

The second stage in the development was to modify the bucket to prevent bed material from piling along the lip. Tubes were placed in the bucket lip through which jets of water flowed to sweep away the deposited material. Results were satisfactory at low discharges, but for the higher flows loose material piled deeply over the tube exits, virtually closing them.

Slots in the bucket lip were then used instead of larger tubes. The slots were found not only to keep the bucket lip free of loose material, but also to provide exits for debris that might find its way into the bucket during unsymmetrical operation of the spillway.

To maintain the effectiveness of the bucket action in dissipating energy, the slots were made just wide enough to prevent deposition at the bucket lip. The first slots tested were 1 foot 9 inches wide, spaced three times that distance apart. The slot bottoms were sloped upward on an 8° angle so that the emerging flow would not scour the channel bottom, and were made tangent to the bucket radius to prevent discontinuities in the surfaces over which the flow passed. The material remaining between the slots then became known as teeth. Three tooth designs, shown in Figure 50, were tested.

Tooth shape, spacing, and pressures. With Tooth Design I, the energy dissipating action of the bucket and the elimination of piled material along the bucket lip were both satisfactory. However, small eddies, formed by the jets leaving the slots, lifted loose gravel to produce abrasive action on the downstream face of the teeth. Therefore, an upward sloping apron was installed downstream from the teeth to help spread the jets from the slots and also to keep loose material away from the teeth. The apron was sloped upward slightly steeper than the slope of the slots, to provide better contact with the jets and thus spread the jet laterally. The apron was found to perform as intended. However, the best degree of slope for the apron and the shortest possible apron length were investigated after the tooth shape and spacing were determined.

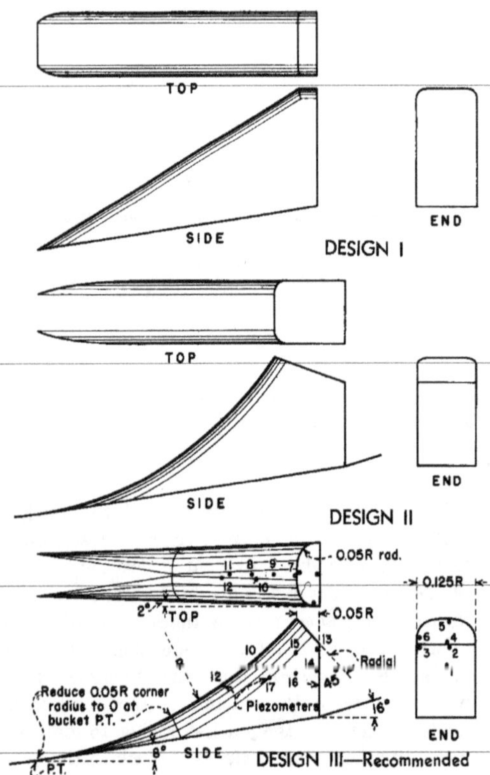

FIGURE 50.—*Tooth shapes tested for slotted bucket.*

The profile of Tooth Design II, Figure 50, was made to conform to the radius of the bucket, eliminating the discontinuity in the flow passing over the teeth. A smoother water surface occurred downstream from the bucket. Pressure measurements showed the necessity of rounding the edges of the teeth. Model radii ranging from 0.1 to 0.3 inch were investigated. The larger radius (12.6 inches prototype) was found to be the most desirable.

Tooth Design III, Figure 50, showed improved pressure conditions on the sides and downstream face of the teeth, when the radius on the tooth edges was increased to 15 inches. Subatmospheric pressures occurred on the downstream face of the teeth at Piezometers 3, 4, and 5, but were above the critical cavitation range.

TABLE 12.—*Pressures on tooth—Design III*

[0.125R width, 0.05R spacing, 1,000 c.f.s. per foot, 77 feet tail water depth]

Piezometer No.	Pressure ft. of water	Piezometer No.	Pressure ft. of water
1	+1 to +16	9	+58
2	+5 to +13	10	+42
3	−2 to +15	11	+68
4	−13 to +16	12	+49
5	−9 to +11	13	+11
6	+8 to +16	14	+13
7	+22	15	+21
8	+62	16	+34
		17	+39

Preliminary tests had shown that pressures on the teeth varied according to the tooth spacing. The most favorable pressures consistent with good bucket performance occurred with Tooth Design III, tooth width 0.125R, and spacing 0.05R at the downstream end. Table 12 shows the pressures in feet of water at the piezometers.

For 1,000 cubic feet per second per foot of width in a 1:42 scale model having a 42-foot-radius bucket, Piezometers 1 through 6 fluctuated between the limits shown. Piezometers 3, 4, and 5 showed subatmospheric values, but since these piezometers are on the downstream face of the teeth, it is unlikely that damage would occur as a result of cavitation. According to the pressure data, significant cavitation should not occur for velocities up to about 75 feet per second; i.e., velocity computed from the difference between headwater and tail water elevations.

Reducing the tooth spacing to 0.035R raised the pressures at Piezometers 3, 4, and 5 to positive values. Pressures on the tooth are shown in Table 13 for a discharge of 1,000 c.f.s. per foot of width in a 1:42 scale model having a 42-foot-radius bucket.

For 0.035R spacing, the teeth should be safe against cavitation for velocities over 75 feet per second. For small buckets, the spaces may be too small for convenient construction. In other respects, the 0.035R tooth spacing is satisfactory.

Apron downstream from teeth. The short apron downstream from the teeth serves to spread the jets from the slots and improve the stability of the flow leaving the bucket. A 16° upward sloping apron was found to be most satisfactory. With a 12° slope, the flow was unstable, intermittently diving from the end of the apron to scour the riverbed. With a 20° slope, longitudinal spreading of the flow was counteracted to some degree by the directional effect of the steep apron.

Two apron lengths, one 10 feet and one 20 feet, were tested to determine the minimum length required for satisfactory operation. The longer apron, 0.5R in length, was found necessary to accomplish lateral spreading of the jets and produce a uniform flow leaving the apron. The 20-foot apron on a 16° slope was therefore adopted for use.

Slotted bucket performance. The slotted bucket thus developed, shown in Figure 47B, operated well over the entire range of discharge and tail water conditions in the sectional model, scale 1:42. The bucket was also tested at a scale of 1:72 on a wide spillway where end effects of the bucket could also be observed and evaluated.

In the 1:72 model, minor changes were made before the bucket was constructed and installed. The bucket radius was changed from 42 to 40 feet, and the maximum discharge was lowered from 277,000 to 247,000 c.f.s. Figure 51 shows the 1:72 model operation for 247,000 c.f.s. (900 c.f.s. per foot of width), erosion after 20 minutes of operation, and erosion after 1½ hours of operation. Performance was excellent in all respects and was better than for any of the solid buckets or other slotted buckets investigated. For all discharges, the water surface was smoother and the erosion of the riverbed was less.

TABLE 13.—*Pressures on tooth—Design III*

[0.125R width, 0.035R spacing, 1,000 c.f.s. per foot, 77 feet tail water depth]

Piezometer No.	Pressure ft. of water	Piezometer No.	Pressure ft. of water
1	+36	9	+62
2	+27	10	+57
3	+30	11	+71
4	+26	12	+63
5	+14	13	+21
6	+27	14	+28
7	+39	15	+40
8	+64	16	+47
		17	+58

Slotted Bucket Generalization Tests

Test equipment. A testing flume and sectional model were constructed, as shown in Figure 52, and used in all subsequent tests. The test flume was 43 feet 6 inches long and 24 inches wide. The head bay was 14 feet deep and the tail bay was 6 feet 3 inches deep and had a 4- by 13-foot glass window on one side. The discharge end of the flume was equipped with a motor-driven tailgate geared to raise or lower the tail water level slowly so that continuous observations could be made.

The sectional spillway model was constructed to fill the flume width with an ogee crest at the top of a 0.7 sloping spillway face. The bucket assembly was made detachable from the spillway face. Four interchangeable buckets having radii of 6, 9, 12, and 18 inches, constructed according to the dimension ratios shown in Figure 47B, were designed so that they could be installed with the bucket inverts located 5 feet below the spillway crest and about 6 inches above the floor of the flume. All flow surfaces were constructed of galvanized sheet metal with smooth joints. The downstream channel was a movable bed molded in pea gravel. The gravel analysis:

	Percent
Retained on ¾-inch screen	6
Retained on ⅜-inch screen	66
Retained on No. 4 screen	25
Retained on Pan	3

Flow was supplied to the test flume through a 12-inch centrifugal pump and was measured by one of a bank of venturi meters permanently installed in the laboratory. Additional water, beyond the capacity of the 12-inch pump, was supplied by two vertical-type portable pumps equipped with two portable 8-inch orifice venturi meters. All venturi meters were calibrated in the laboratory. Water surface elevations were measured with hook gages mounted in transparent plastic wells.

Verification of the Slotted Bucket

General. The generalization tests began by first verifying and then attempting to improve the performance of the slotted bucket. The performance of the slotted bucket with the teeth removed was evaluated, and the performance of the slotted and solid buckets was compared.

Maximum discharge 900 c.f.s. per foot of width. Bucket invert El. 3,040, Tail water El. 3,114

Recommended slotted bucket 1:72 Scale Model

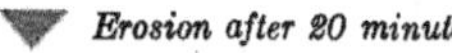

Erosion after 20 minutes

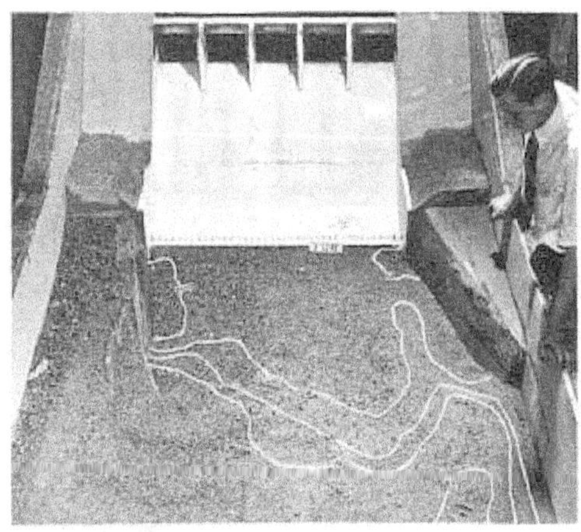

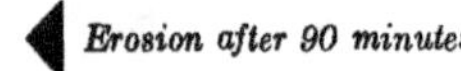

Erosion after 90 minutes

FIGURE 51 *Erosion test on Angostura Dam spillway.*

To determine whether practical modification could be made to improve performance, a 12-inch radius slotted bucket was used. The Angostura type shown in Figure 47 and Figure 53 was tested first to establish a performance standard with which to compare modified buckets. Since little bed erosion occurred with this bucket, improvements in bucket performance were directed toward reducing wave action in the downstream channel. Each modification was subjected to a standard test of 3 c.f.s. per foot of bucket width, with the tail water 2.3 feet above the bucket invert, Figure 48B. This was judged to be bucket capacity at a normal tail water. The movable bed was molded level, just below the bucket apron lip, at the start of each test.

Investigations were undertaken of four modifications of the bucket teeth, of the bucket with teeth removed, and of a solid bucket. The modifications tested are shown in Figure 53. Tooth Modifications I, III, and IV proved to be of no value. Tooth Modification II was an improvement, but was not considered to be of practical use for large buckets.

Tooth Modification I. The teeth were extended in height along the arc of the bucket radius from 45° to 60°, as shown in Figure 53. For the standard test, the bucket performed much the same as the original. However, a boil occurred about 6 inches farther upstream and was slightly higher. Waves were also slightly higher.

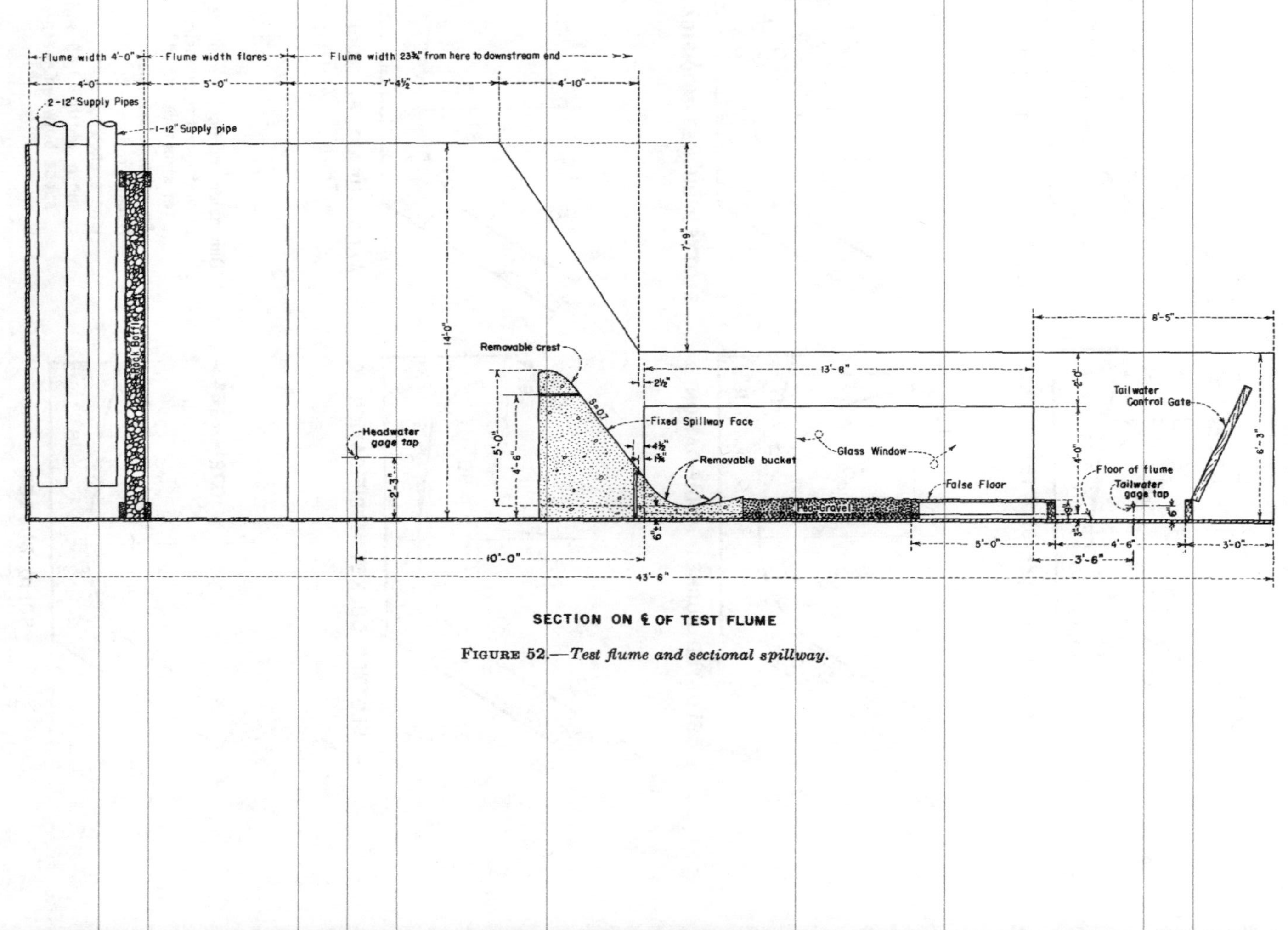

SECTION ON ℄ OF TEST FLUME

FIGURE 52.—*Test flume and sectional spillway.*

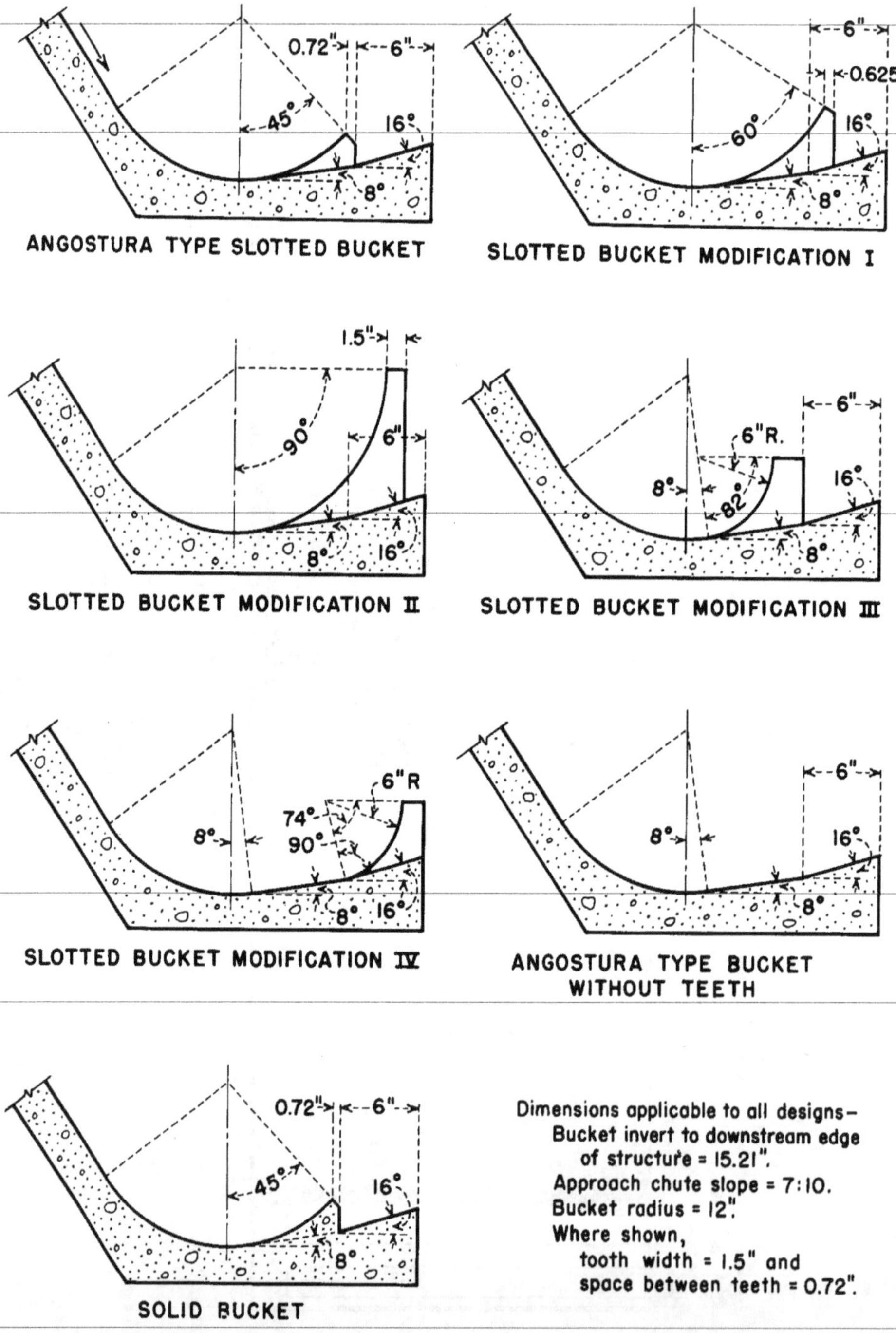

0.72"
6"
45°
16°
8°
ANGOSTURA TYPE SLOTTED BUCKET
6"
0.625'
60°
16°
8°
SLOTTED BUCKET MODIFICATION I
1.5"
90°
6"
8°
16°
SLOTTED BUCKET MODIFICATION II
6"
6"R.
8°
82°
16°
8°
SLOTTED BUCKET MODIFICATION III
6"R
74°
90°
8°
8°
16°
SLOTTED BUCKET MODIFICATION IV
6"
8°
16°
8°
ANGOSTURA TYPE BUCKET
WITHOUT TEETH
0.72"
6"
45°
16°
8°
SOLID BUCKET
Dimensions applicable to all designs–
Bucket invert to downstream edge
of structure = 15.21".
Approach chute slope = 7:10.
Bucket radius = 12".
Where shown,
tooth width = 1.5" and
space between teeth = 0.72".

FIGURE 53.—*Slotted bucket modifications tested.*

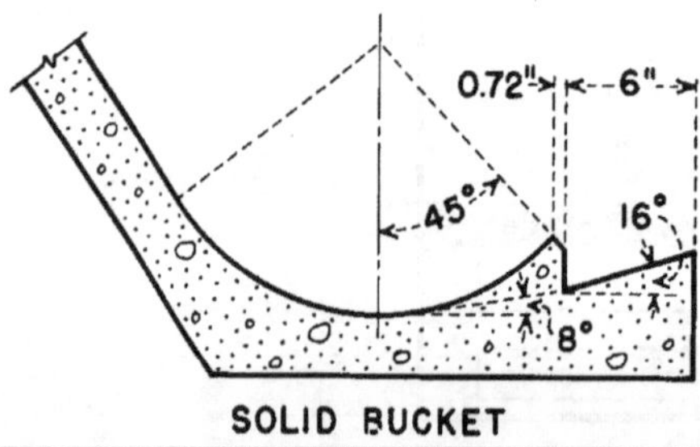

Tooth Modification II. The teeth were extended in height along the arc of the bucket radius to an angle of 90°, as shown in Figure 53. It was realized that the teeth would be too tall to be structurally stable in any but a small bucket, but the trend in performance was the primary purpose in making the test.

Performance was excellent for the standard test. A large portion of the flow was turned directly upward to the water surface where it rolled back into the bucket. The tail water depth in the bucket was about the same as the depth downstream. Only a slight boil could be detected over the teeth. The flow passing between the teeth provided uniform distribution of velocity from the channel bed to the water surface in the channel downstream. The downstream water surface was smooth and the channel bed was not disturbed. The bucket also performed well for high and low tail water elevations. In fact, the range of tail water depths for which the bucket operated satisfactorily was greater than for any other slotted bucket tested. The teeth are suggested for possible use in small buckets.

Tooth Modification III. In the third modification, a radius, half that of the bucket radius, was used as shown in Figure 53 to extend the teeth to a height of 90°. This modification was made to determine whether the height of the teeth, or the 90° curvature of the teeth, provided the improved performance.

Tests showed that the shorter teeth were not effective in lifting flow to the surface. Flow passed over and through the teeth to form a high boil downstream similar to the first modification.

Tooth Modification IV. The teeth from Modification III were placed on the apron at the downstream end of the bucket, as shown in Figure 53. Performance tests showed that the teeth turned some of the flow upward but the performance was no better than for the Angostura design.

Slotted bucket with teeth removed. Tests were made to indicate the value of the teeth and slots in dissipating the energy of the spillway flow. The bucket without teeth is shown in Figure 53. Operation was satisfactory for flows up to 2 c.f.s. per foot of width, about two-thirds maximum capacity of the bucket. For larger discharges, the flow leaving the bucket was unstable and the water surface was rough. For a few seconds, the boil would be quite high then suddenly would become quite low. However, erosion of the riverbed was negligible for all flows.

The tests indicated that the primary function of the teeth is to stabilize the flow and reduce water surface fluctuations in the channel downstream. The tests also suggested that should the teeth in a prototype slotted bucket deteriorate over a period of time, the degree of deterioration could be evaluated from the appearance of the surface flow. Discharges up to about half maximum would be satisfactory if the teeth were entirely gone.

Solid bucket. The solid bucket, shown in Figure 53, was tested to compare the action with that of a slotted bucket. The performance was similar to that shown in Figure 48A and described previously. These tests confirmed the earlier conclusion that a solid bucket may not be desirable when loose material can be carried into the bucket, when the high boil would create objectionable waves, or when a deep erosion hole located from 1 to 3 bucket radii downstream from the bucket lip would be objectionable.

Bucket Size and Tail Water Limits

General. The investigation to determine the minimum bucket size and tail water limits for a range of structure sizes, discharges, and overfall height was accomplished by testing 6-, 9-, 12-, and 18-inch-radius buckets. Each bucket was tested over a range of discharges and tail water elevations with the bed molded in two different positions. For each test, the head on the spillway was measured and recorded. The relationship between head and discharge on the spillway is shown in Figure 54.

Lower and upper tail water limits. Testing began with the bed molded slightly below the apron lip at a distance of approximately 0.05 of the bucket radius, R. For each discharge, q in cubic feet per second per foot of width, the tail water depth was lowered slowly until the flow swept out of the bucket, as shown in Figure 55A. The sweepout depth considered to be too low for proper bucket performance was a limiting tail water depth and was recorded in Tables 14 to 17 (line 2) and plotted in Figure 56. Tail water depth is the difference in elevation between the bucket invert and the tail water surface measured at the tail water gage shown in Figure 52. Figure

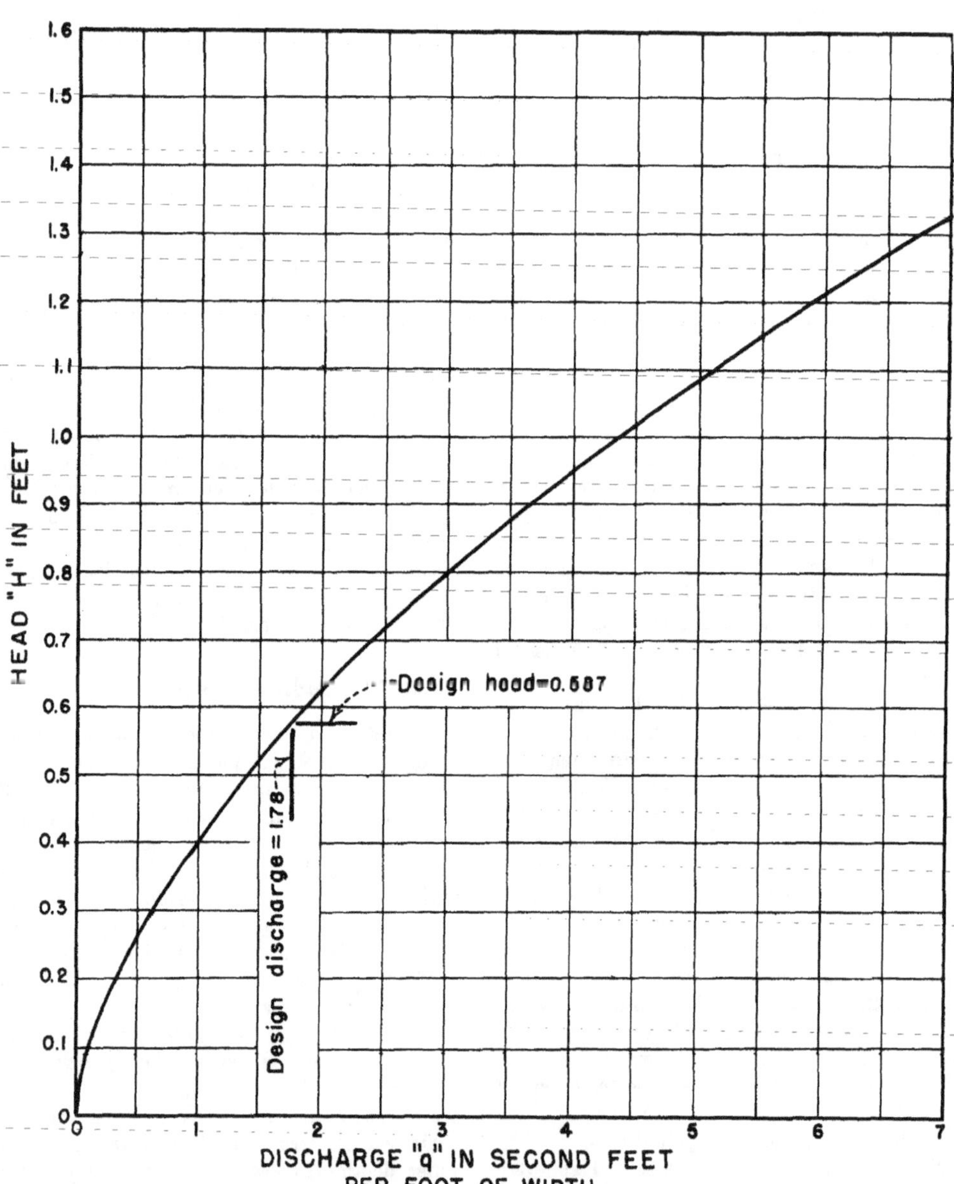

FIGURE 54.—*Discharge calibration of the 5-foot model spillway.*

55B shows the 6-inch bucket operating with tail water depth just safely above sweepout. The tail water depth just safely above the depth required for the sweepout will henceforth in this monograph be called the lower or minimum tail water limit.

At the sweepout depth, the flow left the bucket in the form of a jet, Figure 55A. The jet scoured

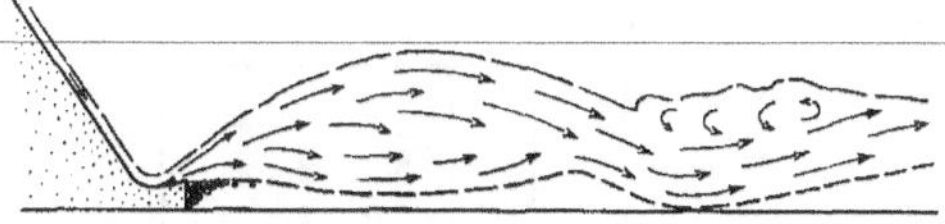

A—*Tail water below minimum. Flow sweeps out.*

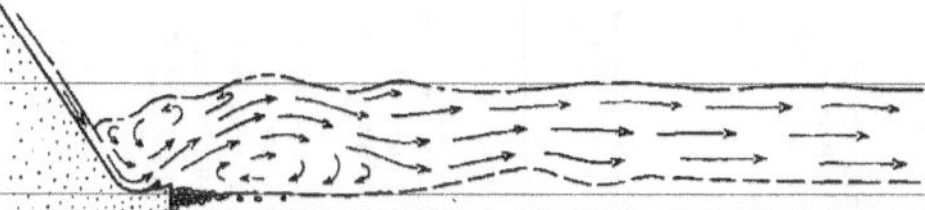

B—*Tail water below average but above minimum. Within normal operating range.*

C—*Tail water above maximum. Flow diving from apron scours channel.*

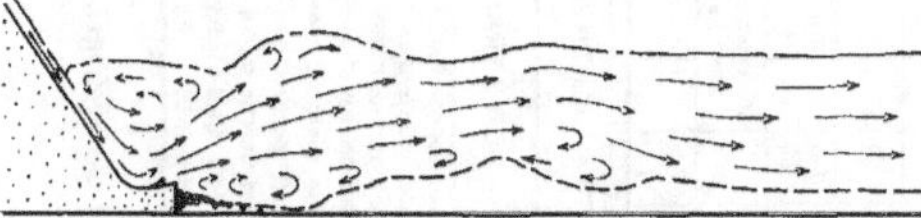

D—*Tail water same as in C. Diving jet is lifted by ground roller. Scour hole backfills similar to B. Cycle repeats. (Bed level 0.3 inch below apron lip at start of test.)*

FIGURE 55.—*Six-inch bucket discharging 1.75 c.f.s. per foot of width (design capacity).*

the channel bed at the point of contact but did not cause excessive water surface roughness downstream. However, a more undesirable flow pattern occurred just before sweepout. An unstable condition developed in the bucket, causing excessive erosion and water surface roughness. Therefore, it is undesirable to design a bucket for both submerged and flip action because of this transition region. The lower tail water limit was found to be from 0.05 to 0.15 foot above the sweepout depth. Only the sweepout depth was actually measured since it was a more definite point. A safe lower limit, T_{min}, was established at the conclusion of all model testing by adding 0.2 foot to the sweepout tail water depth.

For each discharge, the upper tail water limit was also investigated. The tail water was raised slowly until the flow dived from the apron lip, as shown in Figures 49 and 55C. When diving occurred, a deep hole was scoured in the channel bed near the bucket. The tail water depth for diving, considered to be too high for proper performance of the bucket, was also recorded in Tables 14 to 16 (line 12) and plotted in Figure 56. The tail water depth just safely below the depth required for diving will henceforth be called the upper or maximum tail water limit.

At the tail water depth required for diving to occur, Figure 55C, it was noted that after 3 or 4 minutes (model time) diving suddenly ceased and the flow rose to the surface as shown in Figure 55D. The changeover occurred only after the movable bed had become sufficiently scoured to allow a ground roller to form beneath the jet and lift the flow from the apron lip to the water surface. The ground roller then moved the deposited gravel upstream into the scoured hole until the riverbed was nearly level with the apron lip. At the same time, the strength of the ground roller was reduced until it was no longer capable of lifting the flow to the water surface and the flow dived again to start another cycle which was repeated over and over. Very little bed material was moved downstream out of reach of the ground roller even after several cycles. Five or six minutes were required for one cycle as a general rule.

When the flow was diving, the water surface was very smooth, but when the flow was directed toward the surface, a boil formed, and a rough downstream water surface was in evidence. In the former case, part of the energy was dissipated on the channel bed; in the latter case, energy was dissipated on the surface.

In approaching the upper tail water limit, diving occurred in spurts not sufficiently long to move bed material. As the depth approached that required for sustained periods of diving; the momentary spurts occurred more often. In the test data recorded in Table 14 and plotted in Figure 56, the tail water depth required to cause sustained diving was used, because it was a definite point. At the conclusion of all model testing, the upper tail water limit, T_{max}, was established by subtracting 0.5 foot from the tail water depth at which sustained diving occurred. In analyzing the data, as is explained later, an additional safety factor was included in the design curves.

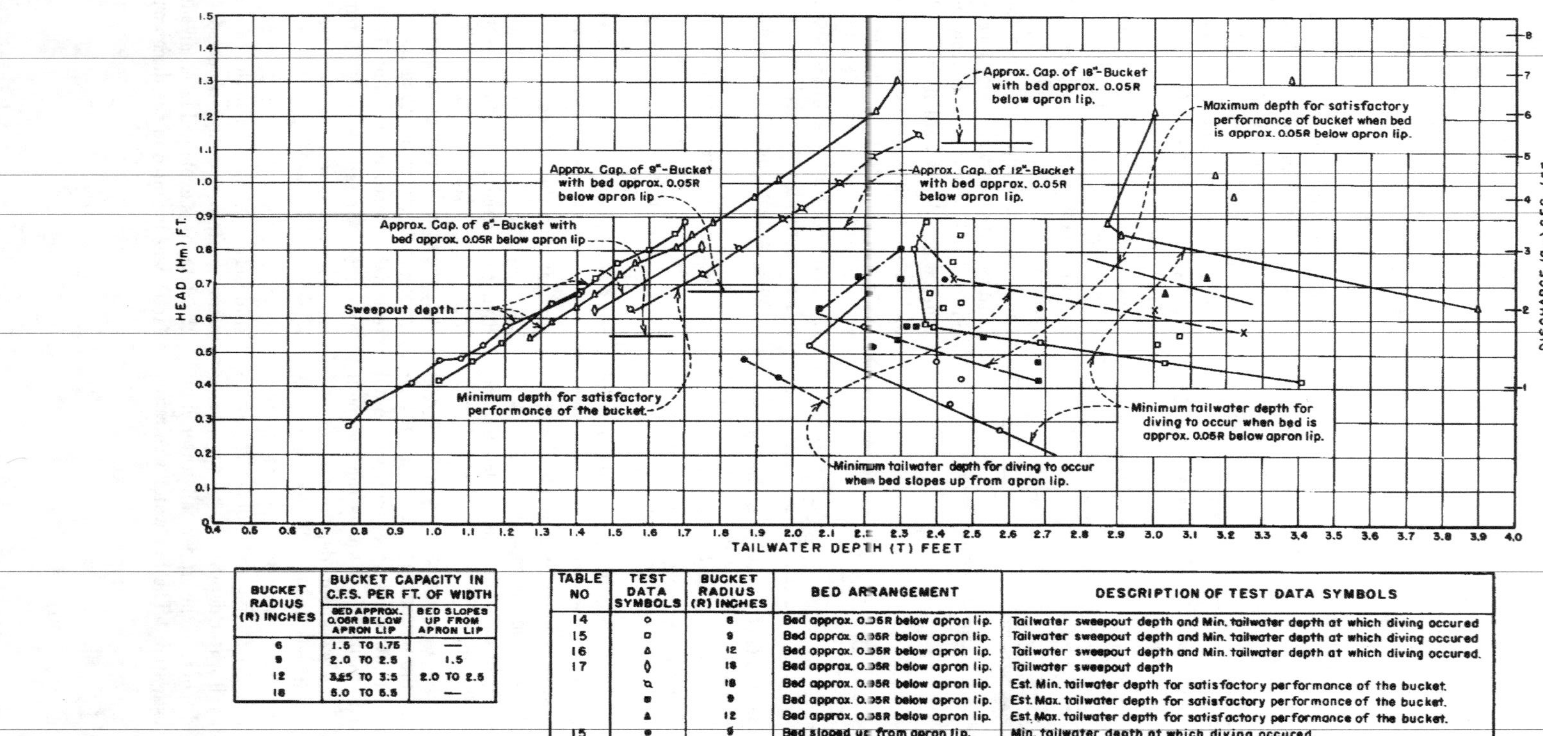

BUCKET RADIUS (R) INCHES	BUCKET CAPACITY IN C.F.S. PER FT. OF WIDTH	
	BED APPROX. 0.05R BELOW APRON LIP	BED SLOPES UP FROM APRON LIP
6	1.5 TO 1.75	—
9	2.0 TO 2.5	1.5
12	3.25 TO 3.5	2.0 TO 2.5
18	5.0 TO 5.5	—

TABLE NO	TEST DATA SYMBOLS	BUCKET RADIUS (R) INCHES	BED ARRANGEMENT	DESCRIPTION OF TEST DATA SYMBOLS
14	○	6	Bed approx. 0.05R below apron lip.	Tailwater sweepout depth and Min. tailwater depth at which diving occured
15	□	9	Bed approx. 0.05R below apron lip.	Tailwater sweepout depth and Min. tailwater depth at which diving occured
16	△	12	Bed approx. 0.05R below apron lip.	Tailwater sweepout depth and Min. tailwater depth at which diving occured.
17	◊	18	Bed approx. 0.05R below apron lip.	Tailwater sweepout depth
	□	18	Bed approx. 0.05R below apron lip.	Est. Min. tailwater depth for satisfactory performance of the bucket.
	■	9	Bed approx. 0.05R below apron lip.	Est. Max. tailwater depth for satisfactory performance of the bucket.
	▲	12	Bed approx. 0.05R below apron lip.	Est. Max. tailwater depth for satisfactory performance of the bucket.
15	●	9	Bed sloped up from apron lip.	Min. tailwater depth at which diving occured.
16	x	12	Bed sloped up from apron lip.	Min. tailwater depth at which diving occured.

FIGURE 56.—*Tail water limits and bucket capacities.*

TABLE 14.—*Data and calculated values for 6-inch-radius bucket*

	Run No.	Bed was approx 0.05R below apron lip at beginning of each run								
		1	2	3	4	5	6	7	8	9
		Sweepout Conditions								
1	H	0. 198	0. 274	0. 352	0. 413	0. 480	0. 481	0. 526	0. 581	0. 678
2	T (sweepout depth)	. 767	. 765	. 826	. 943	1. 023	1. 081	1. 139	1. 203	1. 403
3	q	. 31	. 54	. 81	1. 03	1. 30	1. 30	1. 50	1. 75	2. 25
4	T_{min}	. 967	. 965	1. 026	1. 143	1. 223	1. 281	1. 339	1. 403	1. 603
5	$\frac{V_1^2}{2g}$	4. 231	4. 309	4. 326	4. 270	4. 257	4. 200	4. 187	4. 178	4. 075
6	V_1	16. 50	16. 65	16. 70	16. 58	16. 56	16. 45	16. 42	16. 41	16. 20
7	D_1	. 019	. 032	. 048	. 062	. 078	. 079	. 091	. 107	. 139
8	$F=\frac{V_1}{\sqrt{gD_1}}$	21. 2	16. 31	13. 36	11. 72	10. 42	10. 31	9. 50	8. 85	7. 65
9	$\frac{T_{min}}{D_1}$	51. 43	29. 78	21. 10	18. 40	15. 57	16. 21	14. 66	13. 16	11. 54
10	$D_1+\frac{V_1^2}{2g}$	4. 245	4. 341	4. 374	4. 332	4. 335	4. 279	4. 278	4. 285	4. 214
11	$\frac{R}{D_1+\frac{V_1^2}{2g}}$	. 12	. 12	. 11	. 12	. 12	. 12	. 12	. 12	. 12
		Diving Flow Conditions								
12	T (diving depth)	2. 565	2. 576	2. 435	2. 464	2. 439	2. 397	2. 043	2. 200	2. 213
13	T_{max}	2. 065	2. 076	1. 935	1. 964	1. 939	1. 897	1. 543	1. 700	1. 713
14	$\frac{V_1^2}{2g}$	2. 133	2. 198	2. 417	2. 449	2. 541	2. 584	2. 983	2. 881	2. 965
15	V_1	11. 72	11. 89	12. 47	12. 55	12. 78	12. 90	13. 86	13. 62	13. 81
16	D_1	. 026	. 045	. 065	. 089	. 102	. 101	. 108	. 128	. 163
17	$F=\frac{V_1}{\sqrt{gD_1}}$	12. 67	9. 84	8. 62	7. 40	7. 06	7. 20	7. 42	6. 70	6. 02
18	$\frac{T_{max}}{D_1}$	77. 92	45. 72	29. 76	21. 62	19. 06	18. 82	14. 26	13. 23	10. 51
19	$D_1+\frac{V_1^2}{2g}$	2. 159	2. 243	2. 486	2. 538	2. 643	2. 685	2. 983	3. 009	3. 128
20	$\frac{R}{D_1+\frac{V_1^2}{2g}}$	. 23	. 22	. 20	. 20	. 19	. 19	. 17	. 17	. 16

R = bucket radius (ft.)

H = ht. of reservoir above the crest (ft.)

T = depth of TW above the bucket invert (ft.)

T_{min} = min. TW depth for good performance (ft.) = sweepout depth + 0.2 ft.

T_{max} = max. TW depth for good performance (ft.) = diving depth − 0.5 ft.

q = discharge per ft. of model crest length (c.f.s.)

X = ht. of crest above bucket invert = 5 ft.

V_1 = velocity of flow entering the bucket computed at TW el. (ft. per sec.)

D_1 = depth of flow entering the bucket computed at TW el. (ft.)

F = Froude number of flow entering bucket computed at TW el.

Maximum capacity of bucket estimated to be 1.5 to 1.75 c.f.s. per ft. of width.

TABLE 15.—*Data and calculated values for 9-inch-radius bucket*

	Run No.	Bed approx. 0.05R below apron lip at beginning of each run					
		1	2	3	4	5	6
		Sweepout Conditions					
1	H	0.419	0.476	0.531	0.642	0.682	0.722
2	T (sweepout depth)	1.02	1.11	1.19	1.33	1.41	1.45
3	q	1.05	1.28	1.52	2.05	2.28	2.50
4	T_{min}	1.22	1.31	1.39	1.53	1.61	1.65
5	$\frac{V_1^2}{2g}$	4.199	4.166	4.141	4.112	4.072	4.072
6	V_1	16.45	16.38	16.33	16.28	16.20	16.20
7	D_1	.064	.078	.093	.126	.141	.154
8	$F=\frac{V_1}{\sqrt{gD_1}}$	11.49	10.34	9.44	8.09	7.62	7.27
9	$\frac{T_{min}}{D_1}$	19.12	16.77	14.93	12.15	11.44	10.69
10	$D_1+\frac{V_1^2}{2g}$	4.262	4.244	4.234	4.238	4.212	4.226
11	$\frac{R}{D_1+\frac{V_1^2}{2g}}$	.18	.18	.18	.18	.18	.18
		Diving Flow Conditions					
12	T (diving depth)	3.40	3.03	3.01	2.46	2.38	2.44
13	T_{max}	2.90	2.53	2.51	1.96	1.88	1.94
14	$\frac{V_1^2}{2g}$	1.519	1.946	2.021	2.682	2.802	2.782
15	V_1	9.89	11.20	11.40	13.14	13.43	13.40
16	D_1	.106	.114	.133	.156	.170	.187
17	$F=\frac{V_1}{\sqrt{gD_1}}$	5.34	5.84	5.49	5.85	5.72	5.46
18	$\frac{T_{max}}{D_1}$	2.733	22.13	18.82	12.56	11.07	10.39
19	$D_1+\frac{V_1^2}{2g}$	1.625	2.060	2.154	2.838	2.972	2.969
20	$\frac{R}{D_1+\frac{V_1^2}{2g}}$	.46	.36	.35	.26	.25	.25

NOTE: See Table 14 for definition of symbols.

Maximum capacity of bucket estimated to be 2.0 to 2.5 c.f.s. per ft. of width.

TABLE 15.—*Data and calculated values for 9-inch-radius bucket*—Continued

	Run No.	Bed approx. 0.05R below apron lip at beginning of each run						
		7	8	9	10	11	12	13
		Sweepout Conditions						
1	H	0. 764	0. 805	0. 852	0. 884	0. 534	0. 578	0. 585
2	T (sweepout depth)	1. 51	1. 60	1. 67	1. 70			
3	q	2. 74	3. 00	3. 30	3. 52	1. 53	1. 73	1. 78
4	T_{min}	1. 71	1. 80	1. 87	1. 90			
5	$\frac{V_1^2}{2g}$	4. 054	4. 005	3. 982	3. 984			
6	V_1	16. 16	16. 06	16. 02	16. 02			
7	D_1	. 170	. 187	. 206	. 220			
8	$F=\frac{V_1}{\sqrt{gD_1}}$	6. 92	6. 56	6. 22	6. 03			
9	$\frac{T_{min}}{D_1}$	10. 08	9. 63	9. 07	8. 64			
10	$D_1+\frac{V_1^2}{2g}$	4. 224	4. 192	4. 188	4. 204			
11	$\frac{R}{D_1+\frac{V_1^2}{2g}}$	. 18	. 18	. 18	. 18			
		Diving Flow Conditions						
12	T (diving depth)	2. 44	2. 32	2. 46	2. 37	2. 68	2. 39	2. 37
13	T_{max}	1. 94	1. 82	1. 96	1. 87	2. 18	1. 89	1. 87
14	$\frac{V_1^2}{2g}$	2. 824	2. 985	2. 892	2. 014	2. 354	2. 688	2. 715
15	V_1	13. 48	13. 87	13. 65	13. 94	12. 31	13. 16	13. 22
16	D_1	. 203	. 216	. 242	. 252	. 126	. 131	. 135
17	$F=\frac{V_1}{\sqrt{gD_1}}$	5. 26	5. 25	4. 89	4. 89	6. 11	6. 41	6. 38
18	$\frac{T_{max}}{D_1}$	9. 54	8. 54	8. 10	7. 40	17. 28	14. 38	13. 89
19	$D_1+\frac{V_1^2}{2g}$	2. 077	3. 201	3. 134	3. 266	2. 480	2. 819	2. 850
20	$\frac{R}{D_1+\frac{V_1^2}{2g}}$	. 25	. 23	. 24	. 23	. 30	. 27	. 26

TABLE 15.—*Data and calculated values for 9-inch-radius bucket*—Continued

	Run No.	Bed approx. 0.05R below apron lip at beginning of each run			Bed slopes up from apron lip			
		14	15	16	17	18	19	20
		Sweepout Conditions						
1	H	0.633	0.54	0.433	0.485	0.527	0.634	0.723
2	T (sweepout depth)							
3	q	2.02	1.56	1.12	1.32	1.50	2.01	2.50
4	T_{min}							
5	$\frac{V_1^2}{2g}$							
6	V_1							
7	D_1							
8	$F=\frac{V_1}{\sqrt{gD_1}}$							
9	$\frac{T_{min}}{D_1}$							
10	$D_1+\frac{V_1^2}{2g}$							
11	$\frac{R}{D_1+\frac{V_1^2}{2g}}$							
		Diving Flow Conditions						
12	T (diving depth)	2.42	3.07	1.96	1.86	2.23	2.69	2.43
13	T_{max}	1.92	2.57	1.46	1.36	1.73	2.19	1.93
14	$\frac{V_1^2}{2g}$	2.713	1.970	2.790	3.125	2.797	2.444	2.793
15	V_1	13.21	11.26	13.84	14.18	13.42	12.55	13.40
16	D_1	.153	.138	.081	.093	.112	.160	.187
17	$F=\frac{V_1}{\sqrt{gD_1}}$	5.95	4.15	8.59	8.19	7.08	5.53	5.46
18	$\frac{T_{max}}{D_1}$	12.55	18.55	18.00	14.60	15.47	13.67	10.34
19	$D_1+\frac{V_1^2}{2g}$	2.866	2.108	3.054	3.218	2.909	2.604	2.980
20	$\frac{R}{D_1+\frac{V_1^2}{2g}}$	.26	.35	.25	.23	.26	.29	.25

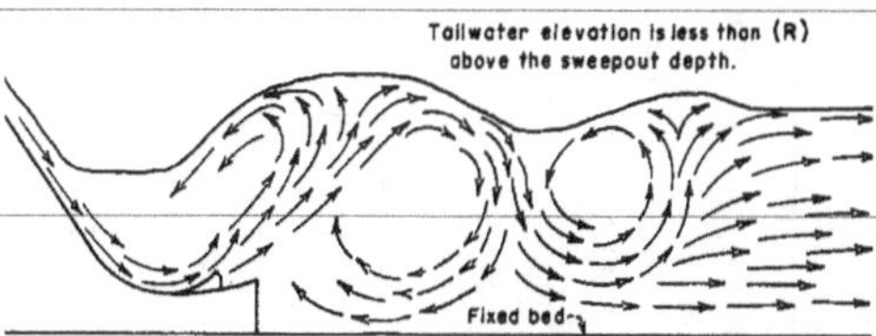

A—*Fixed bed below bucket invert. Desirable tail water depth*

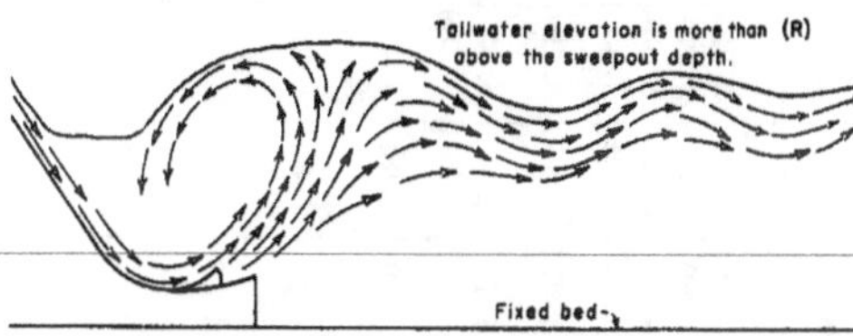

B—*Fixed bed below bucket invert. Less desirable tail water depth*

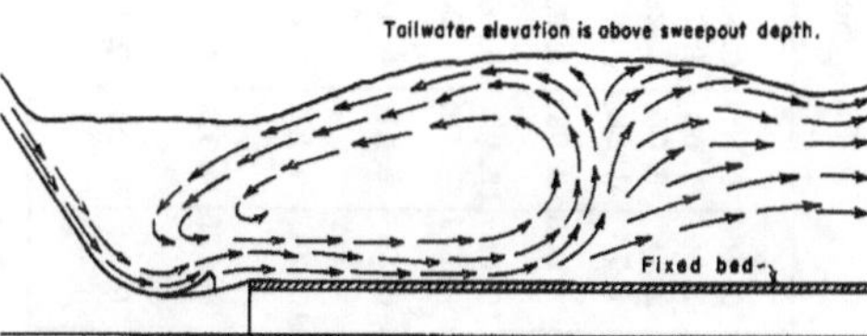

C—*Fixed bed at apron lip level*

NOTE: Bucket radius (R) is 6 inches. Design discharge, 1.75 c.f.s. per foot of width.

FIGURE 57.—*Flow currents for various arrangements of fixed beds.*

It was difficult to obtain consistent results for the tail water depth at which diving occurred, because the upper tail water limit was affected by the shape and elevation of the channel bed with respect to the apron lip. Since it was difficult to maintain the bed shape during the starting of a run, the gravel was removed from the model in anticipation that the upper tail water limit could be determined from observations of the flow pattern.

The gravel was removed completely so that the floor of the test flume was the channel bed. This arrangement proved unsuccessful, since diving did not occur. However, this test showed that excellent performance occurred, Figure 57A, when the tail water depth above the bucket invert was less than the bucket radius. With deeper tail water, the performance was not as good but was still satisfactory, Figure 57B.

The channel bed, represented by a wood floor at the elevation of the apron lip, produced flow currents that followed along the floor for quite some distance before rising to the surface, Figure 57C. The flow followed along the floor for a greater distance with higher tail water. No other changes in flow pattern occurred at high tail water elevations, and again no upper limit could be found.

Testing was continued with the gravel bed molded level slightly below the apron lip. It was necessary to reshape the bed before each test to obtain reasonable upper limit results. Despite every effort, consistent upper limit results were difficult to obtain. Testing showed that it was important that the channel bed be below the apron lip elevation to prevent the diving flow pattern from occurring at a much lower tail water elevation. Therefore, the bed was maintained at approximately 0.05R, or 0.3 of an inch, below the apron lip of the bucket at the beginning of each test. However, in testing the larger radius buckets a sloping bed was included in the investigation. Upper and lower tail water limits were also determined with the bed sloping 16° upward from the apron lip to approximately 6 inches above the lip, since this type of installation may be desirable in order to reduce excavation costs in many instances. Tests showed that sweepout occurred at the same depth, but diving occurred at a much lower tail water depth. For the 9-inch bucket, diving occurred at about the same tail water depth as for the 6-inch bucket with bed level 0.05R below the lip. For the 12-inch bucket it occurred at about the same tail water depth as for the 9-inch bucket with bed level 0.05R below the lip. Thus, the effect of the sloping bed was to reduce the operating range between minimum and maximum tail water depth limits by lowering the upper tail water limit.

Depths for sweepout and diving were difficult to determine precisely for the larger buckets. In fact, for the 18-inch bucket, the sustained diving condition could not be reached at any discharge, even when the tail water was raised to crest elevation. However, the tendency to dive was present, and momentary diving occurred, but in no case was it sustained.

TABLE 16.—*Data and calculated values for 12-inch-radius bucket*

Run No.	Bed was approx. 0.05R below apron lip at beginning of each run												Bed slopes up from apron lip			
	1	2	3	4	5	6	7	8	9	10	11	12	13	14	15	16
	Sweepout Conditions															
1 H	0.543	0.592	0.637	0.679	0.729	0.765	0.811	0.850	0.887	0.961	1.02	1.221	0.565	0.651	0.723	0.839
2 T (sweepout depth)	1.27	1.33	1.40	1.45	1.52	1.56	1.68	1.72	1.78	1.89	1.96	2.23				
3 q	1.58	1.82	2.03	2.25	2.53	2.75	3.05	3.28	3.54	4.06	4.48	6.08	1.67	2.00	2.50	3.21
4 T_{min}	1.47	1.53	1.60	1.65	1.72	1.76	1.88	1.92	1.98	2.09	2.16	2.43				
5 $\frac{V_1^2}{2g}$	4.073	4.062	4.037	4.029	4.009	4.005	3.931	3.930	3.907	3.871	3.860	3.791				
6 V_1	16.20	16.17	16.12	16.11	16.07	16.06	15.92	15.91	15.86	15.79	15.77	15.63				
7 D_1	.099	.112	.126	.140	.157	.171	.192	.206	.223	.257	.284	.389				
8 $F=\frac{V_1}{\sqrt{gD_1}}$	9.10	8.51	8.01	7.60	7.14	6.84	6.41	6.18	5.93	5.49	5.22	4.42				
9 $\frac{T_{min}}{D_1}$	14.91	13.60	12.71	11.81	10.93	10.28	9.81	9.31	8.87	8.13	7.60	6.25				
10 $D_1+\frac{V_1^2}{2g}$	4.172	4.175	4.163	4.169	4.166	4.176	4.123	4.136	4.133	4.128	4.144	4.180				
11 $\frac{R}{D_1+\frac{V_1^2}{2g}}$	0.24	0.24	0.24	0.24	0.24	0.24	0.24	0.24	0.24	0.24	0.24	0.24				

		Diving Flow Conditions															
12	T (diving depth)	3.90	4.00	3.95	3.95	3.95	3.95	------	2.91	2.87	3.22	3.17	3.00	3.25	3.00	2.45	2.35
13	T_{max}	3.45	3.50	3.40	3.45	3.45	3.45	------	2.41	2.37	2.72	2.67	2.50	2.75	2.50	1.95	1.85
14	$\frac{V_1^2}{2g}$	1.093	1.092	1.237	1.229	1.279	1.315	------	2.440	2.517	2.241	2.35	2.721	1.815	2.131	2.773	2.889
15	V_1	8.39	8.39	8.92	8.90	9.07	9.20	------	12.54	12.72	12.01	12.30	12.23	10.81	11.71	13.36	13.64
16	D_1	.188	.217	.228	.253	.279	.299	------	.262	.278	.338	.364	.460	.154	.171	.187	.235
17	$F=\frac{V_1}{\sqrt{gD_1}}$	3.42	3.17	3.29	3.11	3.02	2.96	------	4.31	4.25	3.64	3.41	3.44	4.86	4.98	5.54	4.96
18	$\frac{T_{max}}{D_1}$	18.35	16.12	14.91	14.63	12.36	11.53	------	9.19	8.52	8.04	7.33	5.54	17.85	14.61	10.42	7.87
19	$D_1+\frac{V_1^2}{2g}$	1.281	1.309	1.465	1.482	1.558	1.614	------	2.702	2.795	2.579	2.714	3.181	1.969	2.302	2.960	3.124
20	$\frac{R}{D_1+\frac{V_1^2}{2g}}$	.78	.72	.68	.67	.64	.62	------	.37	.36	.39	.37	.31	.51	.43	.34	.32

NOTE: See Table 14 for definition of symbols.
Maximum capacity of bucket estimated to be 3.25 to 3.50 c.f.s. per ft. of width.

TABLE 17.—*Data and calculated values for 18-inch-radius bucket*

	Run No.	Bed was approx. 0.05R below apron lip at beginning of each run							
		1	2	3	4	5	6	7	8
		Sweepout Conditions							
1	H	0.631	0.734	0.804	0.898	0.926	1.001	1.083	1.150
2	T (sweepout depth)	1.45	-------	1.78	-------	-------	-------	-------	-------
3	q	2.00	2.56	2.99	3.61	3.80	4.35	4.98	5.48
4	T_{min}	1.65	1.85	1.98	2.07	2.15	2.23	2.32	2.45
5	$\frac{V_1^2}{2g}$	3.981	3.884	3.824	3.828	3.776	3.771	3.763	3.700
6	V_1	16.02	15.86	15.70	15.70	15.27	15.68	15.67	15.44
7	D_1	.125	.161	.190	.230	.249	.277	.318	.355
8	$F=\frac{V_1}{\sqrt{gD_1}}$	7.98	6.94	6.33	6.76	5.39	5.24	4.88	4.56
9	$\frac{T_{min}}{D_1}$	13.22	11.46	10.39	9.00	8.64	8.03	7.30	6.70
10	$D_1+\frac{V_1^2}{2g}$	4.106	4.045	4.014	4.058	4.025	4.043	4.081	4.055
11	$\frac{R}{D_1+\frac{V_1^2}{2g}}$	.37	.37	.37	.37	.37	.37	.37	.37
		Diving Flow Conditions							
12	T (diving depth)	No Data Taken.							
13	T_{max}								
14	$\frac{V_1^2}{2g}$								
15	V_1								
16	D_1								
17	$F=\frac{V_1}{\sqrt{gD_1}}$								
18	$\frac{T_{max}}{D_1}$								
19	$D_1+\frac{V_1^2}{2g}$								
20	$\frac{R}{D_1+\frac{V_1^2}{2g}}$								

NOTE: See Table 14 for definition of symbols.

Maximum capacity of bucket estimated to be 5.0 to 5.5 c.f.s. per ft. of width.

A—Flow is about to dive from apron lip—maximum tail water limit has been exceeded

B—Flow is diving from the apron, lip—maximum tail water limit has been exceeded

FIGURE 58.—Nine-inch bucket discharging 1.5 c.f.s.

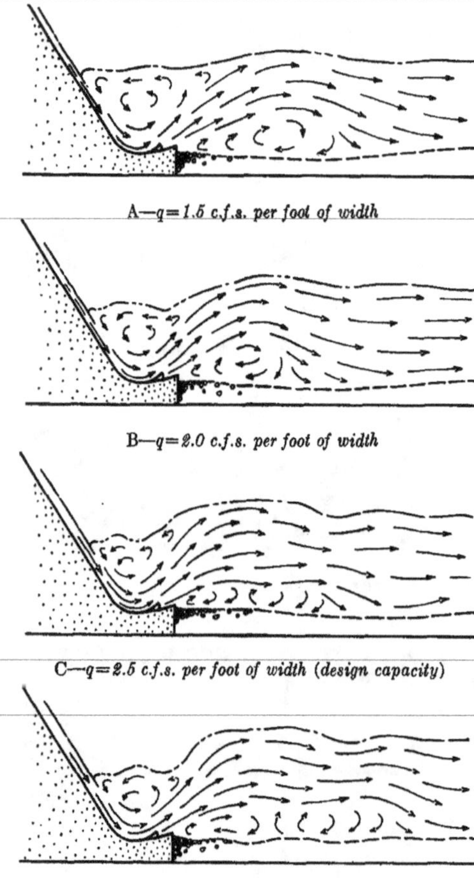

A—$q=1.5$ *c.f.s. per foot of width*

B—$q=2.0$ *c.f.s. per foot of width*

C—$q=2.5$ *c.f.s. per foot of width (design capacity)*

D—$q=3.0$ *c.f.s. per foot of width*

(Bed level 0.5 inch below apron lip at start of test)

FIGURE 59.—*Nine-inch bucket discharging—tail water depth=1.85 feet.*

Maximum capacity. As the discharge capacity of the bucket was approached, the difference between the upper and lower tail water limits became smaller. The maximum capacity of the bucket was judged from its general performance and by the range of useful tail water elevations between the upper and lower tail water limits, Figure 56. The maximum capacity of the 6-inch bucket was found to be 3 to 3.5 c.f.s. or 1.5 to 1.75 c.f.s. per foot of bucket width. The performance of the bucket for 1.75 c.f.s. with normal tail water elevation is shown in Figure 55B.

The maximum capacity of the 9-inch bucket was determined to be 2 to 2.5 cubic feet per foot of width. Discharges of 1.5 to 3 cubic feet per second with a normal tail water depth of 1.85 feet are shown in Figure 59.

Figure 61 shows the performance of the 12-inch bucket for unit flows ranging from 2.5 to 4 c.f.s. with normal tail water depth of 2.3 feet. The maximum capacity of the bucket was determined to be from 3.25 to 3.5 cubic feet per second.

The performance of the 18-inch bucket is shown in Figure 62 for unit discharges ranging from 3 to 5.5 cubic feet per second with normal tail water depths. The capacity of the bucket was determined to be 5 to 5.5 cubic feet per second.

Larger and smaller buckets. Increasing difficulties in determining bucket capacity and tail water depth limits for near capacity flows made it inadvisable to test larger buckets on the 5-foot spillway. In addition, maximum tail water depths would either have submerged the crest or closely approached that condition, and it was not intended at this time to investigate a bucket downstream from a submerged spillway crest.

It was unnecessary to test smaller buckets because very few, if any, prototype structures would use a bucket radius smaller than one-tenth the height of the spillway. A short radius bend is usually avoided on high structures where velocities are also high. Therefore, the available data were analyzed and, with some extrapolation, found to be sufficient.

Water Surface Characteristics

Figure 60 shows water surface characteristics for the 9- and 12-inch buckets. To aid in defining water surface profiles, measurements were made for a range of flows with the tail water at about halfway between the upper and lower limits.

Data Analysis

Safety factor. At the conclusion of the testing, the data for the four buckets were surveyed and the margins of safety, between sweepout depth and minimum tail water depth and between maximum tail water depth and the diving depth, were definitely established. An ample margin of safety for the lower limit was 0.2 foot and for the upper limit 0.5 foot. These values were sufficient

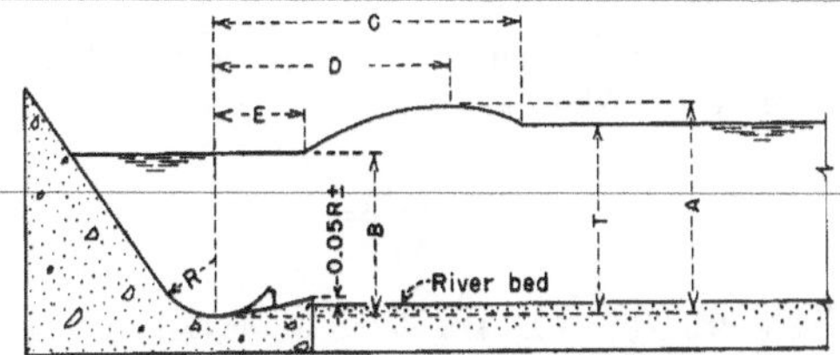

(Crest elevation to bucket invert "x" = 5 feet)

9-INCH BUCKET (R)

Q-cfs	q-cfs/ft.	T-ft.	A	B	C	D	E
3.0	1.50	1.85	25	19	45	25	5
3.0	1.50	2.40	32	26	46	27	1
3.50	1.75	1.85	26	19	45	25	5
4.00	2.00	1.85	27	19	46	25	5
4.50	2.25	1.85	28	19	48	26	6
5.00*	2.50	1.85	28	18	50	32	6
5.50	2.75	1.85	29	17	51	31	6
6.00	3.00	1.85	30	16	52	32	6

12-INCH BUCKET (R)

Q-cfs	q-cfs/ft.	T-ft.	A	B	C	D	E
5.0	2.5	2.30	32	23	62	35	14
6.0	3.0	2.30	33	22	62	37	11
7.0*	3.5	2.30	33	21	68	37	9
8.0	4.0	2.30	35	19	70	37	6
12.0	6.0	—	36	7	90	40	1

NOTE: Dimensions A, B, C, D, and E are in inches.
*Design capacity.

FIGURE 60.—*Average water surface measurements.*

for both the level and sloping movable beds previously described and are included in items T_{min} and T_{max} of Tables 14, 15, 16, and 17.

Evaluation of variables. To generalize the design of a bucket from the available data, it is necessary to determine the relation of the variables shown in Figure 63. The available data are shown in Tables 14 through 17 and are plotted in Figure 56.

Figure 56 shows that, for a given height of structure having a particular overfall shape and spillway surface roughness, the sweepout depth, T_s, and minimum tail water depth limit, T_{min}, are functions of the radius of the bucket, R, and the head on the crest, H. The height of structure may be expressed as the height of fall, h, from the spillway crest to the tail water elevation. The overfall shape and H determine the discharge per foot of spillway width, and may be expressed as q. Since the spillway surface roughness and the spillway slope had negligible effect on flow in the model, they were not considered in the analysis of model data. However, prototype effects are discussed in a subsequent section of this monograph. By actual test, it was found that the elevation or shape of the movable bed did not affect the minimum tail water limits. Therefore,

$$T_{min} \text{ or } T_s = f(h, R, \text{ and } q)$$

Similarly, the maximum tail water depth limit, T_{max}, is a function of the same variables, but since the slope and elevation of the movable bed

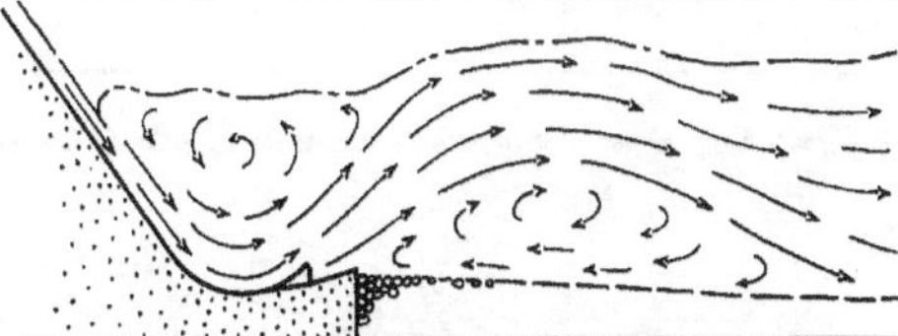

A—*q=2.5 c.f.s. per foot of width*

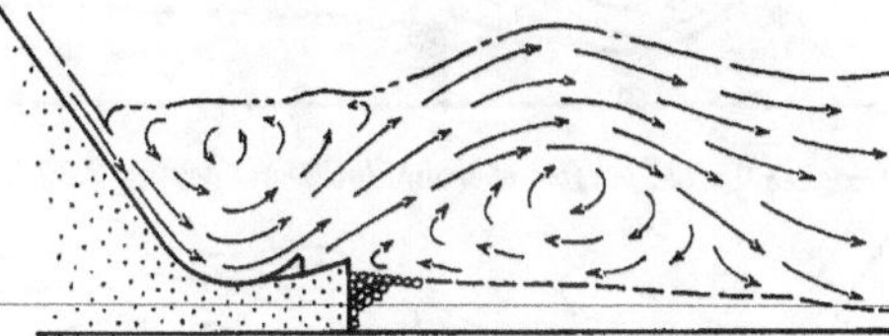

B—*q=3.0 c.f.s. per foot of width*

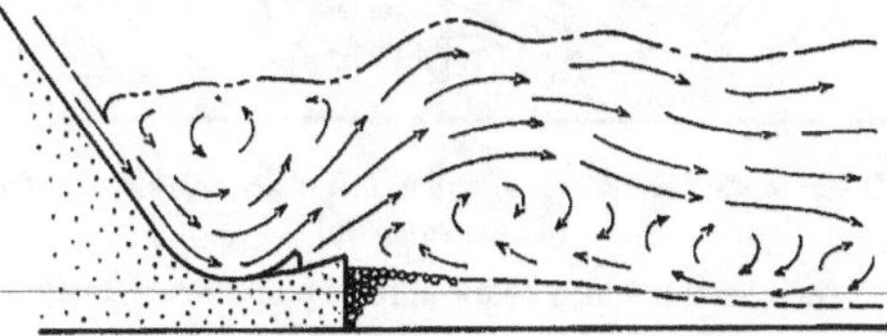

C—*q=3.5 c.f.s. per foot of width (design capacity)*

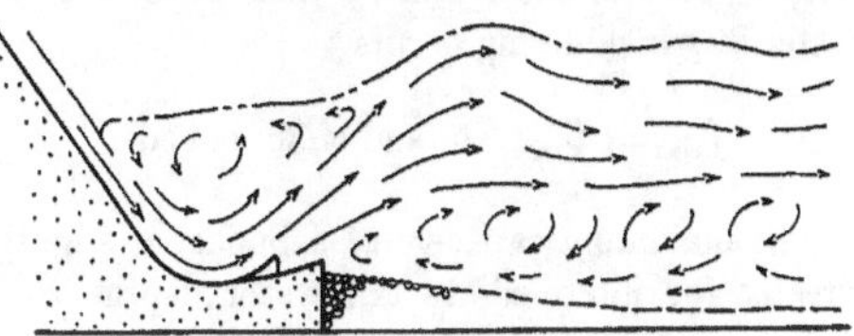

D—*q=4.0 c.f.s. per foot of width*

(Bed level 0.6 inch below apron lip at start of test)

FIGURE 61.—*Twelve-inch bucket discharging. Tail water water depth=2.30 feet.*

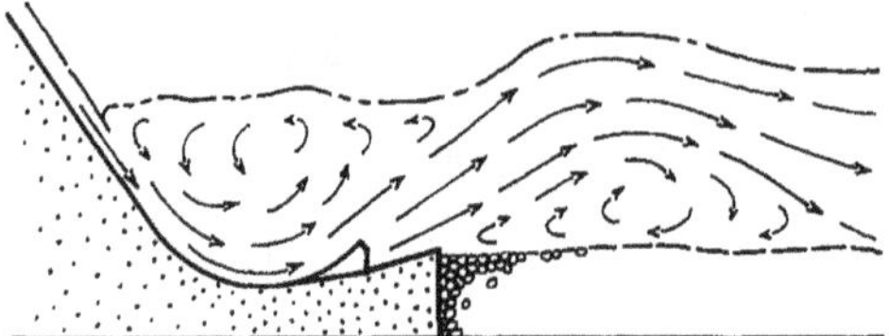

A—$q=3.0$ *c.f.s. per foot of width, tail water depth=2.30 feet*

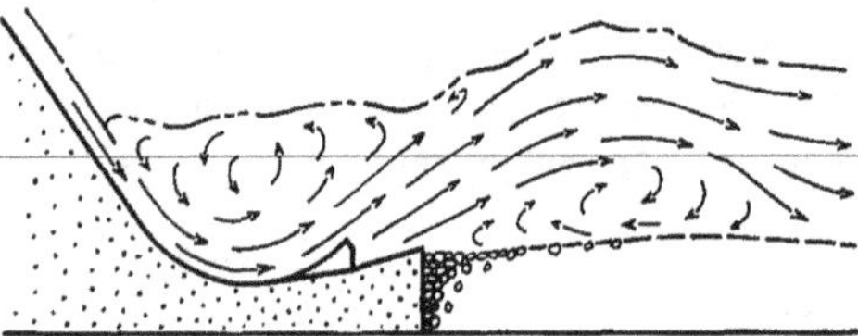

B—$q=3.5$ *c.f.s. per foot of width, tail water depth=2.30 feet*

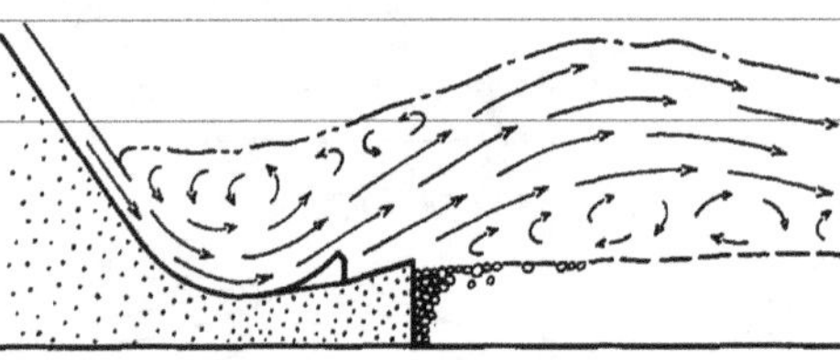

C—$q=4.0$ *c.f.s. per foot of width, tail water depth=2.30 feet*

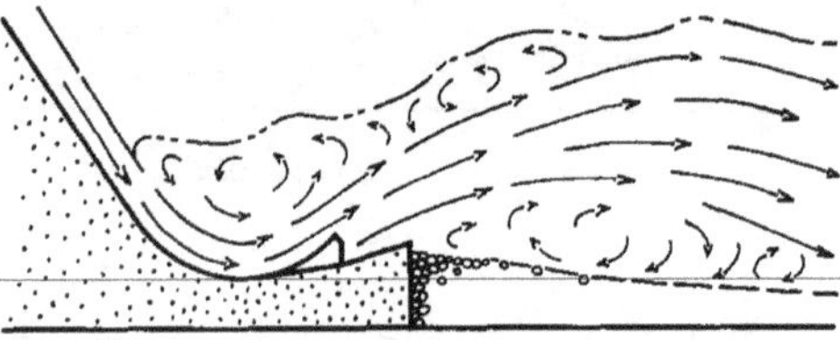

D—$q=5.5$ *c.f.s. per foot of width, tail water depth=2.45 feet (design capacity)*

(Bed level 0.9 inch below apron lip at start of test)

FIGURE 62.—*Eighteen-inch bucket performance.*

with respect to the apron lip does affect the tail water at which diving occurs,

$$T_{max}=f\ (h,\ R,\ q,\ \text{and channel bed}).$$

The maximum capacity of a bucket is slightly greater for intermediate tail water depths than for the extremes. However, the bucket is expected to operate over a range of tail water depths; therefore, the minimum bucket radius is a function of only h and q.

$$R_{min}=f\ (h\ \text{and}\ q)$$

The Froude number is a function of velocity and depth of flow and may be expressed

$$F=\frac{V_1}{\sqrt{gD_1}}$$

in which V_1 and D_1 are at tail water elevation, as shown in Figure 63. Since V_1 and D_1 are functions of h and q, they may be replaced by the Froude number F. Substituting, then

$$T_{min}\ \text{and}\ T_s=f\ (R,\ F)$$

$$T_{max}=f\ (R,\ F\ \text{and channel bed})$$

and

$$R_{min}=f\ (F)$$

Numerical values for the Froude number were computed from the available test data in the tables for points on the spillway face at the tail water elevation. At these points, all necessary information for computing velocity and depth of flow can be determined from the available test data which include headwater elevation, discharge, and tail water elevation. Since the Froude number expresses a ratio of velocity to depth and is dimensionless, a numerical value expresses a prototype as well as a model flow condition. To express T_{min}, T_{max}, and R_{min} as dimensionless numbers so that they may also be used to predict prototype flow conditions, T_{min} and T_{max} were divided by D_1; R_{min} was divided by $D_1+\frac{V_1^2}{2g}$, the depth of flow plus the velocity head at tail water elevation on the spillway face. These dimensionless ratios and the Froude number, computed from test data, are shown in Tables 14, 15, 16, and 17. In computing the tabular values, frictional resistance in the 5-foot model was considered to be negligible.

To provide data that are useful for determining the minimum bucket radius for a given Froude number, the bucket radius dimensionless ratio

$$\frac{R}{D_1+\frac{V_1^2}{2g}}$$

is plotted against the Froude number in Figure 64, using only maximum capacity discharge values. The maximum capacity discharge values are plotted for both the sweepout and diving tail water elevations, since the Froude number and

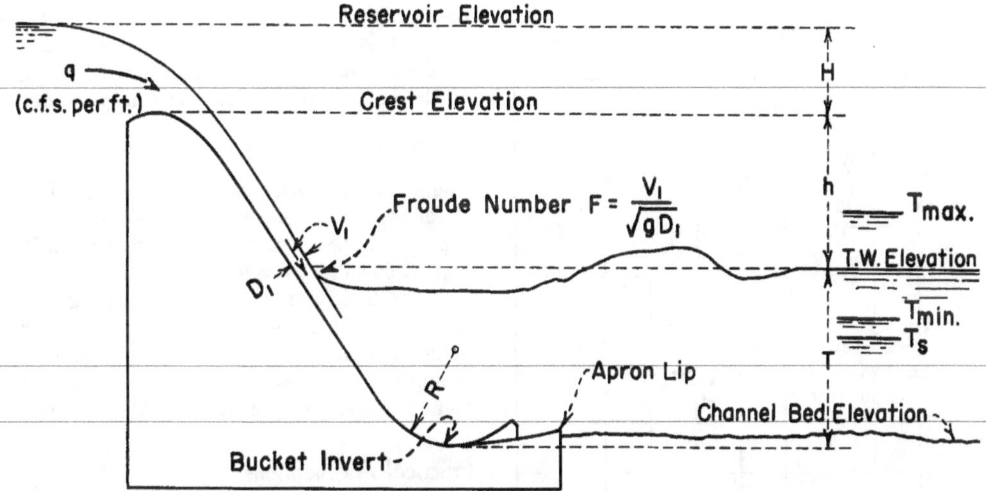

FIGURE 63.—*Definition of symbols.*

bucket radius ratio both vary with tail water elevation. For example, the maximum capacity of the 6-inch bucket is q=1.5 to 1.75 c.f.s. In Columns 7 and 8 of Table 14, data from lines 8 and 11 and lines 17 and 20 were plotted on Figure 64, and since each discharge has two tail water limits, four points may be plotted. The two points obtained for each discharge were connected by a dashed line to indicate the trend in bucket size for constant discharge and varying heights of fall to the tail water surface. Eight dashed lines were thereby obtained for the four buckets. A single envelope curve was then drawn, shown as the solid line to the right of the preliminary lines, to indicate the minimum bucket radius. The solid line, therefore, includes a factor of safety which is measured by the distance between the solid line and the test points.

To provide data useful for determining tail water depth limits for a given Froude number, the dimensionless ratios for tail water depth limits, $\frac{T_{min}}{D_1}$ and $\frac{T_{max}}{D_1}$ for each test point in Tables 14 through 17, were plotted versus the Froude number in Figure 65, and each point was labeled with the computed value of the bucket radius ratio. Then, curves were drawn through both the minimum and maximum tail water depth limits having the same bucket radius ratio values. The upper four curves are for the minimum tail water limit and apply to any bed arrangement. The 10 lower curves apply to the maximum tail water limitation and have two sets of labels, one for the sloping bed and one for the level bed. Two curves are shown for each value of the bucket radius ratio for the upper tail water limit. The upper or solid line curves have an extra factor of safety included because of the difficulty in obtaining consistent upper tail water limit values. The lower or dashed line curves are a strict interpretation of the values in Tables 14 through 17, including the safety factor incorporated into the data as previously explained in the discussion of lower and upper tail water limits.

The curves of Figure 65 may be used directly to determine minimum and maximum tail water limits for a given Froude number and bucket ratio. However, a version of the same data that is simpler and easier to use is given in Figures 66 and 67, which were obtained by cross-plotting the curves of Figure 65. Figure 66 contains a family of curves to determine $\frac{T_{min}}{D_1}$ values in terms of the Froude number and

$$\frac{R}{D_1+\frac{V_1^2}{2g}}.$$

Figure 67 contains similar curves to determine $\frac{T_{max}}{D_1}$ and includes the extra factor of safety dis-

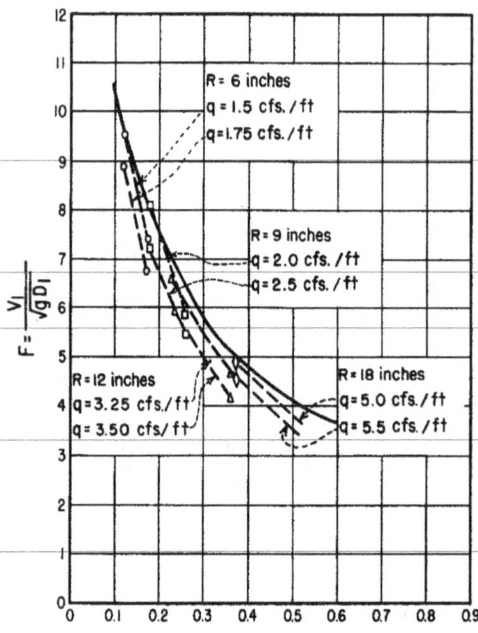

MINIMUM ALLOWABLE $\frac{R}{D_1 + V_1^2/2g}$

EXPLANATION

- o For bucket radius (R) = 6 inches
- □ For bucket radius (R) = 9 inches
- △ For bucket radius (R) = 12 inches
- ◇ For bucket radius (R) = 18 inches

Bed level approximately 0.05R below lip of apron.

FIGURE 64.—*Minimum allowable bucket radius.*

cussed for Figure 65. The two abscissa scales in Figure 67 differentiate between the sloping bed and the level bed used in the tests.

The tail water sweepout depth, T_s, in Tables 14 through 17 was also expressed as a dimensionless ratio $\frac{T_s}{D_1}$ and plotted versus the Froude number in Figure 68, and a curve for each bucket size was drawn. These curves were then cross-plotted in Figure 69 to provide more convenient means for determining the sweepout depth for any installation. The difference between sweepout depth indicated by the curves and the depth to be expected in the prototype indicates the margin of safety.

To aid in determining approximate water surface profiles in and downstream from the bucket, the data of Figure 60 and values scaled from photographs of other bucket tests were analyzed and plotted. Refinement of the curves obtained resulted in the curves of Figure 70. The height of the boil above the tail water may be determined from the Froude number and the ratio $\frac{R}{X}$, where R is the bucket radius and X is the height of the spillway from crest to bucket invert. The depth of the water in the bucket, dimension B in Figures 60 and 70, was found to remain fairly constant over most of the design operating range, about 80 to 85 percent of the dimension T. For minimum recommended tail water, the percentage dropped to 70 percent, and with high tail water the value increased to approximately 90 percent.

Practical Applications

Sample problems. To illustrate the use of the methods and charts given in this monograph, a step-by-step procedure for designing a slotted bucket is presented. Discharge data, height of fall, etc., from Grand Coulee Dam spillway will be used in the example so that the resulting slotted bucket may be compared with the solid bucket individually determined from model tests and now in use at Grand Coulee Dam. The calculations are summarized in Table 18.

For maximum reservoir elevation 1291.65, the spillway discharge is 1 million c.f.s. Since the spillway crest is at elevation 1260, the head is 31.65 feet. The width of the bucket is 1,650 feet, making the discharge per foot 606 c.f.s. The tail water in the river is expected to be at elevation 1011 for the maximum flow. The theoretical velocity head of the flow entering the basin is the difference between tail water elevation and reservoir elevation, or 280.65 feet. Then, the theoretical velocity, V_T, entering the tail water is 134.4 feet per second; $V_1 = \sqrt{2g(H+h)}$. See Figure 63.

The actual velocity is less than theoretical at this point, because of frictional resistance on the spillway face. Using Figure 71, the actual velocity is found to be 91 percent of theoretical. Figure 71 is believed to be reasonably accurate, but since only a limited amount of prototype data were available to develop the chart, information obtained from it should be used with caution. The actual velocity, V_A, in this example is 91 percent

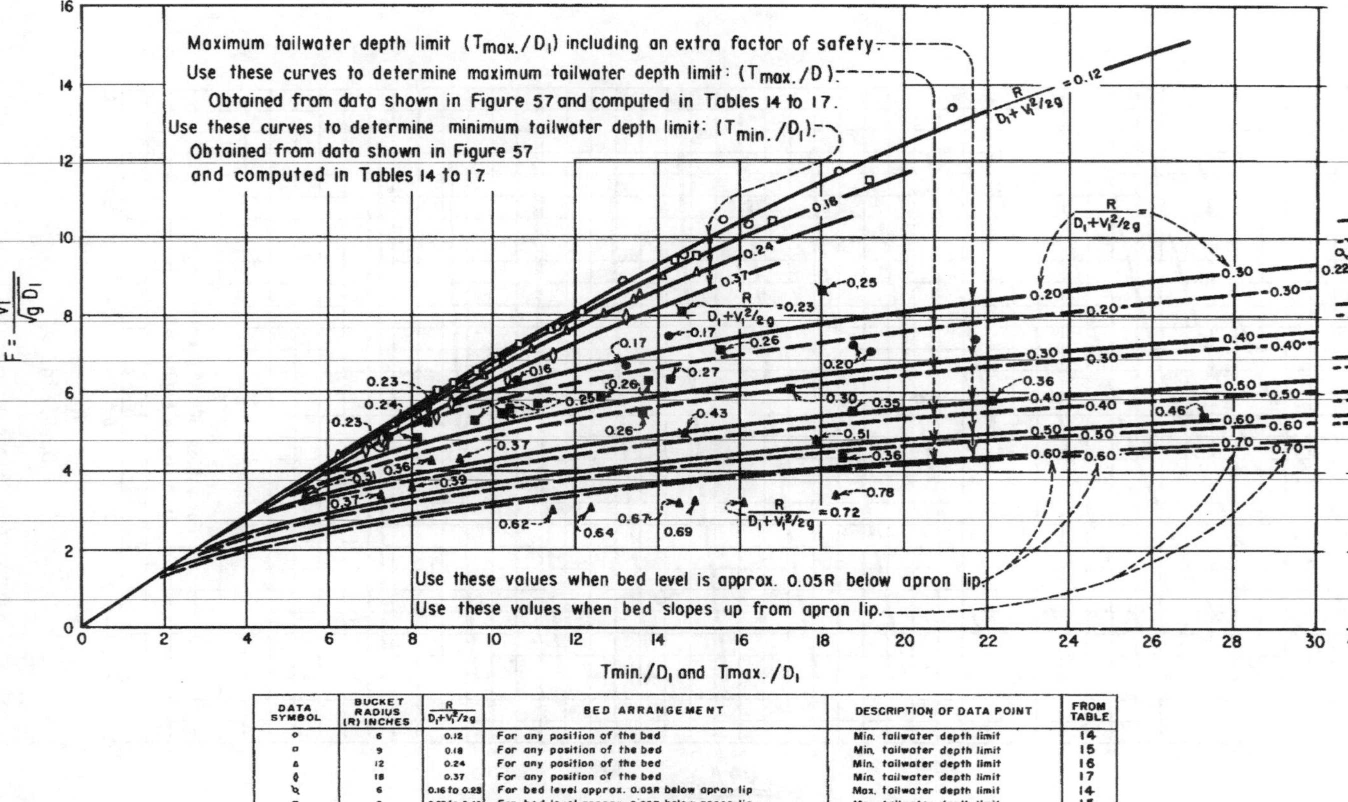

DATA SYMBOL	BUCKET RADIUS (R) INCHES	$\frac{R}{D_1+V_1^2/2g}$	BED ARRANGEMENT	DESCRIPTION OF DATA POINT	FROM TABLE
○	6	0.12	For any position of the bed	Min. tailwater depth limit	14
□	9	0.18	For any position of the bed	Min. tailwater depth limit	15
△	12	0.24	For any position of the bed	Min. tailwater depth limit	16
◊	18	0.37	For any position of the bed	Min. tailwater depth limit	17
⌀	6	0.16 to 0.23	For bed level approx. 0.05R below apron lip	Max. tailwater depth limit	14
■	9	0.23 to 0.46	For bed level approx. 0.05R below apron lip	Max. tailwater depth limit	15
▲	12	0.31 to 0.68	For bed level approx. 0.05R below apron lip	Max. tailwater depth limit	16
●	9	0.25	For bed sloping up from apron lip	Max. tailwater depth limit	15
x	12	0.43 to 0.51	For bed sloping up from apron lip	Max. tailwater depth limit	16

FIGURE 65.—*Dimensionless plot of maximum and minimum tail water depth limits.*

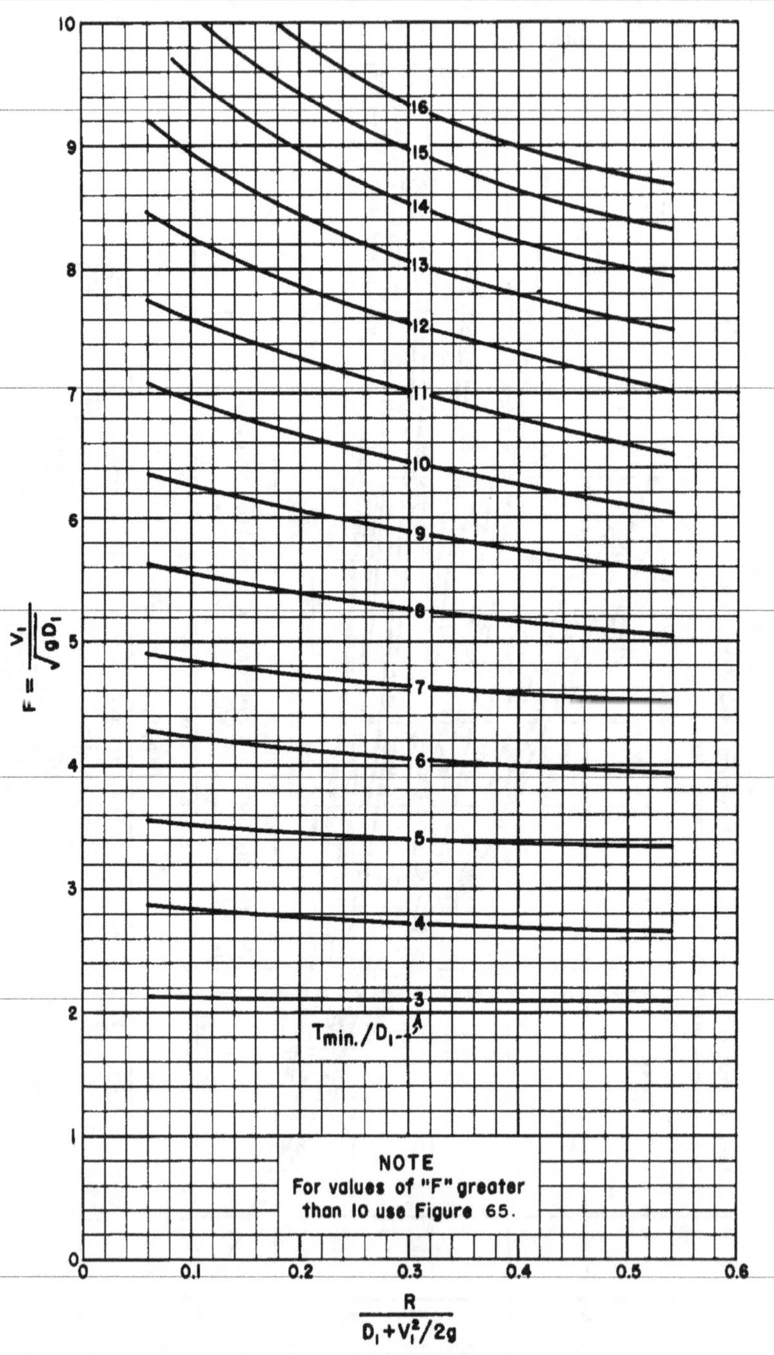

FIGURE 66.—*Minimum tail water limit.*

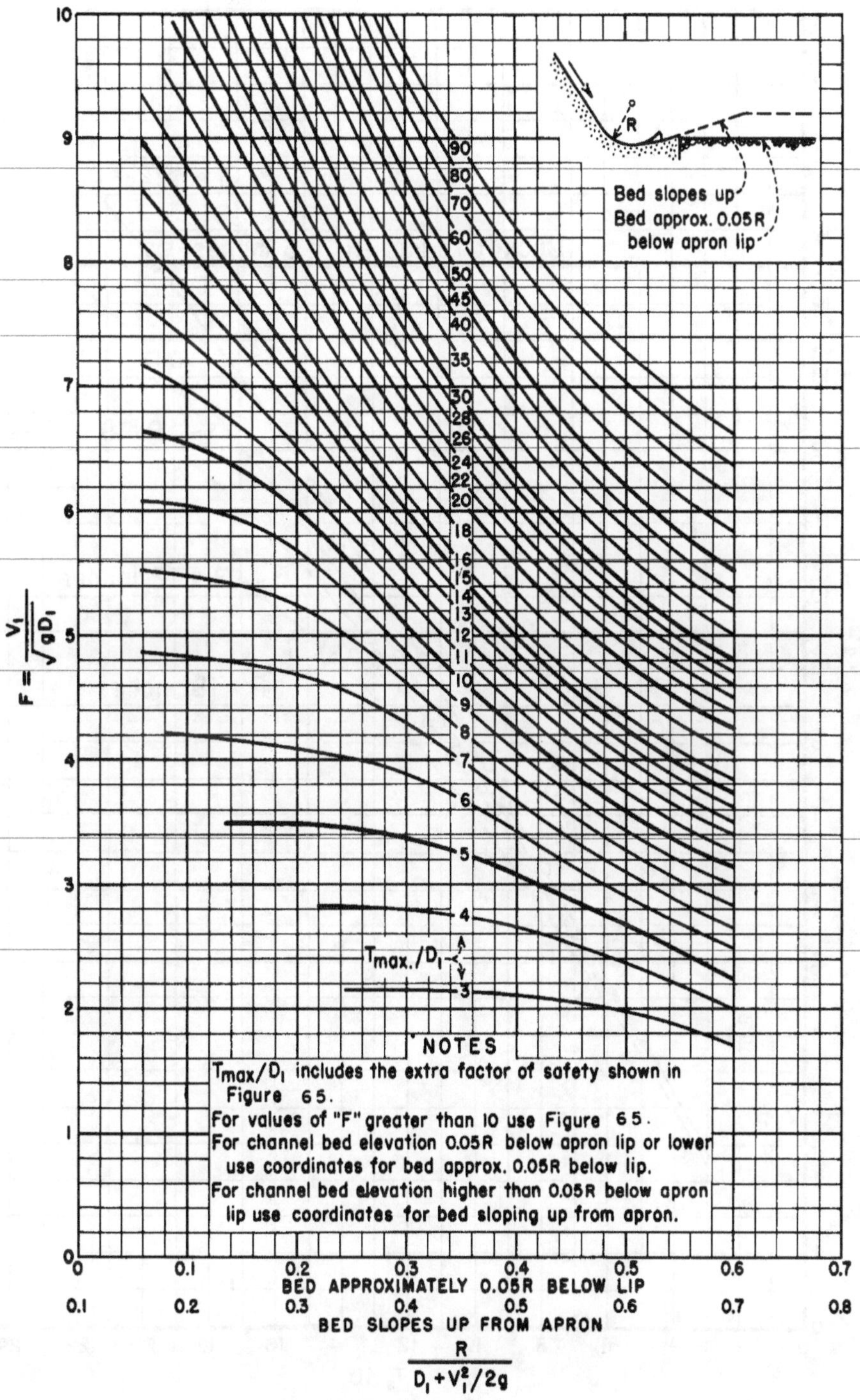

FIGURE 67.—*Maximum tail water limit.*

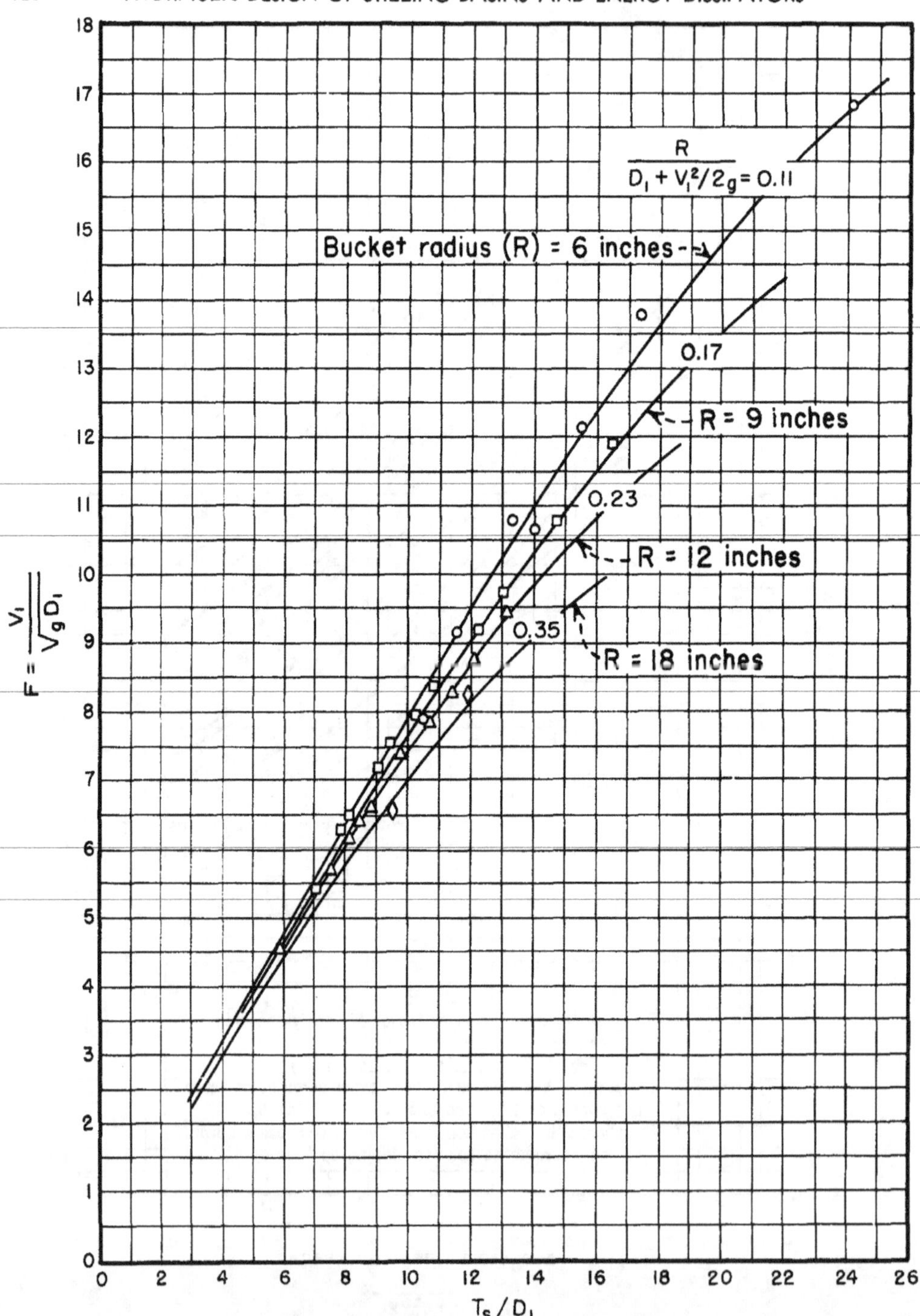

NOTE: Bed arrangement not critical for sweepout condition.

FIGURE 68.—*Tail water depth at sweepout.*

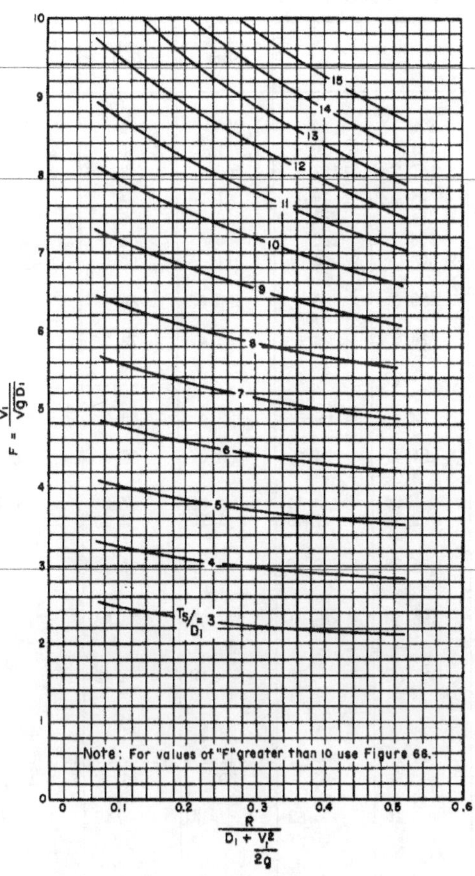

FIGURE 69.—*Tail water sweepout depth.*

of 134.4, or 122.4 feet per second. The corresponding depth of flow D_1 on the spillway face is $\frac{q}{V_1}$ or 4.95 feet. Having determined D_1 and V_1, the Froude number is computed to be 9.7.

Entering Figure 64 with Froude number 9.7, the dimensionless ratio for the minimum allowable bucket radius is found to be 0.12 from the solid line curve, from which the radius is computed to be 28.5 feet. In round numbers, a 30-foot bucket radius probably would be used. This is smaller than the 50-foot radius of the solid type bucket that was actually used at Grand Coulee. For the 30-foot radius, the dimensionless ratio would be 0.13. Entering Figure 66 with the dimensionless ratio and the Froude number, $\frac{T_{min}}{D_1}$ is found to be 14.7, from which T_{min} is 73 feet. Similarly, from Figure 67, $\frac{T_{max}}{D_1}$ for the bed elevation below the apron lip is found to be 23, from which T_{max} is 114 feet.

From Figure 69, the sweepout dimensionless depth ratio is 12.6, from which the sweepout depth is 63 feet. Thus, the minimum tail water depth limit of 73 feet provides 10 feet of margin against flow sweeping out of the bucket at the maximum discharge.

Tail water elevation 1011 at Grand Coulee provides 111 feet of tail water depth above riverbed elevation 900. Therefore, the bucket invert should be set no lower than 3 feet below riverbed elevation or more than 38 feet above. In the latter position, there would be no bed scour, and the water surface would be as smooth as possible. However, this location may not be practical, and it may be necessary to set the bucket on bedrock so that the invert is more than 3 feet below the riverbed.

The data in Figure 60 and the curves of Figure 70 may be used to obtain an approximate water

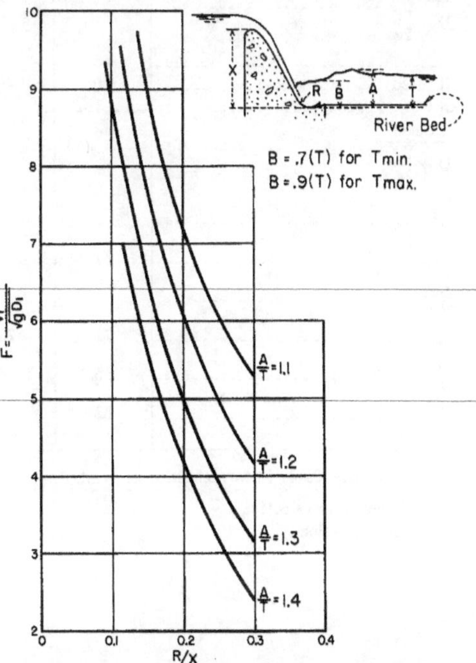

FIGURE 70.—*Water surface profile characteristics for slotted buckets only.*

TABLE 18.—*Examples of bucket design procedures*

	Angostura Dam				Grand Coulee Dam	Trenton Dam	Missouri Diversion Dam
Q (1,000 c.f.s.)	247	180	100	40	1, 000	133	90
HW el.	3, 198. 1	3, 191. 0	3, 181. 5	3, 170. 4	1, 291. 65	2, 785	2, 043. 4
Crest el.	3, 157. 2	3, 157. 2	3, 157. 2	3, 157. 2	1, 260	2, 743	2, 032
H	40. 9	33. 8	24. 3	13. 2	31. 65	42	11. 4
L	274	274	274	274	1, 650	266	644
q	901	657	365	146	606	500	140
TW el.	3, 114	3, 106	3, 095	3, 084	1, 011	2, 700. 6	2, 018. 3
$\frac{V_1^2*}{2g}$	84. 1	85. 0	86. 5	86. 4	280. 65	84. 4	25. 1
V_1*	73. 6	74	74. 6	74. 5	134. 4	73. 7	40. 2
$\frac{V_1**}{V_1*}$	. 98	. 98	. 97	. 93	. 91	----------	. 98
V_1**	72. 2	72. 5	72. 4	69. 3	122. 4	66. 3	39. 4
$\frac{V_1^2**}{2g}$	80. 9	81. 6	81. 4	74. 6	233. 0	68. 3	24. 1
D_1	12. 48	9. 06	5. 04	2. 11	4. 95	7. 54	3. 55
$\sqrt{D_1}$	3. 53	3. 01	2. 24	1. 45	2. 23	2. 75	1. 88
$F=\frac{V_1}{\sqrt{gD_1}}$	3. 61	4. 25	5. 68	8. 42	9. 70	4. 25	3. 70
$D_1+\frac{V_1^2}{2g}$	93. 38	90. 66	86. 44	76. 71	237. 95	75. 84	27. 65
$\frac{R}{D_1+\frac{V_1^2}{2g}}$	. 60	. 43	. 30	. 16	. 12	. 43	. 49
R	56	39	26	12	28. 5	33	14
R (used)	40	40	40	40	----------	----------	----------
R (rec)	----------	----------	----------	----------	30	35	12. 5
$\frac{R \text{ (used)}}{D_1+\frac{V_1^2}{2g}}$	. 43	. 44	. 46	. 52	. 13	. 46	. 45
$\frac{T_{min}}{D_1}$	5. 4	6. 5	9. 1	15. 3	14. 7	6. 5	5. 6
T_{min}	67	59	46	32. 5	73	49	20
$\frac{T_{max}}{D_1}$	5. 7	7. 9	17. 6	100	23	13. 0	8. 9
T_{max}	71	72	89	210	114	98	32
$\frac{T_s}{D_1}$	5. 0	6. 0	8. 2	14. 4	12. 6	6. 0	5. 2
T_s	62	54	41	30	63	45	18

NOTE: See Table 14 for definition of symbols.

*Theoretical velocity.

**Actual velocity.

surface profile if the bucket invert is placed near channel bed elevation 900.

For $F=9.7$ and $\frac{R}{X}=\frac{30}{360}=0.093$; $\frac{A}{T}=1.3$. The maximum height of boil A is then 1.3×111, or 144 feet above the bucket invert and 33 feet above the tail water. In the bucket, the depth of water B would be 90 percent of 111 feet, or approximately 100 feet. The maximum difference (A−B) would be about 44 feet for the design tail water. The length and location of the boil may be estimated from the data in Figure 60.

Another solution would be to use a larger bucket radius. For a 50-foot radius bucket, which is the radius of the solid bucket actually used at Grand Coulee, the tail water depth limits are 78 and 183 feet, and sweepout depth is 67 feet. Thus, the bucket invert can be placed below the riverbed, but the apron lip should be set about 0.05R, or 2.5 feet above riverbed elevation. If the 50-foot radius bucket is placed below riverbed elevation so that the bed slopes upward from the apron lip, the ratio $\frac{T_{max}}{D_1}$ is 20.5, Figure 67, and the upper depth limit is computed to be only 101 feet. In this case, the flow from the apron might scour the channel bed because the tail water depth above the bucket invert is greater than the maximum limit, and a still larger bucket radius would be required.

If the invert of the 50-foot radius bucket is placed at bed elevation to provide 111 feet of tail water, $\frac{R}{X}=0.14$ and $\frac{A}{T}=1.1$.

The height of the boil then would be about 122 feet above the bucket invert, or 11 feet above tail water. The water in the bucket, 80 percent of 111, would be 89 feet deep. The 50-foot bucket would provide a smoother water surface profile than the 30-foot bucket as is shown by comparing the 11-foot high boil with the 33-foot high boil.

Before adopting a design, all factors which might affect the tail water range should be investigated, i.e., large or sudden increases in spillway discharge and effects of discharges from outlet works or powerplant. Tail water elevations for flows less than maximum should also be examined. If V_1 is more than 75 feet per second, pressures on the teeth should be investigated on a hydraulic model.

Discharges for maximum and less than maximum design were investigated for the Angostura installation in Table 18, using the methods presented in this monograph. These computations show that the bucket radius obtained for the maximum flow is larger than necessary for the smaller flows and that the tail water depth range for satisfactory performance is greater for smaller flows than for the maximum flow.

The Angostura analysis in Table 18 shows, too, that the bucket radius determined from the Angostura model study is smaller than the radius shown in the table, indicating that the methods presented in this monograph provide a factor of safety. This is a desirable feature when hydraulic model studies are not contemplated. On the other hand, hydraulic model studies make it possible to explore regions of uncertainty in particular cases and help to provide the absolute minimum bucket size consistent with acceptable performance.

Other examples in Table 18 include an analysis using the data from Trenton Dam spillway. Although Trenton Dam spillway utilizes a hydraulic jump stilling basin, the data were ideal for an example. This spillway utilizes a long flat chute upstream from the energy dissipator. Friction losses are considerably higher than would occur on the steep spillways for which Figure 71 was drawn. Other means must therefore be used to obtain V_1 and D_1 for the bucket design. In this example, actual velocity measurements taken from a model were used. If frictional resistance is neglected in the velocity computations, the minimum tail water limit would be higher, providing a greater factor of safety against sweepout. But the maximum tail water limit would also be higher, which reduces the factor of safety against flow diving.

Tail water requirements for bucket versus hydraulic jump. In general, a bucket-type dissipator requires a greater depth of tail water than a stilling basin utilizing the hydraulic jump. This is illustrated in Table 19, where pertinent data from Table 18 are summarized to compare the minimum tail water depth necessary for a minimum radius bucket with the computed conjugate tail water depth for a hydraulic jump. Line 6 shows T_{min} for the buckets worked out in the section Practical Applications. Line 7 shows the conjugate tail water depth required for a hydraulic jump for the same Froude number and D_1 determined from the equation $\frac{D_2}{D_1}=1/2(\sqrt{1+8F^2}-1)$.

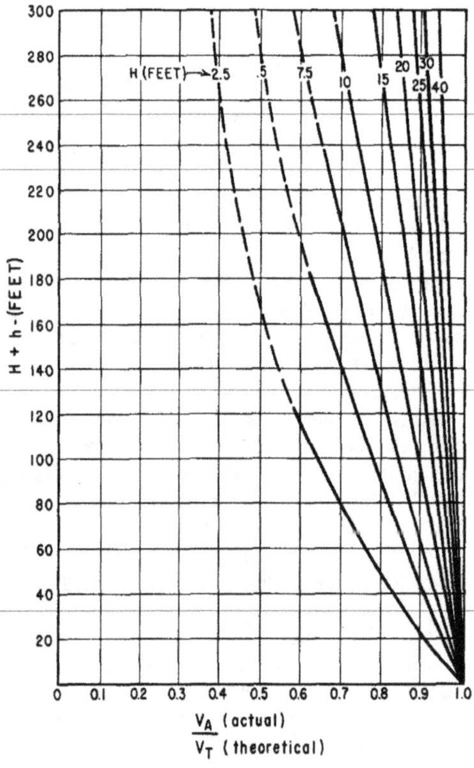

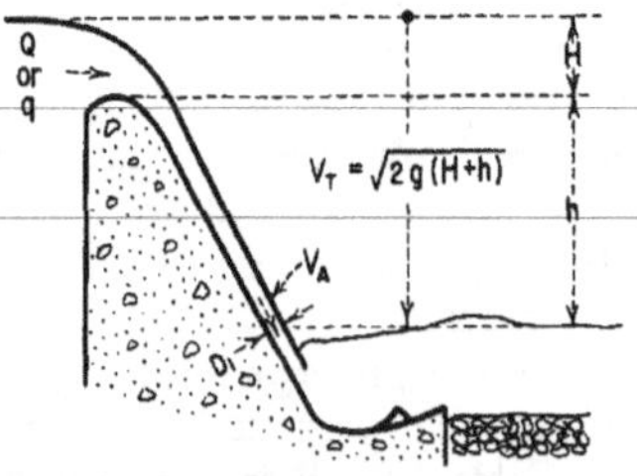

FIGURE 71.—*Curves for determination of velocity entering bucket for steep slopes—0.8:1 to 0.6:1.*

Recapitulation of Bucket Design Procedure

The slotted bucket, Figure 47B, may be used as an energy dissipator at the base of an overfall. Tests showed the slotted bucket to be superior to the solid bucket in all respects. Wherever practicable, the higher teeth recommended in Design Modification II, Figure 53, should be used.

A simplified version of the seven steps required to design a bucket is given below:

1. Determine Q, q (per foot of bucket width), V_1, D_1; compute Froude number from $F=\frac{V_1}{\sqrt{gD_1}}$ for maximum flow and intermediate flows. In some cases V_1 may be estimated from Figure 71.
2. Enter Figure 64 with F to find bucket radius parameter $\frac{R}{D_1+\frac{V_1^2}{2g}}$ from which minimum allowable bucket radius, R, may be computed.
3. Enter Figure 66 with $\frac{R}{D_1+\frac{V_1^2}{2g}}$ and F to find $\frac{T_{min}}{D_1}$ from which minimum tail water depth limit T_{min}, may be computed.
4. Enter Figure 67 as in Step 3 above to find maximum tail water depth limit, T_{max}.
5. Set bucket invert elevation so that tail water curve elevations are between tail water depth limits determined by T_{min} and T_{max}. Keep apron lip and bucket invert above riverbed, if possible. For best performance, set bucket so that the tail water depth is nearer T_{min}. Check setting and determine factor of safety against sweepout from Figure 69 using methods of Step 3.
6. Complete the design of the bucket, using Figure 47 to obtain tooth size, spacing, dimensions, etc.
7. Use Figures 60 and 70 to estimate the probable water surface profile in and downstream from the bucket. The sample calculations in Table 18 may prove helpful in analyzing a particular problem.

TABLE 19.—*Comparison of tail water depths required for bucket and hydraulic jump*

		Angostura Dam	Angostura Dam	Angostura Dam	Angostura Dam	Grand Coulee Dam	Grand Coulee Dam	Trenton Dam	Missouri Diversion Dam [1]
1	Q in thousands c.f.s	247	180	100	40	1,000	1,000	133	90
2	V_1 ft/sec	72	72	73	70	122.4	122.4	66	39
3	D_1 ft	12.5	9.1	5.0	2.1	5.0	5.0	7.6	3.6
4	F	3.6	4.3	5.7	8.5	9.7	9.7	4.2	3.7
5	T_{max} ft	71	72	89	210	114	183	98	32
6	T_{min} ft	67	59	46	32	73	78	49	20
7	T_{conj} ft	57	52	38	24	66	66	40	16
8	Bucket radius	47	39	26	12	30	50	35	12.5

[1] Proposed diversion dam on the Missouri River Basin project.

NOTE: If a larger than minimum bucket radius is used, the required minimum tail water depth becomes greater, as shown for the two Grand Coulee bucket radii.

Section 8

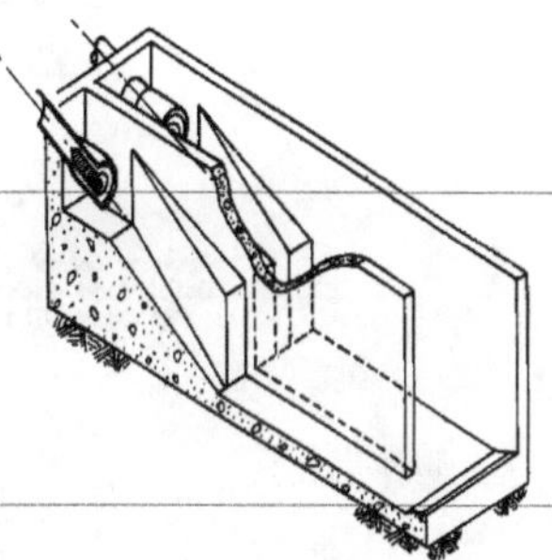

Hydraulic design of hollow-jet valve stilling basins (Basin VIII)

THE hollow-jet valve stilling basin, about 50 percent shorter than a conventional basin, is used to dissipate hydraulic energy at the downstream end of an outlet works control structure. To reduce cost and save space, the stilling basin is usually constructed within or adjacent to the powerhouse structure as shown in Figures 72 and 73.

The hollow-jet valve, Figure 74, controls and regulates the flow. Regardless of the valve opening or head, the outflow has the same pattern, an annular or hollow jet of water of practically uniform diameter throughout its length, Figure 75. The stilling basin is designed to take advantage of the hollow-jet shape; solid jets cannot be used in this basin.

The hollow-jet valve was developed by the Bureau of Reclamation in the early 1940's to fill a need for a dependable regulating valve. A complete 6-inch-diameter hydraulic model and a sectional 12-inch-diameter air model aided the design, and were tested in the Bureau of Reclamation Hydraulic Laboratory. To evaluate the valve characteristics at greater than scale heads, a 24-inch-diameter valve was tested at Hoover Dam under heads ranging from 197 feet to 349 feet.

Piezometer pressure measurements, thrust determinations on the valve needle, and rates of discharge were studied in both field and laboratory tests. It was found that the hydraulic characteristics of the larger valves could be predicted from the performance of the smaller model valves. From these tests and from investigations of prototype valves up to 96 inches in diameter, the valve has been proved to be a satisfactory control device.

Cavitation damage, found on a few of the many prototype valves in use, was minor in nature and was caused by local irregularities in the body casting and by misalinement of the valve with the pipe. These difficulties have been eliminated by careful foundry and installation practices. On one installation, damage that occurred on the cast

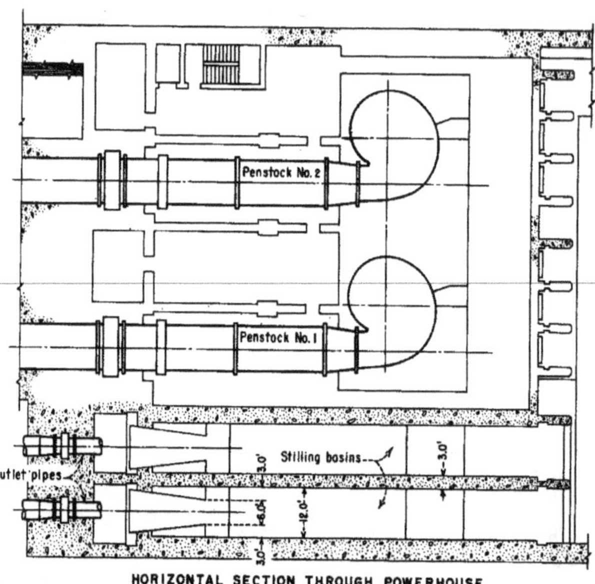

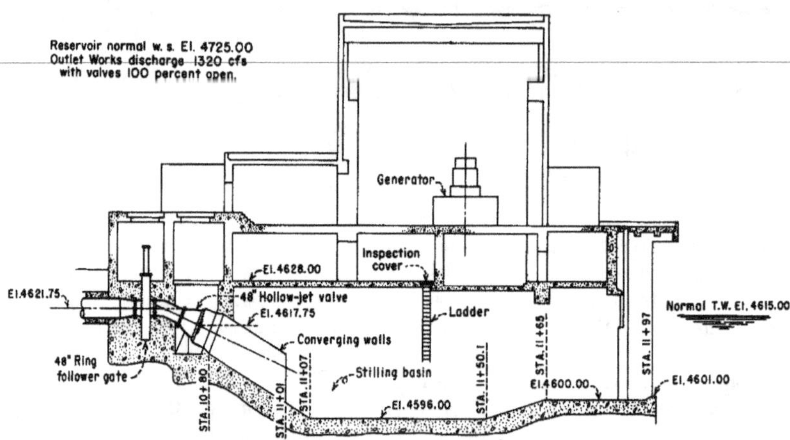

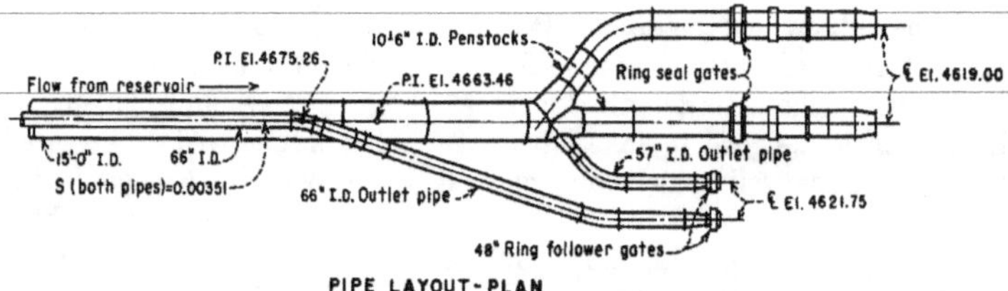

Figure 72.—*Boysen Dam outlet works stilling basin and arrangement of powerplant.*

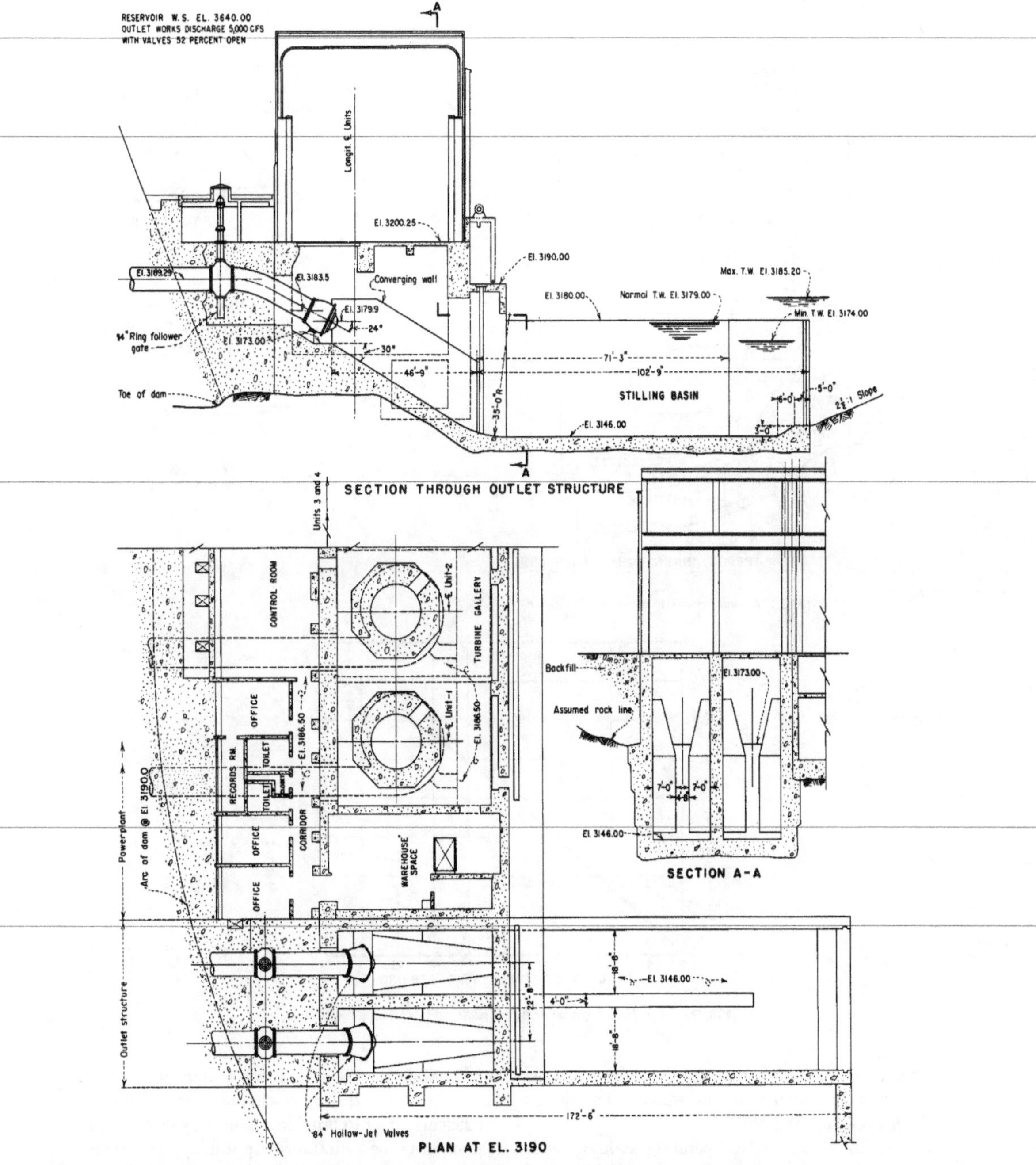

FIGURE 73.—*Yellowtail Dam proposed outlet works stilling basin and powerplant.*

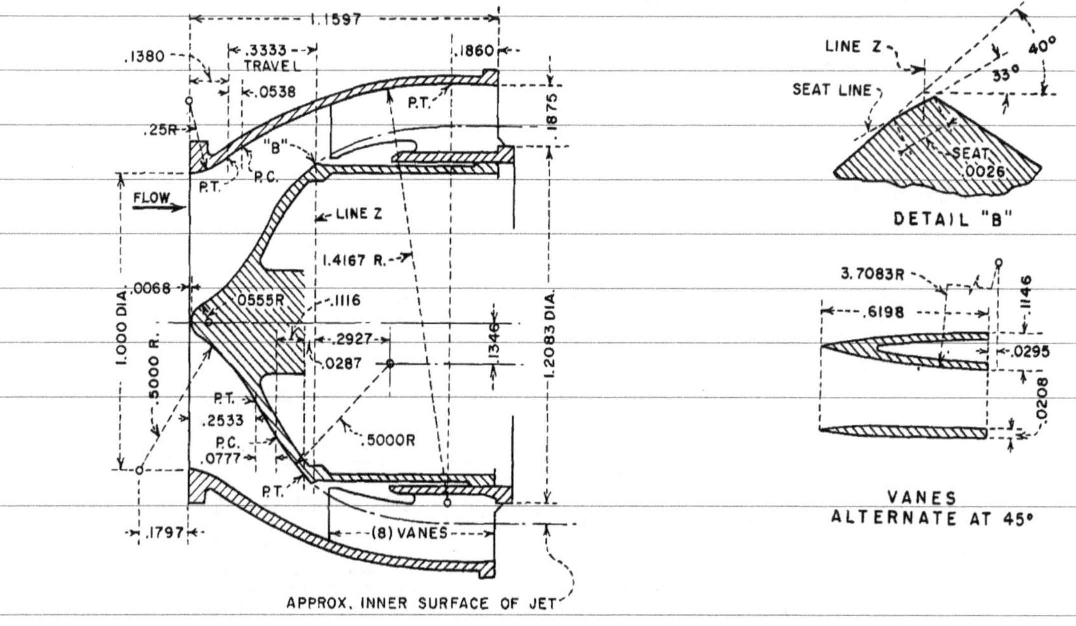

NOTE: All dimensions in terms of diameter

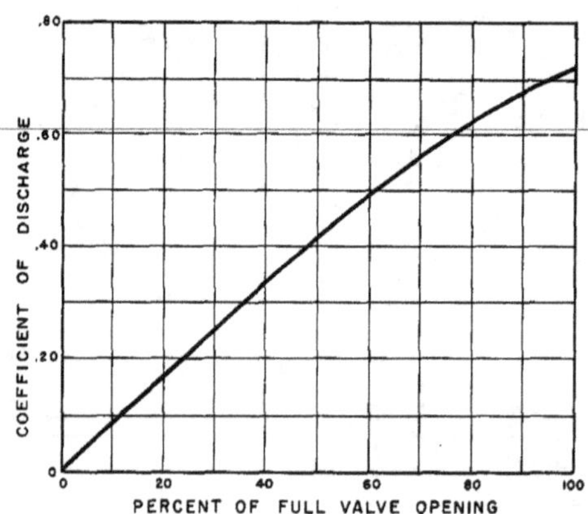

FIGURE 74.—*Hollow-jet valve dimensions and discharge coefficients.*

iron valve support vanes may have been caused by abrasive sediment in the water. The design itself is cavitation free.

Because a large valve operating at high heads can discharge flows having an energy content of up to 150,000 horsepower, a stilling basin is usually required downstream from the valve. In early designs, the valve was discharged horizontally onto a trajectory-curved floor which was sufficiently long to provide a uniformly distributed jet entering the hydraulic jump stilling pool. This resulted in an extremely long structure, twice or more the length of the basin recommended herein. When two valves were used side by side, a long,

TABLE 20.—*Comparison of basin dimensions* [1 2 3]

Basin Dimensions	Boysen	Falcon, U.S.	Falcon, Mexico	Yellowtail	Trinity	Navajo
(1)	(2)	(3)	(4)	(5)	(6)	(7)
Valve diameter, in ft	4	6	7.5	7	7	6
Head at valve, in ft	86	81.5	81.9	380	315	217
Design Q, in c.f.s	660	1,460	2,285	2,500	3,835	2,340
Coefficient C	.70	.70	.70	.41	.70	.70
Percentage valve open	100	100	100	52	100	100
Depth D, in ft	16.2	21.	24.7	31.5	38.5	30
	19	22.5	25.2	32.6	38	[5] 35
Depth D_s, in ft	13.6	17.4	20.2	25.9	31.5	24.6
	14	17.5	19.5	25.6	31.8	24
Length L, in ft	60.4	74.4	86.2	104	129	103
	58	73.9	94	102.8	123	[5] 110
Width W, in ft	10.2	14.7	18	19.2	19.6	16.2
	12	16.2	16.2	18.7	18.9	[5] 18.0
End sill height	3	3	3.1	3.9	4.8	([5])
	4	3	3	3	5	([5])
End sill slope	[4] 3.3:1	2:1	2:1	2:1	2:1	([5])
Converging wall height	3.0d	4.5d	3.9d	3.1d	3.5d	3.4d
Converging wall gap	.50 W	.52 W	.65 W	.25 W	.25 W	.23 W
Center wall length	[4] 1.5 L	.5 L	.4 L	.7 L	.3 L	.5 L
Channel slope	([4])	4:1	4:1	2.5:1	2:1	[5] 6:1

[1] Upper values in each box were calculated from Figs. 82 through 86, lower values in each box were developed from individual model studies.
[2] Valve tilt 24°; inclined floor 30° in all cases.
[3] See Figs. 72, 73, 77, 78, 79, 80, and 82.
[4] Special case, for structural reasons.
[5] Special case, for diversion flow requirements (dentated sill used and basin size increased).

costly dividing wall was also required. Hydraulic model tests showed that the basin length could be reduced more than 50 percent by turning the hollow-jet valves downward and using a different energy dissipating principle in the stilling basin. The first stilling basin of this type was developed for use at Boysen Dam, a relatively low-head structure. Basins for larger discharges and higher heads were later developed from individual hydraulic models of the outlet works at Falcon, Yellowtail, Trinity, and Navajo Dams. It became apparent at this time that generalized design curves could be determined to cover a wide range of operating heads and discharges. Therefore, a testing program was initiated to provide the necessary data. A brief description of the individual model tests made to develop the basin type is given in the following section. Table 20 gives a summary of basin dimensions, valve sizes, test heads, and discharges for these structures.

Development of Basin Features

Boysen Dam. In the Boysen Dam model studies, a series of basic tests was made to determine the optimum angle of entry of a hollow jet into the tail water. For flat angles of entry, the jet did not penetrate the pool but skipped along the tail water surface. For steep angles, the jet penetrated the pool but rose almost vertically to form an objectionable boil on the water surface. When the valves were depressed 24° from the horizontal, Figure 72, and a 30° sloping floor was placed downstream from the valve to protect the underside of the jet from turbulent eddies, optimum performance resulted. The submerged path of the valve jet was then sufficiently long that only a minimum boil rose to the surface. The size and intensity of the boil were further reduced when converging walls were placed on the 30° sloping floor to protect the sides of the jet until it was fully submerged. The converging walls have another function, however; they compress the hollow jet between them to give the resulting thin jet greater ability to penetrate the tail water pool. Sudden expansion of the jet as it leaves the converging walls, plus the creation of fine-grain turbulence in the basin, accounts for most of the energy losses in the flow. Thorough breaking up of the valve jet within

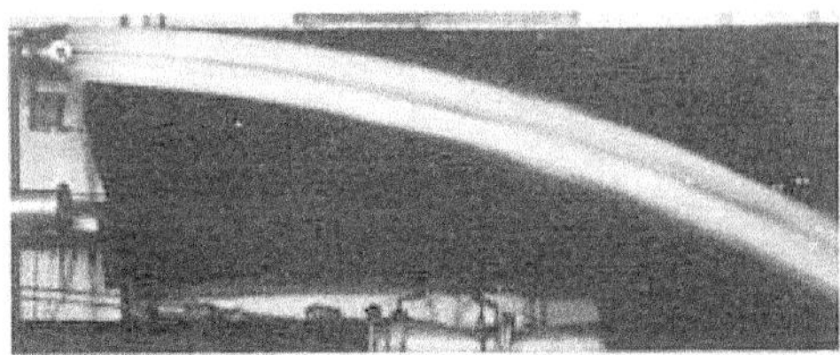

(a) *Valve fully open*

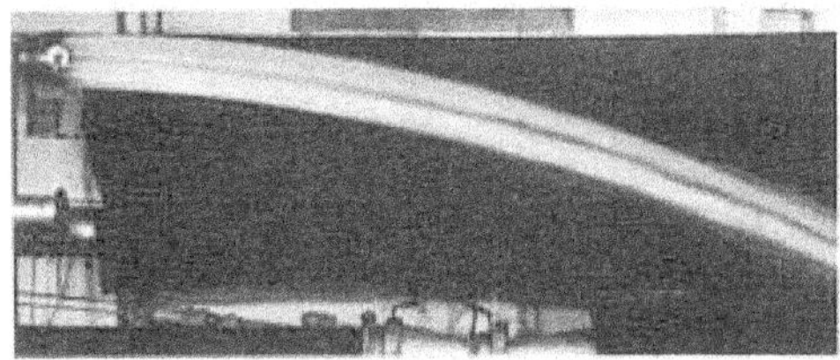

(b) *Valve 50 percent open*

FIGURE 75.—*Six-inch hollow-jet valve discharging.*

the basin and good velocity distribution over the entire cross section of the flow account for the low velocities leaving the basin. Figure 76 shows the performance of a hollow-jet basin both with and without the converging walls.

Pressures on the inside face and downstream end of the converging walls were measured to determine whether low pressures which might induce cavitation were present. The lowest pressure, measured on the end of the wall, was 3 feet of water above atmospheric; therefore, cavitation should not occur. Pressures measured on the sloping floor, and under and near the impinging jet, were all above atmospheric. Maximum pressures did not exceed one-fourth of the total head at the valve.

Scour downstream from the end sill was mild and prototype wave heights were only 0.5 foot in the river channel. A vertical traverse taken near the end sill showed surface velocities to be about 5 feet per second, decreasing uniformly to about 2 feet per second near the floor.

Falcon Dam. In the Falcon Dam tests, two separate basins were developed, one for the United States outlet works and one for the Mexican outlet works, Figures 77 and 78. In these tests, the basic concepts of the Boysen design were proved to be satisfactory for greater discharges. In addition, it was confirmed that dentils on the end sill were not necessary and that the center dividing wall need not extend the full length of the basin. A low 2:1 sloping end sill was sufficient to provide minimum scour and wave heights. Maximum pressures on the floor beneath the impinging jet were found to be about one-third of the total head at the valve, somewhat greater than found in the Boysen tests, but still not excessive.

Yellowtail Dam. In the Yellowtail Dam model studies, the head and discharge were both considerably higher than in the Boysen and Falcon tests. Because of the high-velocity flow from the valves, it was found necessary to extend the converging walls to the downstream end of the sloping floor, Figure 73, and to reduce the wall gap to about one quarter of the basin width. These refinements improved the stilling action within the basin, Figure 76(c), and made it possible to further reduce the basin length. Scour was not excessive, and the water surface in the downstream channel was relatively smooth. Pressures on the converg-

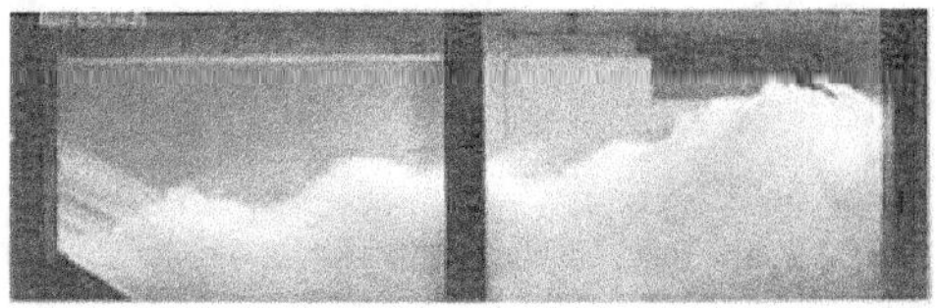

(a) *Stilling action without converging walls*

(b) *Stilling action with short converging walls*

(c) *Stilling action with recommended converging walls*

FIGURE 76.—*Hollow-jet valve stilling basin with and without converging walls.*

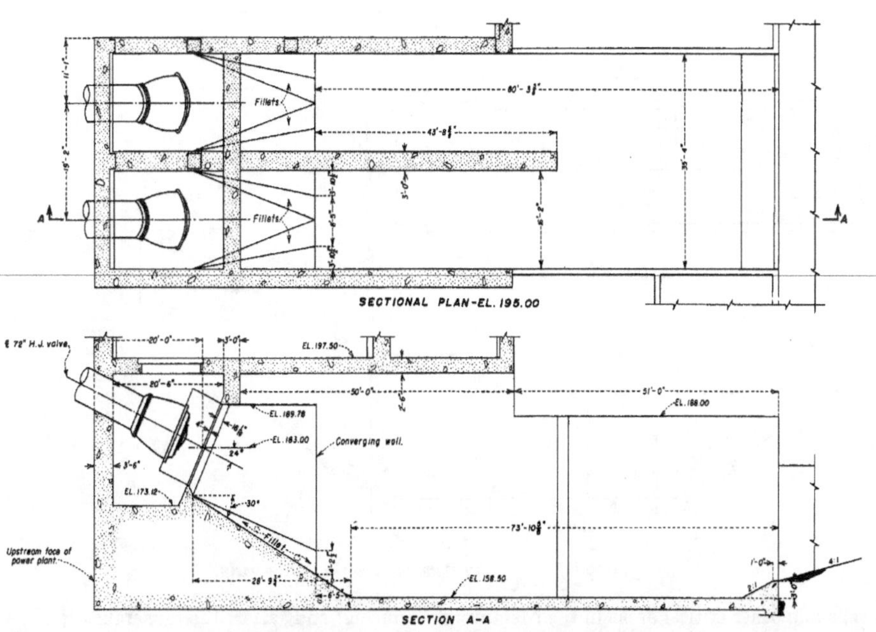

FIGURE 77.—*United States outlet works, Falcon Dam.*

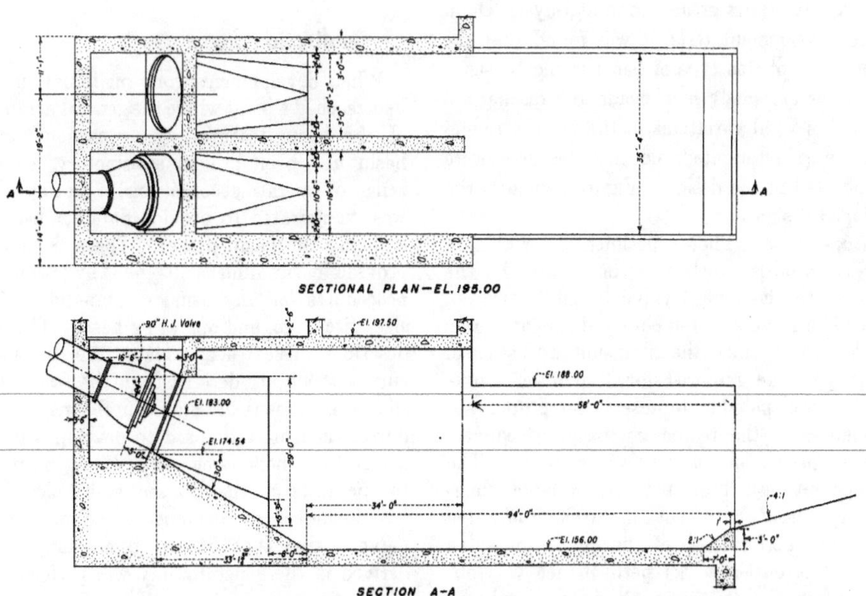

FIGURE 78.—*Mexican outlet works, Falcon Dam.*

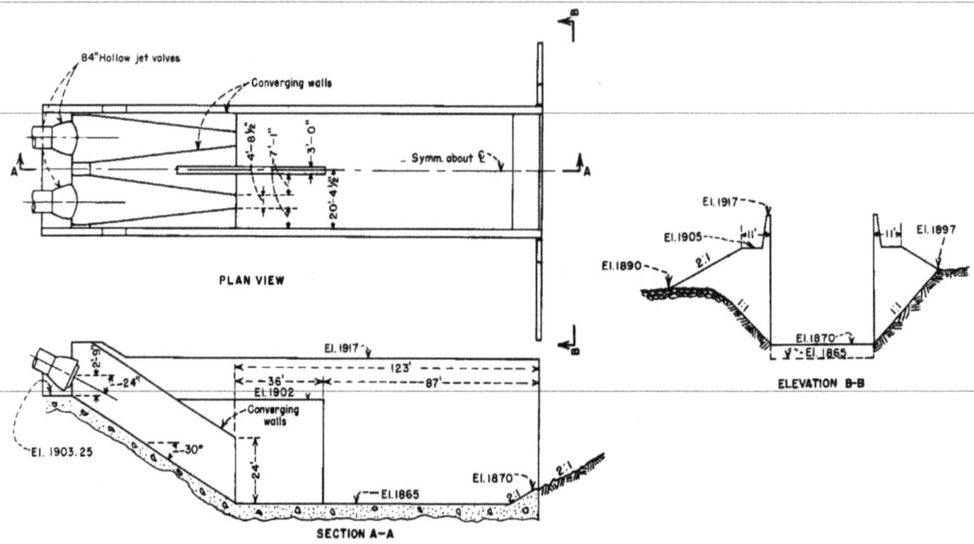

Figure 79.—*Trinity Dam outlet works stilling basin.*

ing walls and other critical areas in the basin were found to be above atmospheric.

Trinity Dam. The Trinity Dam outlet works developed a head almost four times greater and a discharge five times greater than at Boysen Dam. In the development tests, it was found that the performance of this type of basin would be satisfactory for extremely high heads and discharges. Although several variations in the basin arrangement were investigated, no new features were incorporated in the design. Figure 79 shows the developed design.

Navajo Dam. The experimental work on the Navajo outlet works was complicated by the fact that the hollow-jet valve basin, Figure 80, first had to serve as a temporary diversion works stilling basin. Since the diversion works basin was larger than required for the outlet works basin, it was possible to insert the proper appurtenances in the temporary basin to convert it to a permanent outlet works basin. The development tests indicated that a larger-than-necessary basin does not in itself guarantee satisfactory performance of the hollow-jet valve basin. Best outlet works performance was obtained when the temporary basin was reduced in size to conform to the optimum size required for the permanent structure. Since the Navajo Dam outlet works model was available both during and after the generalization tests, the model was used both to aid in obtaining the generalized data and to prove that the design curves obtained were correct.

Generalization Study

When development work on individual basins had reached a point where the general arrangement of the basin features was consistent, and the basin had been proved satisfactory for a wide range of operating conditions, a testing program was inaugurated to provide data for use in generalizing the basin design. These tests were to provide basin dimensions and hydraulic design procedures for any usual combinations of valve size, discharge, and operating head. This section describes these tests, explains the dimensionless curves which are derived from the test data, and shows, by means of sample problems, the procedures which may be used to develop a hydraulic design for a hollow-jet valve stilling basin. Prototype tests on the Boysen and Falcon Basins are included to demonstrate that hollow-jet valve basins that fit the dimensionless curves derived in the general study will perform as well in the field as can be predicted from the model tests.

Test equipment. The outlet works stilling basin model shown in Figure 81 was used for the generali-

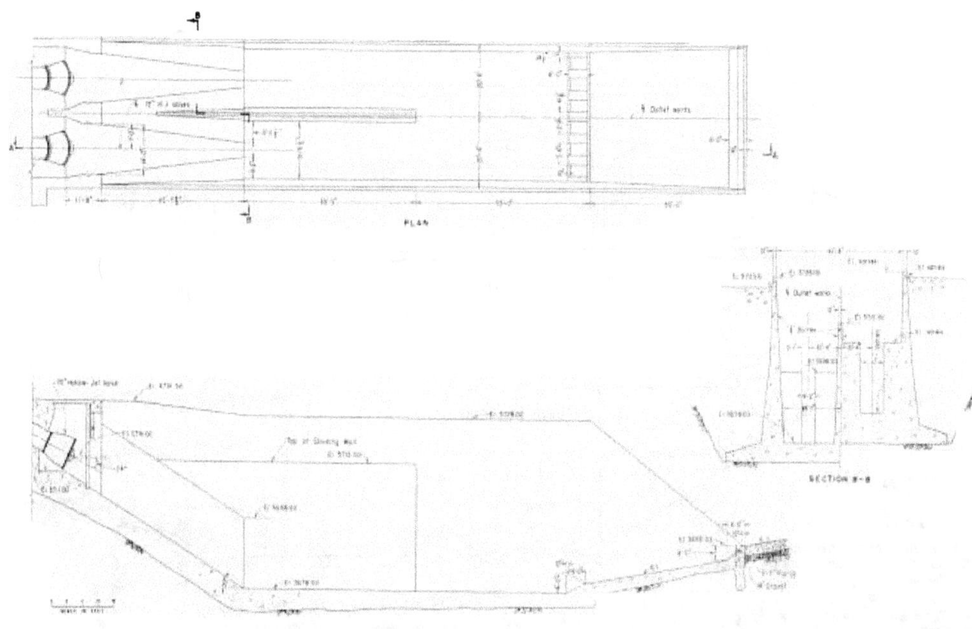

Figure 80—*Navajo Dam outlet works stilling basin.*

Figure 81.—*Hollow-jet valve stilling basin model used for generalization tests.*

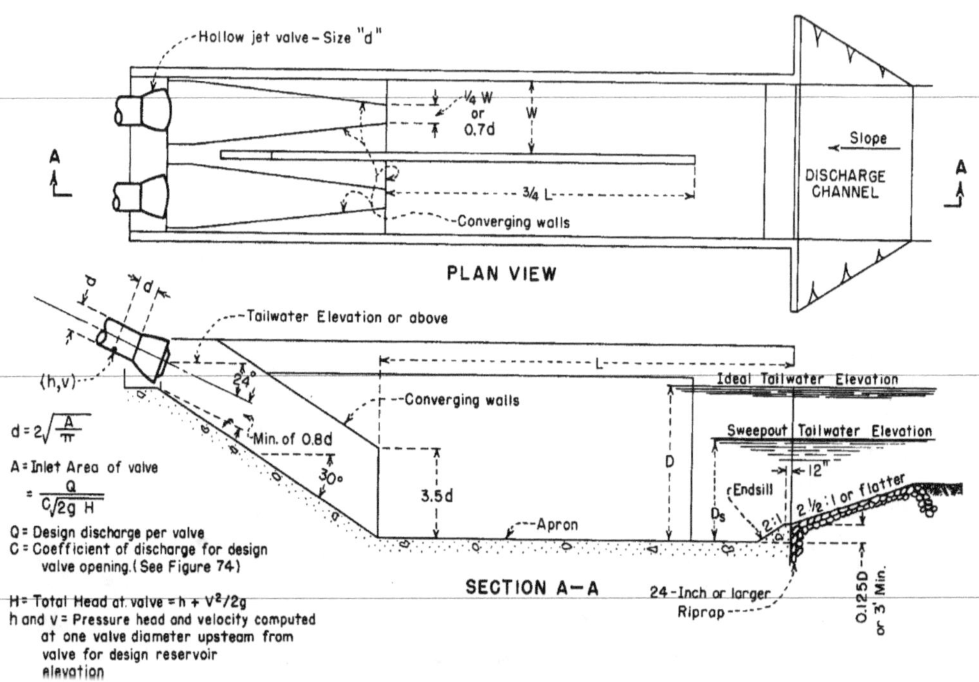

FIGURE 82.—*Generalized design.*

zation tests. The glass-walled testing flume contained two stilling basins separated by a dividing wall. The right-hand basin, having the glass panel as one wall, was operated singly to determine the basin length, width, and depth requirements; both basins were used to study the performance with and without flow in an adjacent basin.

The glass panel permitted observation of the stilling action and the flow currents within and downstream from the basin. The length, width, and depth of the basin were varied by inserting false walls or by moving the basin within the test box. The tail box contained an erodible sand bed to represent the discharge channel bed.

The test valves were exact models of a prototype valve in that the flow surfaces were exactly reproduced, and could be opened and closed to any partial opening. The models were 3-inch valves machined from bronze castings.

The pressure head at each model valve was measured, using a piezometer located in the 3-inch supply pipe one diameter upstream from the valve flange. Calibrated Venturi meters permanently installed in the laboratory measured discharges. The tail water elevation in the discharge channel was controlled with a hinged tailgate in the tail box and tail water elevations were determined visually from a staff gage on the tail box wall located approximately 62 valve diameters downstream from the valves.

Preliminary procedures. The investigation began with tabulating the important dimensions of the Boysen, Falcon, Yellowtail, and Trinity outlet works basins and expressing them in dimensionless form, as shown in Table 20. Based on these dimensions, a model was constructed as shown in Figure 82, using the 3-inch valve dimension to establish the absolute model size. More weight was given to the Yellowtail and Trinity basins because they were developed for higher heads and contained refinements in the converging wall design which improved the basin performance at high heads. Also, the latter basins had been model tested over a greater operating range than were the earlier low-head basins.

To provide practical discharge limits for the tests, the 3-inch model was assumed to represent an 84-inch prototype valve, making the model scale 1:28. Discharges of 2,000 to 4,000 c.f.s. with one valve open 100 percent were considered to be the usual design discharges for a valve of this size. To produce these discharges, heads of 100 feet to 345 feet of water at the valve would be required.

Initial tests were made with the stilling basin apron longer than necessary and with no end sill in place. For a given discharge, the ideal depth of tail water was determined from visual inspection of the stilling action as it occurred over a range of tail water elevations. For each ideal tail water determination, the minimum length of concrete apron was estimated after an inspection of the flow currents in the model had indicated where an end sill should be placed in the prototype. Confirming tests were then conducted successively on a representative group of basins having the apron lengths previously determined and having an end sill at the end of the apron. Preliminary values were then adjusted as necessary to obtain final ideal tail water depths and apron lengths. In the latter tests, the height of the valve above the maximum tail water elevation was adjusted to simulate a typical prototype installation. Similar tests were then made with the valve open 75% and 50%. Finally, a series of tests was made to determine the ideal width of stilling basin and the range of widths over which satisfactory performance could be expected.

Preliminary tests. In a typical test, the desired discharge was set by means of the laboratory venturi meters and passed through the hollow-jet valve or valves opened 100 percent. The tail water elevation was adjusted to provide the best energy dissipating action in the basin. The optimum value, tail water depth D in Figure 82, was judged by the appearance and quality of the stilling action in the basin and on the smoothness of the tail water surface.

For discharges of 2,000 to 4,000 c.f.s. it was found that the tail water could be raised or lowered about 3 feet (0.1 foot in the model) from the ideal tail water elevation without adversely affecting the basin performance. Increasing the tail water depth beyond this margin reduced the efficiency of the stilling action and allowed the jet to flow along the bottom of the basin for a greater distance before being dissipated. This also produced surges in the basin and increased the wave heights in the discharge channel. Decreasing the tail water depth below the 3-foot margin moved the stilling action downstream in the basin and uncovered the valve jets at the end of the converging walls. This increased the flow velocity entering the discharge channel and increased the tendency to produce bed scour. Uncovering of the stilling action also produced objectionable splashing at the upstream end of the basin. If the tail water depth was decreased further, the flow swept through the basin with no stilling action having occurred. The latter tail water depth was measured and recorded as the sweepout depth, D_s. These tests were made with the dividing wall extended to the end of the basin, since this provided the least factor of safety against jump sweepout. With a shorter dividing wall, sweepout occurs at a tail water elevation slightly less than D_s.

With the ideal tail water depth set for a desired flow, the action in the basin was examined to determine the ideal length, L, of the basin apron, Figure 82. The apron length was taken to the point where the bottom flow currents began to rise from the basin floor of their own accord, without assistance from an end sill, Figure 76(c). The water surface directly above and downstream from this point was fairly smooth, indicating that the stilling action had been completed and that the paved apron and training walls need not extend farther. In the individual model studies that preceded the generalized tests, it had been found that when the basin was appreciably longer than ideal, the ground roller at the end sill carried bed material from the discharge channel over the end sill and into the basin. If this action should occur in a prototype structure the deposited material would swirl around in the downstream end of the basin and cause abrasive damage to the concrete apron and end sill. It had also been found that scour tendencies in the discharge channel were materially increased if the basin was appreciably shorter than ideal. Therefore, the point at which the currents turned upward from the apron, plus the additional length required for an end sill, was determined to be the optimum length of apron. At this point, the scouring velocities were a minimum and any scouring tendencies would be reduced by the sloping end sill to be added later.

Practical difficulties were experienced in determining the exact length of apron required, however. Surges in the currents flowing along the basin floor caused the point of upturn to move upstream and downstream a distance of 1/4 to 1/2 D in a period of 15 to 20 seconds in the model. An average apron length was therefore selected in the preliminary test.

The depth D, sweepout depth D_s, and length L were then determined for the range of discharges possible with the hollow-jet valve first open 75 percent, and finally 50 percent, using the testing methods described in the preceding paragraphs.

Partial openings were investigated because the valve size is often determined for the minimum operating head and maximum design discharge. When the same quantity is discharged at higher heads, the valve opening must be reduced. It may be necessary, therefore, to design the basin for maximum discharge with the valves opened less than 100 percent. When the relation between head and discharge through the valve is changed materially, the minimum required basin dimensions will be affected. The data for the partially opened valves are also useful in indicating the basin size requirements for discharges greater or less than the design flow conditions.

Final tests and procedures. The final tests were made to correct or verify the dimensions obtained in the preliminary tests and to investigate the effect of varying the basin width. Scour tendencies were also observed to help evaluate the basin performance. D, D_s, and L for the three valve openings are functions of the energy in the flow at the valve. The energy may be represented by the total head, H, at the valve, Figure 82. Therefore, to provide dimensionless data which may be used to design a basin for any size hollow-jet valve, D, D_s, and L values from the preliminary tests were divided by the valve diameter d, and each variable was plotted against H/d. The resulting curves, similar to those in Figures 83, 84, and 85, were used to obtain dimensions for a group of model basins which were tested with the end sill at the end of the apron and with the valves placed the proper vertical distance above the tail water. For each model basin, a 3:1 upward sloping erodible bed, composed of fine sand, was installed downstream from the end sill. The bed was kept sufficiently low that it did not interfere with tail water manipulation, even when the tail water was lowered for the sweepout tests. Test procedure was essentially as described for the preliminary tests.

Basin depth and length. The preliminary depth curves for both ideal tail water depth and sweepout tail water depth needed but little adjustment. The preliminary basin lengths were found to be too long for the high heads and too short for the lower heads, although both adjustments were relatively minor. The adjusted and final curves are shown in Figures 83, 84, and 85.

It was observed that a longer apron than that indicated by Figure 85 was necessary when the tail water depth exceeded the tail water depth limit in Figure 83. As the stilling action became drowned, the action in the basin changed from fine-grain turbulence to larger and slower moving vertical eddies. The bottom flow currents were not dissipated as thoroughly or as quickly and were visible on the apron for a greater distance, thereby increasing the necessary length of basin. The action is similar to that observed in hydraulic jumps which are drowned by excessive tail water depths. A moderate amount of drowning is tolerable, but it is important that the ideal tail water depth be maintained within stated limits if the best performance is desired. The tail water depth limits—0.1 foot above and below the ideal depth—expressed in dimensionless form is 0.4 d. If this limit is exceeded, a model study is recommended.

Basin width. To determine the effect of basin width, tests on several basins were made in which only the basin width was varied. It was found that the width could be increased to 3.0 times the valve diameter before the action became unstable. The width could be decreased to 2.5 times the valve diameter before the stilling action extended beyond the ideal length of basin. However, the H/d ratio and the valve opening were found to affect the required basin width as shown for 100 percent, 75 percent, and 50 percent valve openings in Figure 86.

Basin width is not a critical dimension but certain precautions should be taken when selecting a minimum value. If the tail water is never to be lower than ideal, as shown by the curves in Figure 83, the basin width may be reduced to 2.5 d. If the tail water elevation is to be below ideal, however, the curve values for width in Figure 86 should be used. In other words, the lower limits for both tail water and basin width should not be

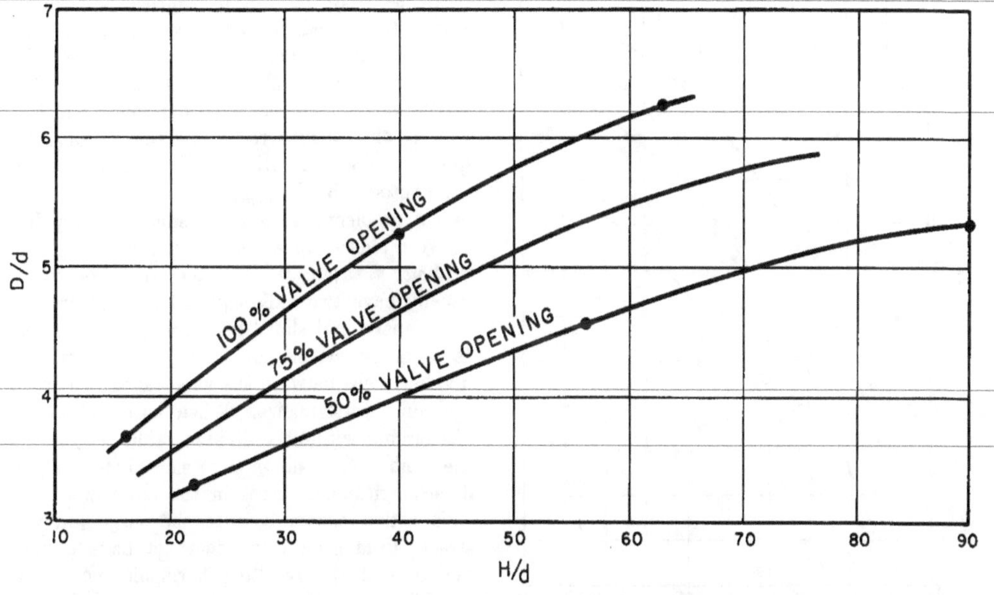

NOTE:

Best hydraulic performance is for ideal depths shown.
Good performance occurs over range of depths 0.4 (d) greater or less than D.
D, H, and d are defined in Figure 82.
"●" represent data points shown in Figures 87 and 88.

FIGURE 83.—*Ideal tail water depth.*

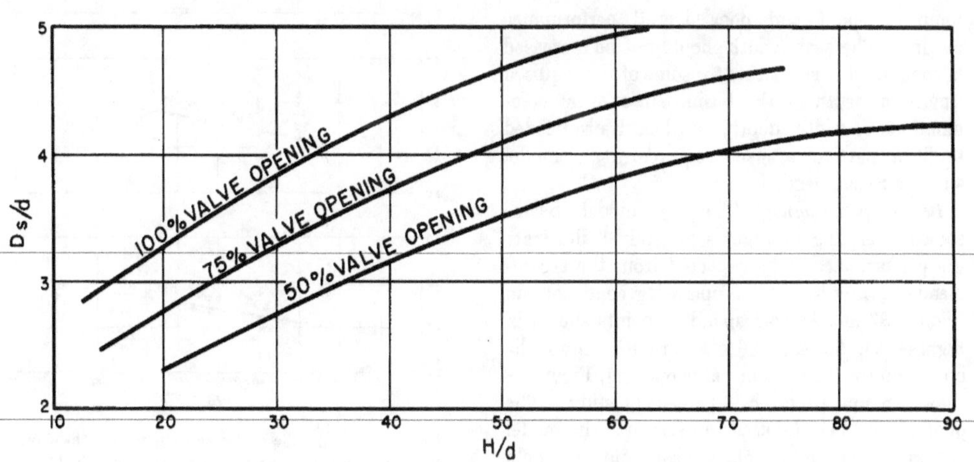

NOTE: D_s is the depth of tail water above the basin apron when the flow from the valve first begins to sweep out of the basin.

H and d are defined in Figure 82.

FIGURE 84.—*Tail water sweepout depth.*

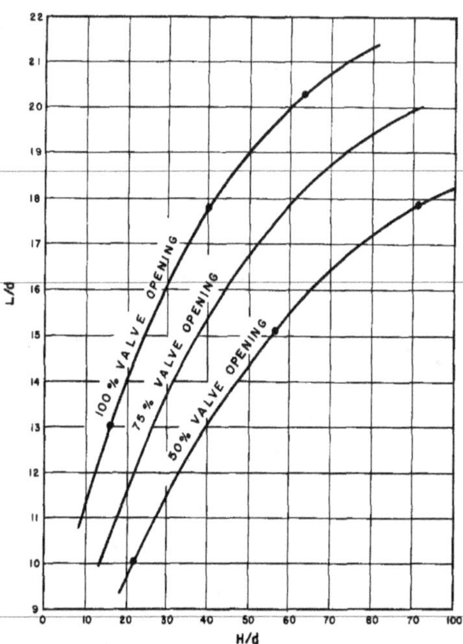

NOTE: H, L, and d are defined in Figure 82.
"●" represent data points shown in Figures 87 and 88.

FIGURE 85.—*Stilling basin length.*

used in the same structure. The combined minimums tend to reduce the safety factor against jump sweepout and poor overall performance results. The basin width should not be increased beyond 3.0 d to substitute for some of the required length or depth of the basin. If unusual combinations of width, depth, and length are needed to fit a particular space requirement, a model study is recommended.

Basin performance. The six model basins shown operating in Figures 87 and 88 illustrate the performance to be expected from the recommended structures. The operating conditions in Figures 87 and 88 correspond to points shown in Figures 83, 85, and 86. Figure 87 shows the operation for 100 percent valve opening; Figure 88 shows the operation for 50 percent opening. The photographs may be used to determine the model appearance of the prototype basin and may help to provide a visual appraisal of the prototype structure. Wave heights, boil heights, or other visible dimensions may be scaled from the photographs (using the scale shown in the photographs) and converted to prototype dimensions by multiplying the scaled distances by the model scale. To determine the model scale, the prototype valve diameter in inches should be divided by 3 (the model valve diameter). To determine which of the six photographs represents the prototype in question, the H/d ratio should be used to select the photograph which most nearly represents the design problem. It is permissible to interpolate between photographs when necessary.

Center dividing wall. Prototype stilling basins usually have two valves placed a minimum distance apart, and alined to discharge parallel jets. It is necessary, without exception, to provide dividing walls between the valves for satisfactory hydraulic performance. When both valves are discharging without a dividing wall, the flow in the double basin sways from side to side to produce longitudinal surges in the tail water pool. This action occurs because the surging downstream from each valve does not have a fixed period, and the resulting harmonic motion at times becomes intense. When only one valve is discharging, conditions are worse. The depressed

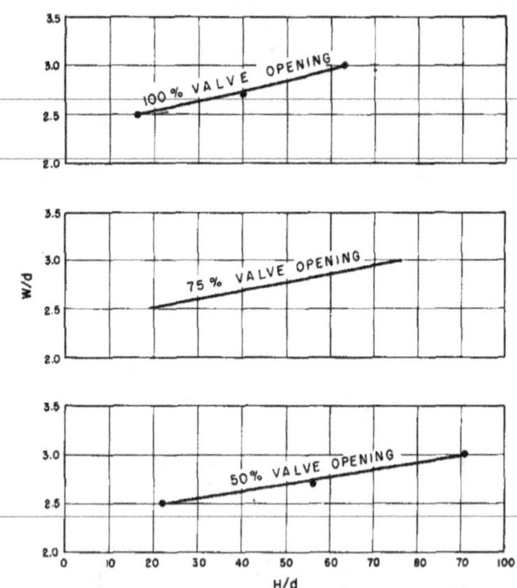

NOTE: Best hydraulic performance is for widths shown. Good performance occurs over range of width 2.5 d to 3.5 d.

W, H, and d are defined in Figure 82.

"●" represent data points shown in Figures 87 and 88.

FIGURE 86.—*Basin width per valve.*

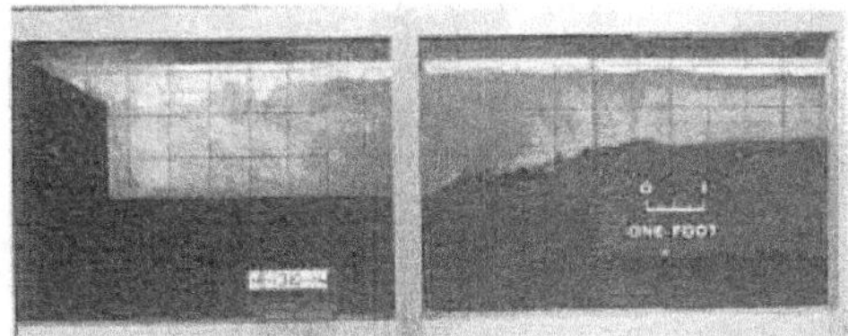

(a) H/d=16, D/d=3.7, L/d=12.9, W/d=2.5

(b) H/d=40, D/d=5.2, L/d=17.8, W/d=2.7

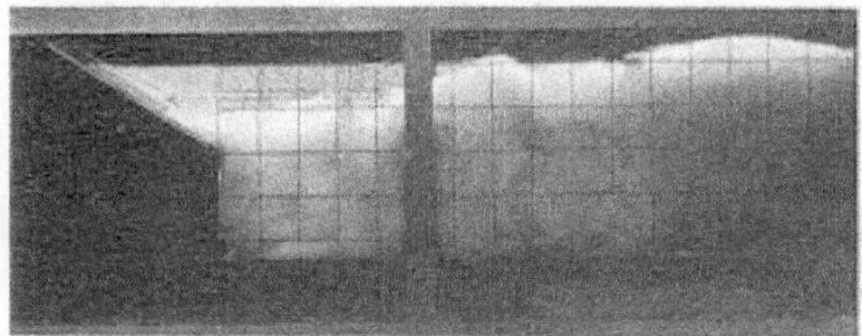

(c) H/d=63, D/d=6.2, L/d=20.3, W/d=3.0

FIGURE 87.—*Hollow-jet valve stilling basin performance, valve 100 percent open.*

water surface downstream from the operating valve induces flow from the higher water level on the nonoperating side. Violent eddies carry bed material from the discharge channel into the basin and swirl it around. This action in the prototype would damage the basin as well as the discharge channel. In addition, the stilling action on the operating side is impaired.

To provide acceptable operation with one valve operating, the dividing wall should extend to three-fourths of the basin length or more. However, if the two adjacent valves discharge equal quantities of flow at all times, the length of the center dividing wall may be reduced to one-half of the basin length. The margin against sweepout is increased, but the stability of the flow pattern is decreased as the dividing wall is shortened. In some installations, a full-length wall may be desirable to help support the upper levels of a powerplant, Figure 72. If other arrangements of the center wall are required, a model study is recommended.

Valve placement. A hollow-jet valve should not operate submerged because of the possibility of cavitation occurring within the valve. However, the valve may be set with the valve top at maximum tail water elevation, and the valve will not be under water at maximum discharge. The valve jet sweeps the tail water away from the downstream face of the valve sufficiently to allow usual ventilation of the valve. However, as a general rule, it is recommended that the valve be placed with its center (downstream end) no lower than tail water elevation.

Riprap size. A prototype basin is usually designed for maximum discharge, but will often be used for lesser flows at partial and full valve openings. For these lesser discharges, the basin will be larger than necessary, and in most respects, the hydraulic performance will be improved. However, at less than design discharge, particularly

(a) H/d=22, D/d=3.3, L/d=10.0, W/d=2.5

(b) H/d=56, D/d=4.5, L/d=15.0, W/d=2.7

(c) H/d=91, D/d=5.3, L/d=17.8, W/d=3.0

FIGURE 88.—*Hollow-jet valve stilling basin performance valve 50 percent open.*

in instances close to the design discharge, the ground roller will tend to carry some bed material upstream and over the end sill into the basin. The intensity of this action is relatively mild over most of the discharge range, and movement of material may be prevented by placing riprap downstream from the end sill. Riprap having 50 percent or more of the individual stones 24 inches to 30 inches or larger in diameter should provide a stable channel downstream from the end sill. The riprap should extend a distance D, or more, from the end sill. If the channel is excavated and slopes upward to the natural river channel, the riprap should extend from the end sill to the top of the slope, or more. The riprap should not be terminated on the slope.

The justification for choosing riprap as described is as follows: Because of the fixed relationships between depth and width of basin, the average velocity leaving the basin will seldom exceed 5 feet per second, regardless of structure size. Surface velocities will therefore seldom exceed 7 to 8 feet per second and bottom velocities 3 to 4 feet per second. To protect against these velocities, stones 10 inches to 12 inches in diameter would be ample. However, the critical velocity for riprap stability is the upstream velocity of the ground roller, which has a curved path and tends to lift the stones out of place. Model tests showed that graded riprap up to 24 inches to 30 inches in diameter was sufficient to provide bed stability.

Application of Results

Problems. Design a stilling basin for (*a*) one hollow-jet valve discharging 1,300 c.f.s., and (*b*) a double basin for two valves discharging 650 c.f.s. each. In both problems, the reservoir is 108 feet above maximum tail water elevation.

One-valve stilling basin design. The valve size should be determined from the equation:

$$Q=CA\sqrt{2gH},$$

in which Q is the design discharge, C is the coefficient of discharge, A is the inlet area to the valve, g is the acceleration of gravity, and H is the usable or total head at the valve with the valve center placed at maximum tail water elevation. In this example, the usable head at the valve is estimated to be 80 percent of the total head of 108 feet, or 86 feet.

From Figure 74, for 100 percent valve opening:

$$C=0.7.$$

Then

$$A=25 \text{ sq. ft.}$$

and

$$d=5.67 \text{ ft.}$$

in which d is the inlet diameter of the valve and also the nominal valve size.

Since nominal valve sizes are usually graduated in 6-in. increments,

$$d=6 \text{ ft.}$$

would be selected. Because the selected valve is larger than required, it would not be necessary to open the valve fully to pass the design flow at the maximum head.

Having determined the valve size and therefore the diameter of the supply conduit, the probable head losses in the system from reservoir to valve may be computed. In this example, the computed losses are assumed to be 20 feet, which leaves 88 feet of head at the valve. Using the equation, C is computed to be 0.61; from Figure 74, the valve opening necessary to pass the design discharge at the design head is 83 percent.

The basin depth, length, and width may be determined from Figures 83, 84, 85, and 86 using the head ratio

$$\frac{H}{d}=\frac{88}{6}=14.67.$$

For 83 percent valve opening, Figure 83 shows the depth ratio

$$\frac{D}{d}=3.4.$$

The depth of the basin is

$$D=20.4 \text{ ft.}$$

therefore, the apron is placed 20.4 feet below the maximum tail water elevation.

For 83 percent valve opening, Figure 85 shows the length ratio

$$\frac{L}{d}=11.2.$$

The length of the basin is

$$L=67 \text{ ft.}$$

For 83 percent valve opening, Figure 86 shows the width ratio

$$\frac{W}{d}=2.5.$$

The width of the basin is

$$W=15 \text{ ft.}$$

The dimensions of other components of the basin may be determined from Figure 82.

The tail water depth at which the flow will sweep from the basin may be determined from Figure 84. For 83 percent valve opening, the depth sweepout ratio

$$\frac{D_s}{d}=2.7.$$

The sweepout depth is

$$D_s=16.2 \text{ ft.}$$

Since 20.4 feet of depth is provided, the basin has a safety factor against sweepout of 4.2 feet of tail water depth. In most installations this is sufficient, but if a greater margin of safety is desired, the apron elevation may be lowered

$$0.4(d)=2.4 \text{ ft.}$$

If greater economy and less margin of safety are desired, the basin floor may be placed 2.4 feet higher to provide only 18 feet of depth.

If the tail water depth from Figure 83 is adopted, the water surface profile will be similar to that shown in Figure 87(a), since the H/d value of 16 in Figure 87(a) is comparable to 14.67 in this example. If tail water depth 2 feet greater or less than the ideal is adopted for the prototype, the water surface profile will be moved up or down accordingly. Water surfaces may be estimated by multiplying the variations shown in Figure 87(a) by the quotient obtained by dividing the prototype valve diameter of 72 inches by the model valve diameter of 3 inches. Wave heights in the downstream channel will be considerably less, as indicated in other photographs showing downstream conditions.

Two-valve stilling basin design. If two valves are to be used to discharge the design flow of 1,300 c.f.s., a double basin with a dividing wall is required. The discharge per valve is 650 c.f.s., and at 100 percent valve opening the valve coefficient is 0.7, Figure 74. The head on the valve is estimated to be 86 feet, as in the first example. From the equation used for one-valve stilling basin design, the inlet area of the valve is found to be 12.48 square feet. A 48-inch valve provides practically the exact area required.

For this example, it is assumed that the computations to determine head losses have been made and that the estimated head of 86 feet at the valves is correct. Therefore, 100 percent valve opening will be necessary to pass the design flow.

Using the methods given in detail in the first example:

$$\frac{H}{d}=21.5$$

$$\frac{D}{d}=4.06, \text{ from Figure 83,}$$

and

$$D=16.2 \text{ ft.}$$

$$\frac{D_s}{d}=3.3, \text{ from Figure 84,}$$

then

$$D_s=13.2 \text{ ft.}$$

The tail water depth for sweepout is therefore 3.0 ft. below the ideal tail water depth. If more or less insurance against the possibility of sweepout is desired, the apron may be set lower or higher by the amount

$$0.4(d)=1.6 \text{ ft.}$$

To aid in determining the apron elevation, the effect of spillway, turbine, or other discharges on the tail water range may need to be considered.

$$\frac{L}{d}=14.4, \text{ from Figure 85,}$$

then

$$L=58 \text{ ft.}$$

$$\frac{W}{d}=2.6, \text{ from Figure 86,}$$

then

$$W=10.4 \text{ ft.}$$

FIGURE 89.—*Boysen Dam: left valve of outlet works basin, discharging 660 c.f.s.—model scale: 1:16.*

Since two valves are to be used, the total width of the basin will be 2(W) plus the thickness of the center dividing wall. The length of the center dividing wall should be three-fourths of the apron length or 43.5 feet long, Figure 82. If it is certain that both valves will always discharge equally, the wall need be only one-half the apron length, or 29 feet long. The hydraulic design of the basin may be completed using Figure 82.

If the tail water depth determined from Figure 83 is adopted, the water surface profile for determining wall heights may be estimated by interpolating between Figure 87 (a) and (b). Water surface variations may be predicted by multiplying values scaled from the photographs by the ratio 48/3.

Prototype Performance

The Boysen Dam and Falcon Dam outlet works stilling basins, Figures 72, 77, and 78, fit the design curves derived from the generalized study quite well, and have been field tested and found to perform in an excellent manner. Table 20 shows the important dimensions of these basins and indicates that the values computed from the design curves of this section are in good agreement with those obtained from the individual model tests.

Boysen Dam. The outlet works basin at Boysen Dam is designed for 1,320 c.f.s. from two 48-in. hollow-jet valves 100 percent open at reservoir elevation 4,725.00. Design tail water elevation at the basin is 4,616.00. The model performance of this basin is shown in Figures 89 and 90.

The prototype tests, Figures 91, 92, and 93, were conducted with the reservoir at elevation 4,723.5 and with the powerplant both operating

FIGURE 90.—*Boysen Dam: outlet works discharging 1,320 c.f.s.—model scale: 1:16.*

Both valves fully open. Reservoir elevation 4,723.5. Dashed lines show the outline of converging walls located beneath spray. Compare with Figure 89.

FIGURE 91.—*Boysen Dam: left valve of outlet works basin discharging 732 c.f.s.—looking upstream.*

Both valves fully open. Reservoir elevation 4723.5. Compare with Figure 89.

FIGURE 92.—*Boysen Dam: left valve of outlet works basin discharging 732 c.f.s.—looking downstream.*

and shut down. The spillway was not operating. The outlet works discharge was measured at a temporary gaging station about 1/2 mile downstream from the dam, using a current meter to determine the discharge. Tail water elevations were read on the gage in the powerhouse.

The prototype performed as well as predicted by the model and was considered satisfactory in all respects. However, the field structure entrained more air within the flow than did the model. This caused the prototype flow to appear more bulky, and "white water" extended farther into the downstream channel than was indicated in the model. A comparison of the model and prototype photographs, Figures 90 and 93, illustrates this difference. Greater air entrainment in the prototype is usually found when making model-prototype comparisons, particularly when the difference between model and prototype velocities is appreciable. In other respects, however, the prototype basin was as good or better than predicted from the model tests.

Both valves fully open and both turbines operating at normal load. Reservoir elevation 4,723.5. Tail water elevati 4,617. Compare with Figure 90.

FIGURE 93.—*Boysen Dam: outlet works discharging 1,344 c.f.s.*

For the initial prototype test, only the left outlet valve was operated; the powerhouse was not operating. At the gaging station, the discharge was measured to be 732 c.f.s. after the tail water stabilized at elevation 4,614.5. (This is a greater discharge than can be accounted for by calculations. It is presumed that valve overtravel caused the valve opening to exceed 100 percent even though the indicator showed 100 percent open.) It was possible to descend the steel ladder, Figure 72, to closely observe and photograph the flow in the stilling basin, Figures 91 and 92. The basin was remarkably free of surges and spray and the energy-dissipating action was excellent. There was no noticeable vibration at the valves or in the basin. The flow leaving the structure caused only slightly more disturbance in the tailrace than the flow from the draft tubes when the turbines were operating at normal load.

Operation of the prototype provided an opportunity to check the air requirements of the structure, which could not be done on the model. With the inspection cover, Figure 72, removed, the basin was open to the rooms above. Air movements through the inspection opening and in the powerplant structure were negligible, which indicated that ample air could circulate from the partially open end of the stilling basin, Figure 92.

When both valves were discharging fully open, the tail water stabilized at elevation 4615. A discharge measurement at the gaging station disclosed that both valves were discharging 1,344 c.f.s. Since the left valve had been found to discharge 732 c.f.s., the right valve was discharging 612 c.f.s.

The reason for the difference in discharge is that the 57-inch-inside-diameter outlet pipe to the left valve is short and is connected to the 15-foot-diameter header which supplies water to the turbines, Figure 72. The right valve is supplied by a separate 66-inch-diameter pipe extending to the reservoir. Therefore, greater hydraulic head losses occur in the right valve supply line, which accounts for the lesser discharge through the right valve. Although it was apparent by visual observation that the left valve was discharging more than the right valve, Figure 93, no adverse effect on the performance of the outlet works stilling basin or on flow conditions in the powerhouse tailrace could be found.

The outlet works basin performance was also observed with the turbines operating and the tail water at about elevation 4617. No adverse effects of the outlet works discharge on powerplant performance could be detected. Flow conditions in the tailrace area were entirely satisfactory, Figure 93. Since the tests were made at normal reservoir level and maximum discharge, the stilling basin was subjected to a severe test.

Falcon Dam. The outlet works basin on the Mexico side at Falcon Dam is designed to accommodate 4,570 c.f.s. from two 90-inch valves or 2,400 c.f.s. from one valve, with the valves 100 percent open and the reservoir at elevation 300. The tail water elevation is 181.2 when the powerplant is discharging 5,400 c.f.s. in conjunction with both valves. The model performance of this basin is shown in Figures 95 and 96.

The outlet works basin on the United States side at Falcon Dam is designed to discharge 2,920 c.f.s. from two 72-inch valves, or 1,600 c.f.s. from one valve, with the valves 100 percent open and the reservoir at elevation 310. Tail water is at elevation 180.8 when two valves are operating and 180.5 when one valve is operating. The model performance of this basin is shown in Figures 97, 98, and 99.

The prototype tests at Falcon, Figures 100, 101, and 102, were conducted at near maximum conditions; the reservoir was at elevation 301.83, and

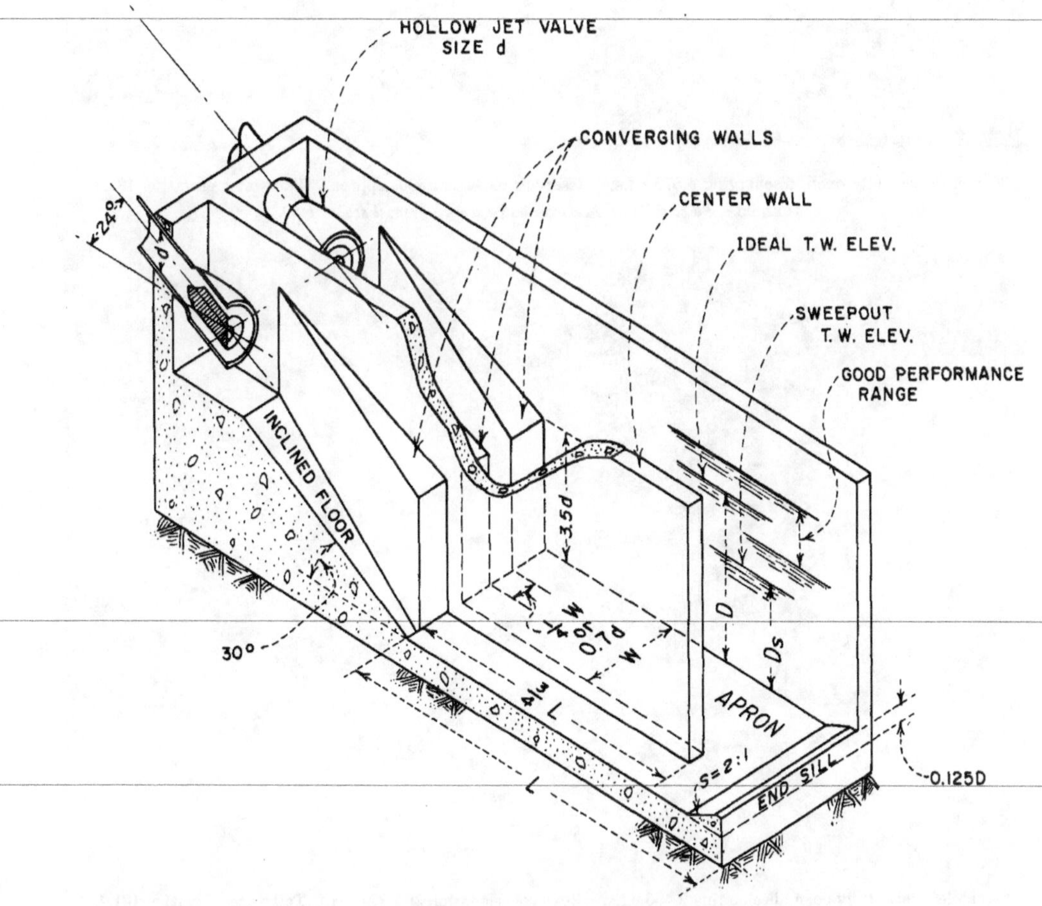

FIGURE 94.—*Developed basin.*

90-inch valves fully open, discharging 4,570 c.f.s. Reservoir elevation 300, approx. Tail water elevation 181.2.

FIGURE 95.—*Falcon Dam: Mexican outlet works—model scale: 1:30.*

90-inch left valve fully open, discharging 2,400 c.f.s. Reservoir elevation 300, approx. Tail water elevation 181.2.

FIGURE 96.—*Falcon Dam: Mexican outlet works—model scale: 1:30.*

72-inch valves fully open, discharging 2,920 c.f.s. Reservoir elevation 310, approx. Tail water elevation 180.8.
FIGURE 97.—*Falcon Dam: United States outlet works—model scale: 1:24.*

72-inch right valve fully open, discharging 1,600 c.f.s. Reservoir elevation 310, approx. Tail water elevation 180.5.
FIGURE 98.—*Falcon Dam: United States outlet works—model scale: 1:24.*

72-inch valves open 100%, discharging 2,920 c.f.s. Reservoir elevation 310, approx. Tail water elevation 180.8.

FIGURE 99.—*Falcon Dam: United States outlet works—model scale: 1:24.*

90-inch left valve 100% open, discharging 2,300 c.f.s., approx. Reservoir elevation 301.83. Tail water elevation 183.0. Compare with Figure 95.

FIGURE 100.—*Falcon Dam: Mexican outlet works.*

72-inch left valve 100 percent open, discharging 1,750 c.f.s., approx. Reservoir elevation 301.83. Tail water elevation 182.7. Compare with Figures 98 and 99.

FIGURE 101.—*Falcon Dam: United States outlet works.*

the valves were 100 percent open. In each outlet works, the valves were operated together and individually. Single-valve operation represents an emergency condition and subjects the stilling basin to the severest test, Figures 100 and 101. All turbines at both powerplants were operating at 72 percent gate and 100 percent load during all tests. The prototype valve discharges were determined from discharge curves based on model test data.

Here, too, more white water was evident in the prototype than in the model. The greater amount of air entrainment in the prototype, evident in the photographs, caused bulking of the flow at the end of the stilling basin and a higher water surface than was observed in the model. However, the prototype tail water is 3 feet to 4 feet higher than shown in the model photograph, and this probably helps to produce a higher water surface boil at the downstream end of the basin by reducing the efficiency of the stilling action. In other respects, the prototype basin performed as predicted by the model.

★ ★ ★

Recapitulation

The schematic drawing, Figure 94, shows the developed basin and the relationships between important dimensions.

A brief description of the seven steps required to design a stilling basin is given below:

1. Using the design discharge Q, the total head at the valve H, and the hollow-jet valve discharge coefficient C from Figure 74, solve the equation $Q=CA\sqrt{2gH}$ for the valve inlet area A and compute the corresponding diameter d which is also the nominal valve size.

2. Use H/d in Figure 83 to find D/d and thus D, the ideal depth of tail water in the basin. Determine the elevation of the basin floor, tail water elevation minus D. It is permissible to increase or decrease D by as much as 0.4 (d).

3. Use H/d in Figure 85 to find L/d and thus L, the length of the horizontal apron.

4. Use H/d in Figure 86 to find W/d and thus W, the width of the basin for one valve.

5. Use H/d in Figure 84 to find D_s/d and thus D_s, the tail water depth at which the action is swept out of the basin. D minus D_s gives the margin of safety against sweepout.

6. Complete the hydraulic design of the basin from the relationships given in Figure 82.

7. Use the H/d ratio to select the proper photograph in Figures 87 and 88 to see the model and help visualize the prototype per-

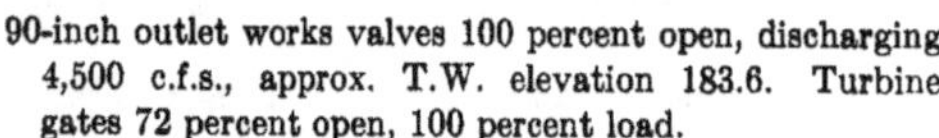
90-inch outlet works valves 100 percent open, discharging 4,500 c.f.s., approx. T.W. elevation 183.6. Turbine gates 72 percent open, 100 percent load.

72-inch outlet works valves 100 percent open, discharging 3,000 c.f.s., approx. T.W. elevation 184.1. Turbine gates 72 percent open, 100 percent load.

FIGURE 102.—*Falcon Dam: Mexican and United States powerplants and outlet works discharging at reservoir elevation 301.83.*

formance of the design. The water surface profile may be scaled from the photograph, using the scale on the photograph. To convert to prototype dimensions, multiply the scaled values by the ratio d (in.)/3.

Stilling basin dimensions calculated in this manner are in close agreement with the dimensions obtained from individual model tests of the basins for Boysen, Falcon, Yellowtail, Trinity, and Navajo Dams, Table 20. Since the Boysen and Falcon basins performed satisfactorily during prototype tests, it is believed that satisfactory future projects may be hydraulically designed from the material presented herein.

Section 9

Baffled apron for canal or spillway drops (Basin IX)

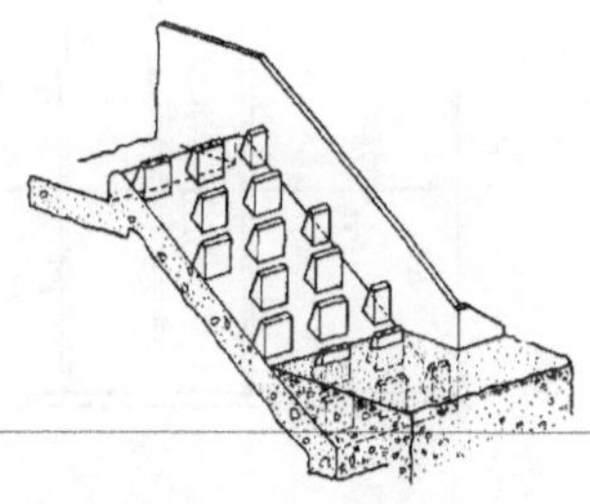

BAFFLED aprons or chutes have been in use on irrigation projects for many years. The fact that many of these structures have been built and have performed satisfactorily indicates that they are practical and that in many cases they are an economical answer to the problem of dissipating energy. Baffled chutes are used to dissipate the energy in the flow at a drop and are most often used on canal wasteways or drops. They require no initial tail water to be effective although channel bed scour is not as deep and is less extensive when the tail water forms a pool into which the flow discharges. The multiple rows of baffle piers on the chute prevent excessive acceleration of the flow and provide a reasonable terminal velocity, regardless of the height of drop. Since flow passes over, between, and around the baffle piers, it is not possible to define the flow conditions in the chute in usual terms. The flow appears to slow down at each baffle pier and accelerate after passing the pier, the degree depending on the discharge and the height of the baffle piers. Lower unit discharges result in lower terminal velocities on the chute.

The chute is constructed on an excavated slope, 2:1 or flatter, extending to below the channel bottom. Backfill is placed over one or more rows of baffles to restore the original streambed elevation. When scour or downstream channel degradation occur, successive rows of baffle piers are exposed to prevent excessive acceleration of the flow entering the channel. If degradation does not occur, the scour creates a stilling pool at the downstream end of the chute, stabilizing the scour pattern. If excessive degradation occurs, it may become necessary to extend the chute.

A number of baffled chutes have been constructed and tested in the field. Some of the existing structures were developed from designs obtained from hydraulic model tests made for the particular structure. Other designs for existing structures were obtained by modifying model-tested designs to the extent believed necessary to account for local changes in topography and flow conditions. The generalized design procedures discussed in this section were obtained from test results on several models of baffled chutes and from one model which was modified as necessary to obtain information of value in designing a chute for any installation.

A study of the existing baffled chutes showed that certain features of the design, such as the 2:1 chute slope, had been utilized in each installation. Thus, when a series of tests to generalize the overchute design was begun, these features were considered to be standard and did not need to be evaluated as variables. However, in a concluding series of tests, the baffle pier row spacing was determined for slopes flatter than 2:1.

Development of Baffled Apron Features

Prior to the generalization tests, individual models were constructed to provide a stilling basin upstream from the baffled chute and to develop the baffled chute and stilling basin as a complete unit. Three models that were tested are described in detail in Hydraulic Laboratory Report No. Hyd-359, "Hydraulic Model Studies of the Outlet Control Structure; Culvert Under Dike; and Wash Overchute at Station 938+00—Wellton-Mohawk Division, Gila Project, Arizona." A fourth study, "Hydraulic Model Studies of

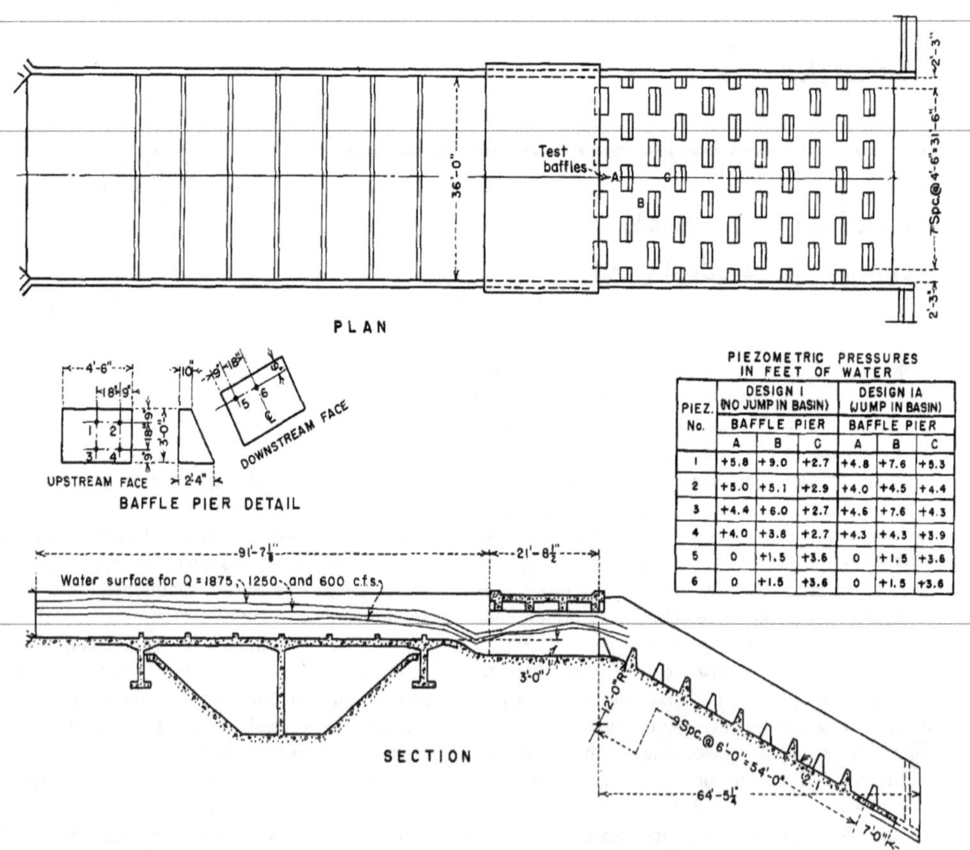

PIEZOMETRIC PRESSURES IN FEET OF WATER

PIEZ. No.	DESIGN I (NO JUMP IN BASIN) BAFFLE PIER			DESIGN IA (JUMP IN BASIN) BAFFLE PIER		
	A	B	C	A	B	C
1	+5.8	+9.0	+2.7	+4.8	+7.6	+5.3
2	+5.0	+5.1	+2.9	+4.0	+4.5	+4.4
3	+4.4	+6.0	+2.7	+4.6	+7.6	+4.3
4	+4.0	+3.8	+2.7	+4.3	+4.3	+3.9
5	0	+1.5	+3.6	0	+1.5	+3.6
6	0	+1.5	+3.6	0	+1.5	+3.6

FIGURE 103.—*Wash overchute, Sta. 938+00, Wellton-Mohawk Canal, Gila project.*

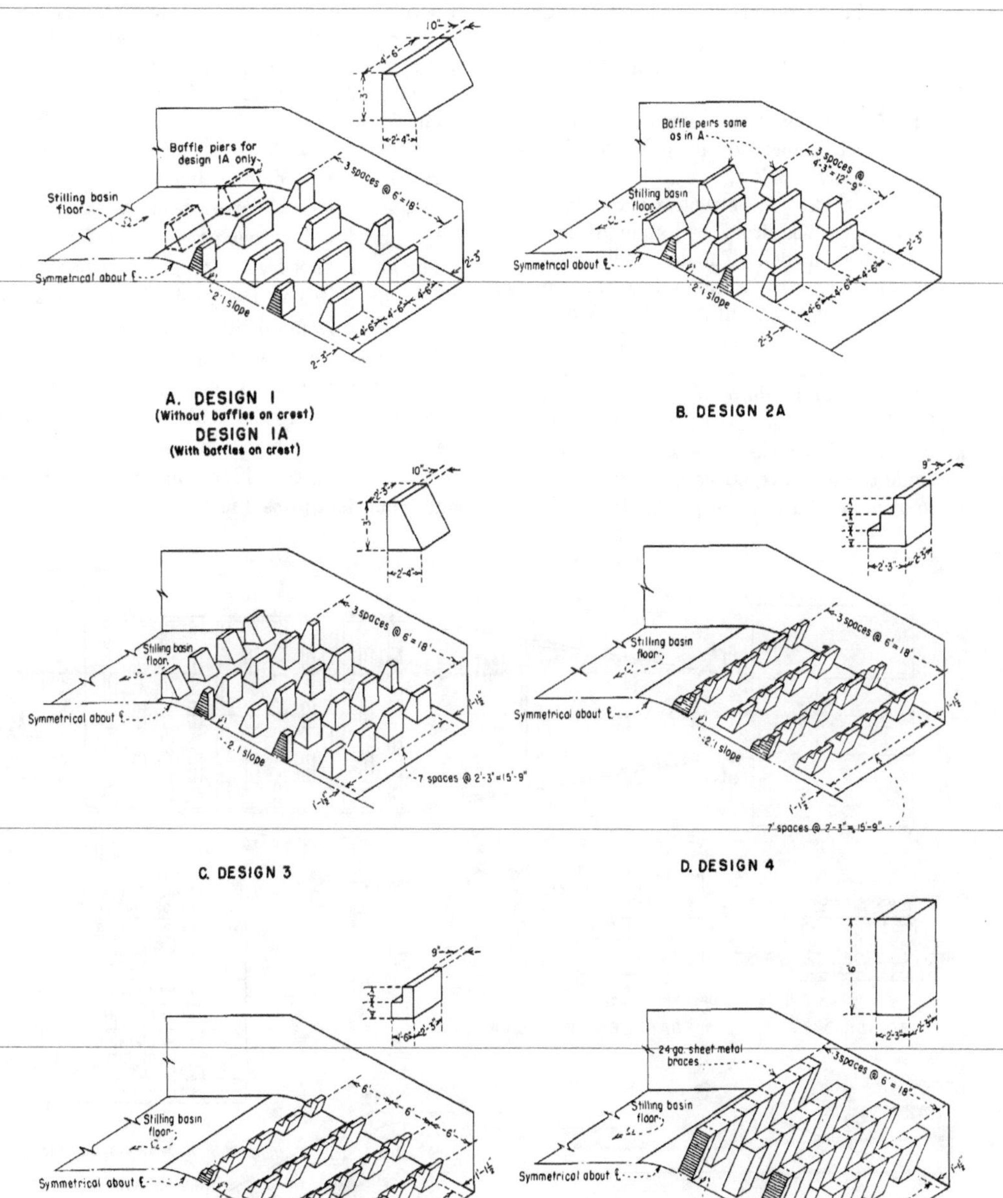

FIGURE 104.—*Wash overchute, Sta. 938+00, Wellton-Mohawk Canal, Gila project, different baffle pier arrangements on 2:1 sloping apron, 1:12 scale model.*

the Check Intake Structure—Potholes East Canal, Columbia River Basin Project, Washington," is the subject of Hydraulic Laboratory Report No. Hyd-411.

A brief summary of the parts of the individual studies which influenced the generalization test procedure is given below.

Wash overchute. The structure shown in Figure 103 was developed from hydraulic model tests on a 1:12 scale model. The design discharge was 1,250 c.f.s. and the chute was 36 feet wide, making the unit discharge about 35 c.f.s. After tests had been made to develop the stilling basin upstream from the chute, six different arrangements of baffle piers on the chute were tested, Figure 104.

For Design 1, the missing row of baffle piers at the top of the chute permitted the flow to continue to accelerate, strike the second row, and jump over the third row of piers. In Design 1A, the top row of baffle piers was in place; the resulting scour depth in the sand bed at the base of the chute was 7 feet, 5 feet less than for Design 1. In Design 2A, the spacing of the rows was reduced from 6 feet to 4 feet 3 inches. This resulted in no apparent difference in the operation of the structure. Scour depth was 7 feet. In Design 3, a greater number of narrow baffle piers was used. These produced a rougher water surface and a scour depth of 8 feet. Stepped face baffle piers were substituted in Design 4. Flow appearance was good and scour depth was 7 feet. For Design 5, the upstream row of baffle piers was reduced to 2 feet in height. Flow appearance was good and scour depth was 5.5 feet. In Design 6, baffle piers 6 feet high and 2 feet square in cross section were used. Flow appearance was poor and scour depth was 9 feet.

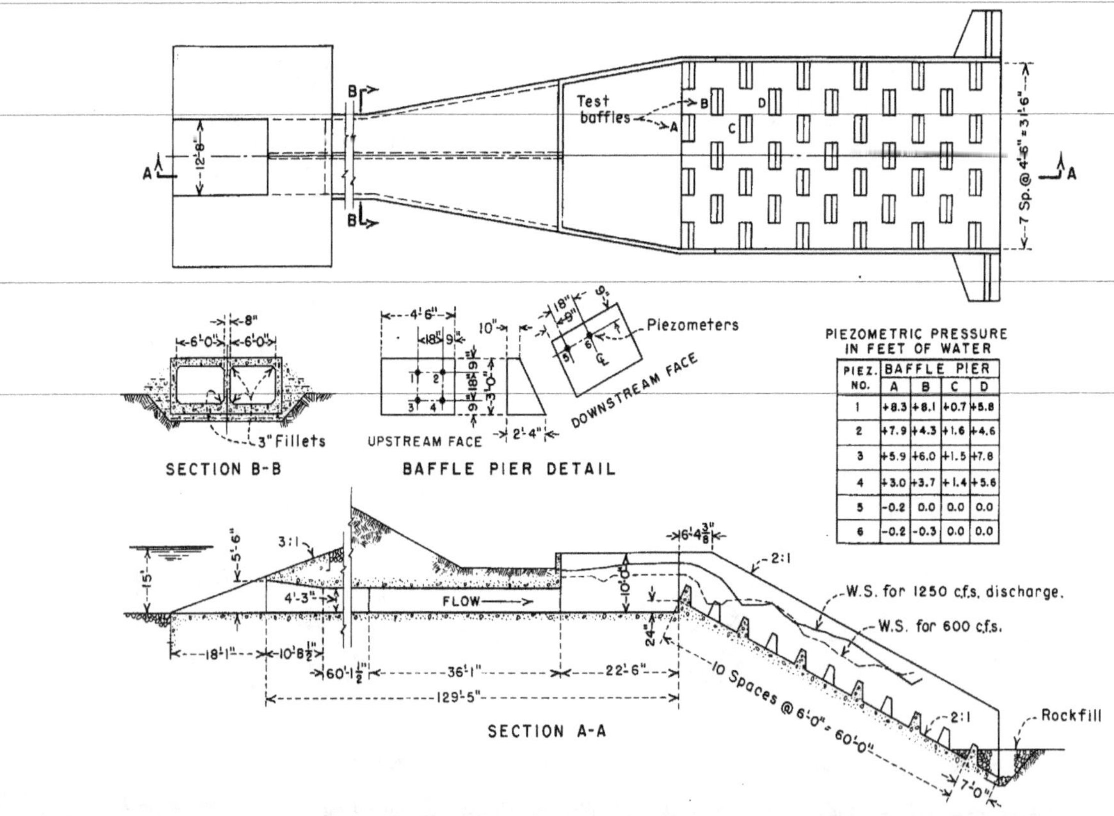

PIEZOMETRIC PRESSURE IN FEET OF WATER

PIEZ. NO.	BAFFLE PIER A	B	C	D
1	+8.3	+8.1	+0.7	+5.8
2	+7.9	+4.3	+1.6	+4.6
3	+5.9	+6.0	+1.5	+7.8
4	+3.0	+3.7	+1.4	+5.6
5	-0.2	0.0	0.0	0.0
6	-0.2	-0.3	0.0	0.0

FIGURE 105.—*Culvert under dike, Gila project.*

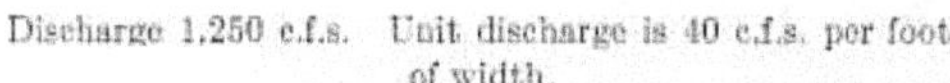

Discharge 1,250 c.f.s. Unit discharge is 40 c.f.s. per foot of width.

Scour pattern for flow in photograph at left.

FIGURE 106.—*Model studies for culvert under dike, Gila project. See details in Figure 105*

Considering all factors, including stilling basin performance, flow appearance, scour depth and extent, and structural problems, it was concluded that the arrangement shown in Figure 103 was most desirable. The piers were 3 feet high and 4 feet 6 inches wide, placed in staggered rows 6 feet apart. Water surface profiles and baffle pier pressures for this arrangement are shown in Figure 103.

Culvert under dike. The culvert structure developed from 1:12 scale hydraulic model tests is shown in Figure 105. The design discharge was 1,250 c.f.s. and the chute width was 31 feet 6 inches, making the unit design discharge approximately 40 c.f.s. After tests had been made to develop the culvert and the stilling basin upstream from the chute, scour tests were made with baffle piers 3, 4, and 5 feet high on the chute. Results of these tests disclosed the depth of scour for the 4- and 5-foot piers to be approximately the same as that obtained for the 3-foot-high piers. Piers 3 feet high provided the best overall performance. The appearance of the design flow and the resulting scour pattern are shown in Figure 106. Water surface profiles and baffle pier pressures for the recommended structure are shown in Figure 105.

Outlet control structure. The outlet control structure stilling basin and baffled chute were developed from 1:24 scale hydraulic model tests on a half model and are shown in Figure 107. The chute width is 140 feet and the design discharge is 7,000 c.f.s., making the unit discharge 50 c.f.s. Tests showed the stilling basin to be adequate for the design flow released through the control notches, Figure 108A. Baffle piers 3 feet high in rows spaced 6 feet apart provided satisfactory flow in the chute. Scour depth was about 5 feet, as shown in Figure 108B.

Check intake structure. A 1:16 scale model was used in this study. Figure 109 shows the developed design which includes the gated control structure, stilling basin, and baffled chute. The chute is 64 feet wide and the discharge is 3,900 c.f.s., making the unit discharge about 61 c.f.s. Baffle piers 4 feet 6 inches high were tested in horizontal rows spaced at intervals of 9 and 6 feet. No differences in the appearance of the flow were apparent for the two spacings, but the scour depth over most of the area was 2 feet less with the larger row spacing. Figure 110 shows the structure in operation and the scour test results.

Figure 111 shows the flow appearance and the resulting scour for a unit discharge of 50 c.f.s. and the 9-foot row spacing. The scour depth is about 1 foot less than for 60 c.f.s. Figure 111 also shows flow conditions for unit discharges of 31 and 16 c.f.s.

Normal versus vertical pier faces. Tests were made to determine the effect of constructing the pier faces vertical rather than normal to the chute, Figure 112. For a unit discharge of 35 c.f.s. there was very little difference in performance between vertical and normal placement. Figure

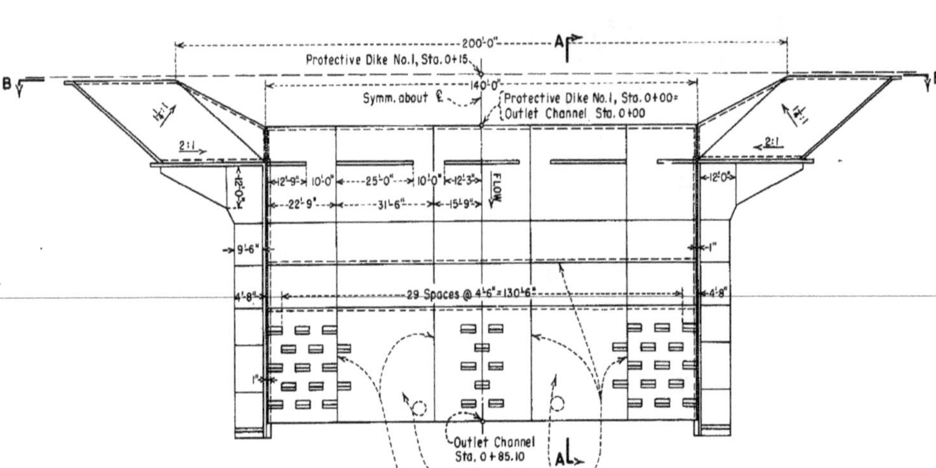

SECTION A-A

HEAD DISCHARGE CURVE

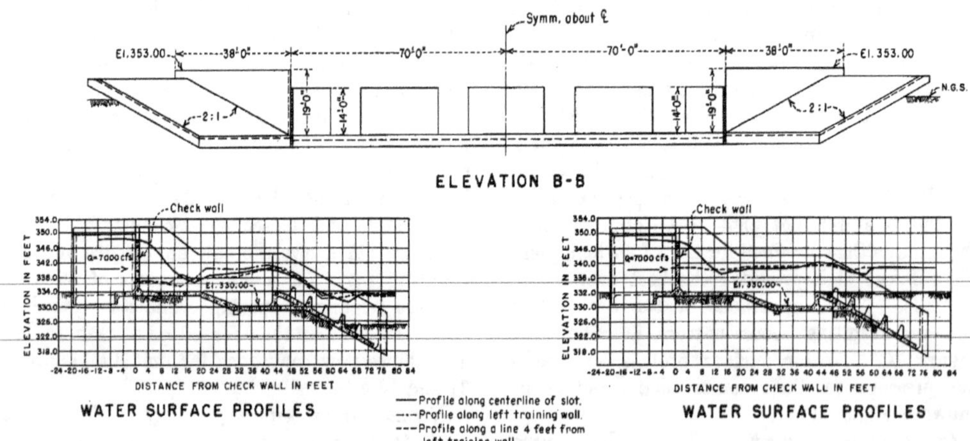

WATER SURFACE PROFILES

—— Profile along centerline of slot.
—·— Profile along left training wall.
- - - Profile along a line 4 feet from left training wall.

WATER SURFACE PROFILES

FIGURE 107.—*Outlet control structure, Gila project.*

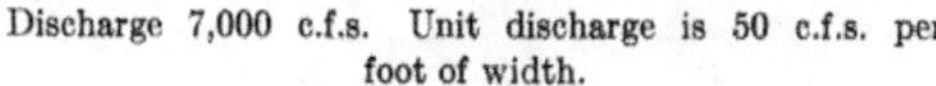

Discharge 7,000 c.f.s. Unit discharge is 50 c.f.s. per foot of width.

Scour pattern for flow in photograph at left.

FIGURE 108.—*Model of outlet control structure, Gila project. See details in Figure 107.*

112 shows that the splash was about 5 feet lower with vertical face piers as indicated by the darker wetted area in the photographs. Figure 112 shows the scour patterns obtained after ½ hour of model operation. There was slightly less scour in the vicinity of the wing wall when normal pier faces were used. The scour pocket (elevation 906) along the wall of symmetry in the model probably would not have occurred if the full width of the model had been built.

The same scour tendencies were prevalent for a unit discharge of 61 c.f.s., Figure 113. There was less overall erosion with the pier faces normal to the slope although the scour depths were the same.

Generalization Tests

The models. A 1:16 scale model of a 171-foot length of the Potholes East Canal between Stations 1367+69 and 1369+40 was used for the generalization tests. Included were a reach of approach canal, the gate control structure upstream from the baffled apron, the 2:1 sloping apron, and approximately 80 feet of outlet channel. To make the model features as large as possible, only one-half of the structure was built and tested, Figure 114. The wall on the right in the photograph is the wall of symmetry and is on the centerline of the full-sized structure. The gate structure, shown in Figure 109, was made removable so that studies could be made for low as well as high velocities at the top of the baffled chute. A painted splashboard was installed along the wall of symmetry to record the height of splash. The paint on the board absorbed the splash and showed the splash area as a darker color. The channel downstream from the baffled chute was molded in sand having a mean diameter of about 0.5 millimeter. Discharges were measured through calibrated venturi meters and velocities were measured with a pitot tube.

On an entirely different model, a series of tests, scale 1:10 to 1:13.5, was conducted to determine the required baffle pier heights and arrangements for chutes constructed on 3:1 and 4.5:1 (flatter) slopes. Testing was started using the chute and baffle pier arrangement recommended for 2:1 sloping chutes. Each variable was investigated in turn and it was determined that only the baffle pier row spacing needed modification. In these tests some of the baffle piers were equipped with an impact tube (piezometer) installed in the upstream face of the pier. The tubes, one in each row on the pier nearest the centerline of the chute, were transparent and were extended

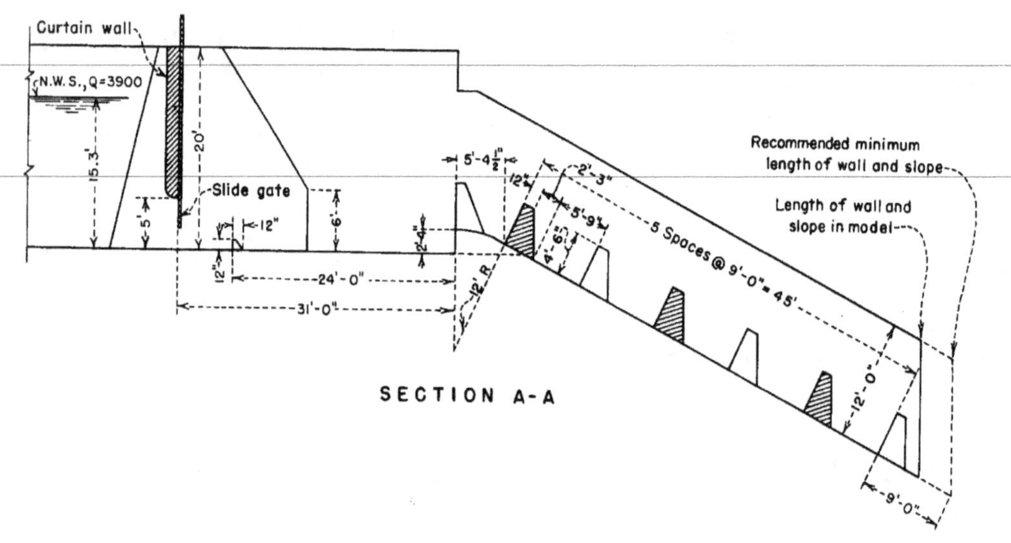

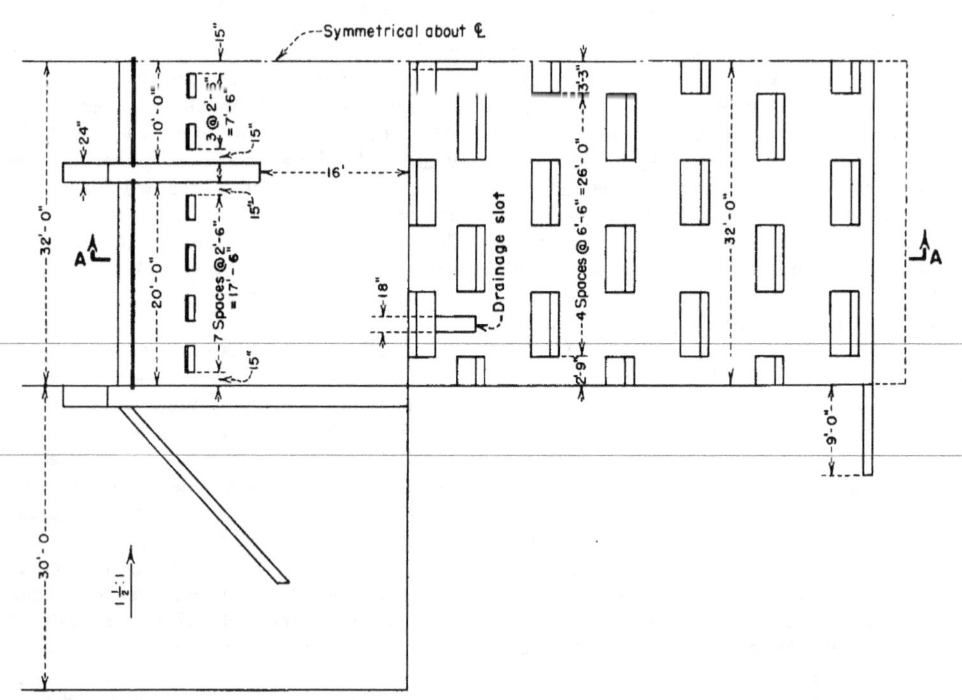

FIGURE 109.—*Check intake structure, Sta. 1369+40, Potholes East Canal, Columbia Basin project, 1:16 scale model.*

Baffle piers 4′6″ high, row spacing 9′0″.

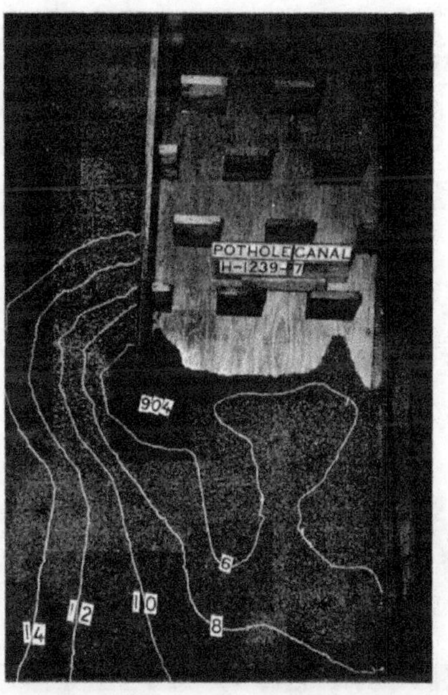

NOTE: Bed was at elevation 914 at start of 30-minute test.

Baffle piers 4′6″ high, row spacing 6′0″.

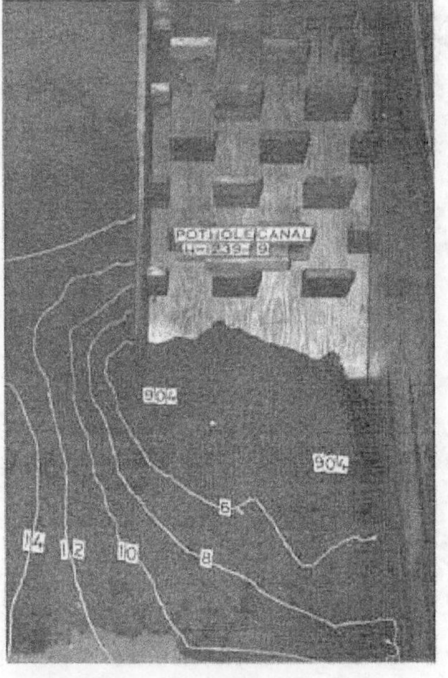

FIGURE 110.—*Model of check intake structure, discharge at 61 c.f.s. per foot of width. See details in Figure 109.*

Flow.

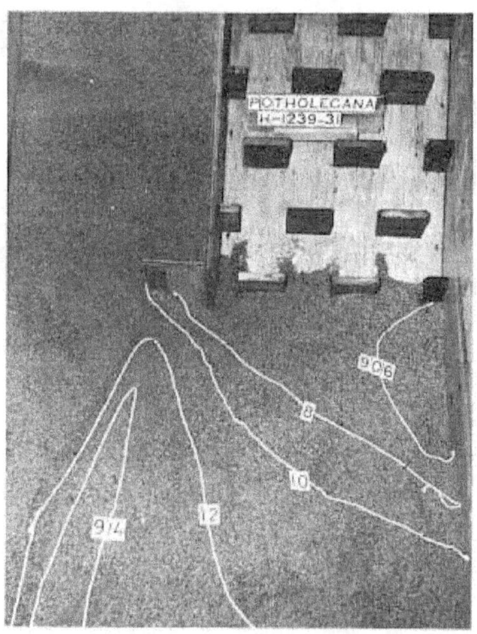

Scour.

Discharge 3,200 c.f.s.—unit discharge 50 c.f.s. per foot width. Baffle piers 4'6" high, row spacing 9'0".

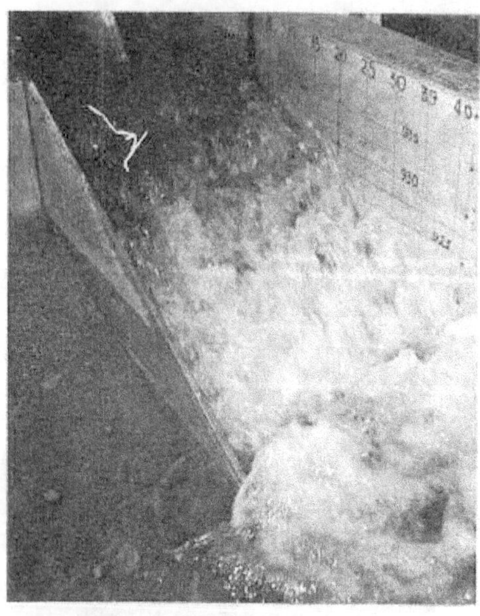

31 c.f.s. per foot width.

16 c.f.s. per foot width.

FIGURE 111.—*Model of check intake structure, Potholes East Canal, discharges at 50 c.f.s., 31 c.f.s., and 16 c.f.s. per foot of width.*

Flow.

Note splash area on wall.

Normal-face piers.

Discharge 35 c.f.s. per foot width.

Flow.

Vertical-face piers.

FIGURE 112.—*Model of check intake structure, Potholes East Canal, tests of various-shaped baffles.*

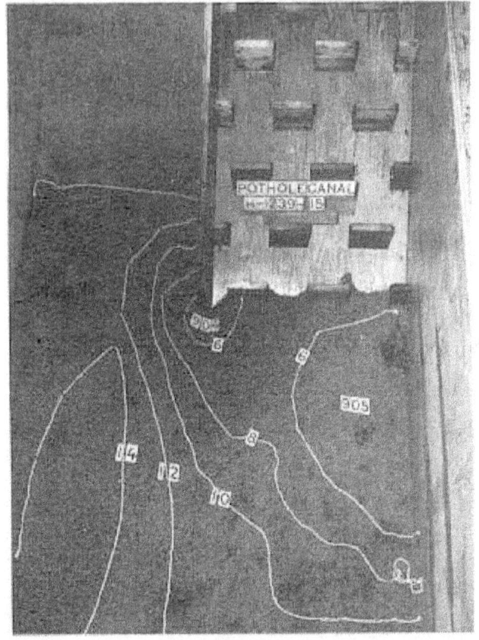

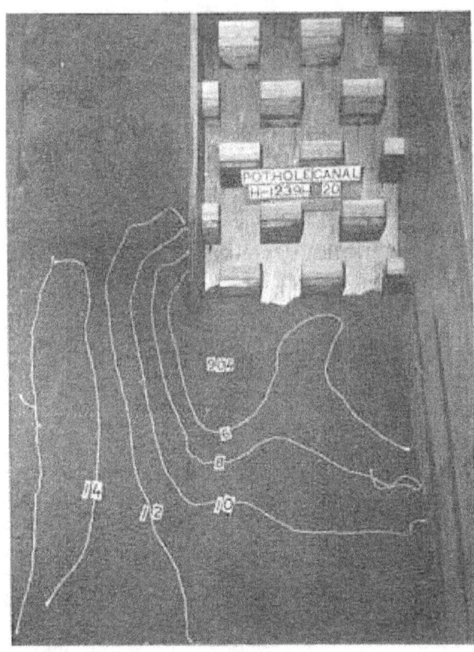

Normal face piers. Vertical-face piers.

NOTE: Bed was at elevation 914 at start of 30-minute test.

Discharge 61 c.f.s. per foot width.

FIGURE 113.—*Model of check intake structure, Potholes East Canal, tests of various-shaped baffles.*

through the pier and bent at right angles to rise above the top of the flowing water surface. The tubes were filled, after the model was operating, with colored water so that the impact pressures on the pier faces could be evaluated visually. These tubes were especially useful in determining the most effective spacing of the baffle pier rows.

Testing procedure. The tests on the 2:1 sloping chute were concerned primarily with the effectiveness of the baffled chute in preventing acceleration of the flow down the chute. This was judged by the appearance or profile of the flow in the chute, the depth and extent of scour in the downstream channel, and by the height of splash shown on the splashboard. For each test, the channel was molded level at the base of the chute at elevation 914 and the model was operated for 30 minutes, after which the erosion in the channel bed was measured. Relative depths were made visible with contour lines of white string. The tailgate in the model was set to provide a tail water depth of 2 feet (elevation 916) in the downstream channel for a discharge of 20 c.f.s. per foot of width of chute. The tailgate setting was not changed for larger discharges; therefore, the tail water depth did not build up as much as it normally would in a field structure. The resulting depths for discharges of 35, 50, and 60 c.f.s. were 2.5, 3.0, and 3.5 feet, respectively. For tests with gate-controlled flow, 15.3 feet of depth was maintained upstream from the gates. For the free flow tests, the gate structure was removed and the normal depth for the particular flow being tested was maintained in the canal. The elevations shown in the drawings and photographs are compatible and apply for a model scale of 1:16.

Four baffle pier heights were included in the original testing program: 3, 4, 5, and 6 feet, measured normal to the 2:1 sloping chute, Figures 115, 116, 117, and 118. Each height was tested

Figure 114.—*Model of check intake structure as used in generalization tests.*

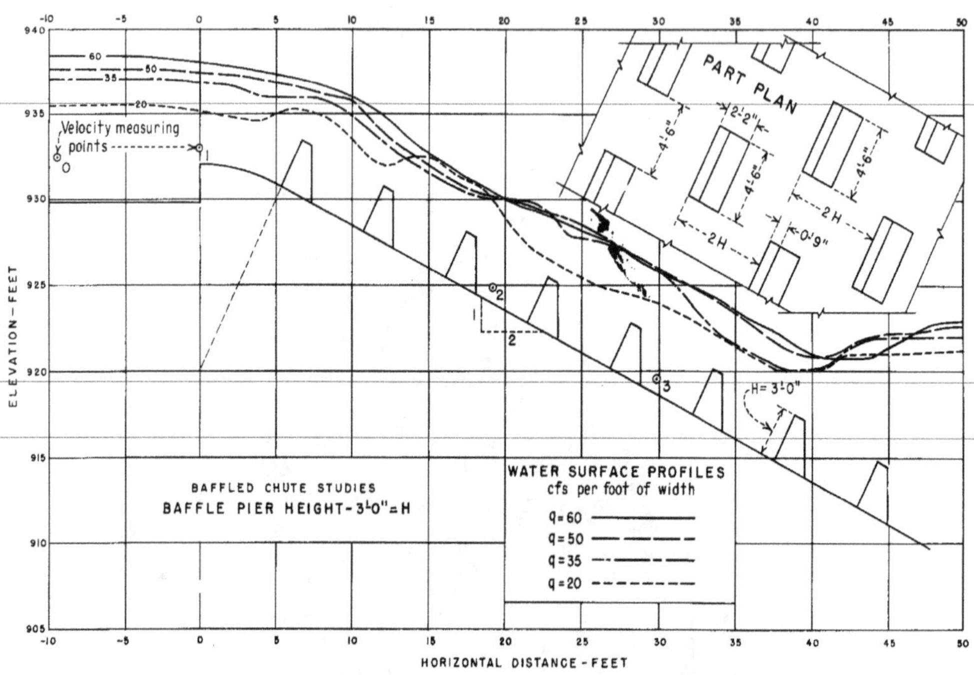

FIGURE 115.—*Baffled chute studies. Baffle pier height, H=3'0''.*

with the spacing between rows fixed at twice the baffle height. The baffle pier widths and spacing within each row were equal to one and one-half times the baffle height. For each baffle pier arrangement, individual tests were made for 20, 35, 50, and 60 c.f.s. per foot of width.

Water surface measurements were made with a point gage and a scale, taking the maximum water surface at each measured point. Since the water surface at any point on the chute varies with respect to time, the profiles obtained are higher than the profiles shown in a photograph of the same test. The measured profiles of Figures 115 to 118 are believed to be more dependable for estimating necessary wall heights than are the photographs in the report, which portray the appearance of the flow at a particular moment.

Velocity measurements were attempted in the locations shown in Figures 115 through 118. At Stations 0 and 1, the flow was smooth and uniform; the data are accurate. On the slope, where turbulence and unsteadiness are characteristic of the flow, only the measurements at Point 3 were considered to be dependable. Even these showed some inconsistencies, but velocity curves for the range of discharges were determined by using general knowledge and judgment to adjust the obviously incorrect measured values. The curves shown in Figure 119 are believed to be reasonably accurate and were found useful in evaluating the height of the baffle piers in terms of general performance. The velocity measurements in other parts of the chute are summarized in the notes of Figure 119.

Test results. For all baffle pier heights and a test discharge of 60 c.f.s. per foot of width, the flow entering the chute had a bottom velocity of about 1.8 feet per second and reached a maximum of 5.5 feet per second at Point 2. At Point 3, the velocity was dependent on the baffle pier height, as shown in Figure 119. The average velocity $V=Q/A$, at the top of the chute was 7 feet per second. For a unit discharge of 20 c.f.s., the initial bottom velocity was about 1.1 feet per second, reached a maximum of about 4.5 feet per second at Point 2, and was reduced at Point 3. The average velocity at the top of the chute was 3 feet per second. The velocities in themselves

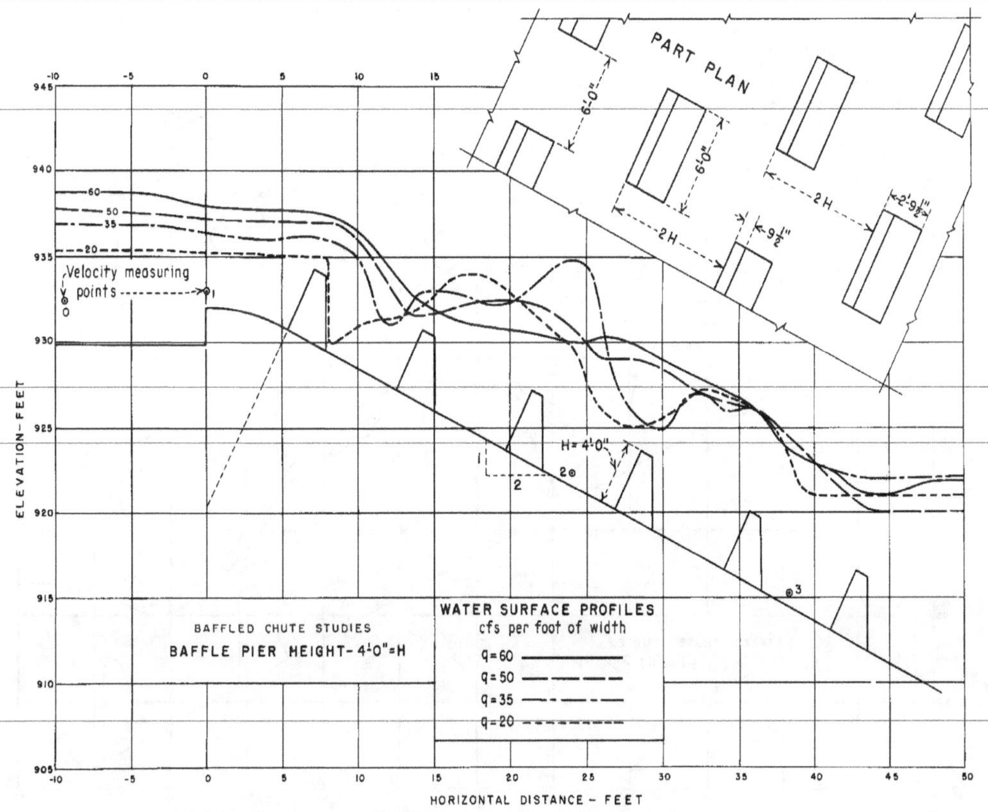

FIGURE 116.—*Baffled chute studies. Baffle pier height, H=4'0".*

are not important in generalizing the design of the baffled chute, but do help the reader to visualize the velocity distribution on the chute. With low baffles and high discharges, the bottom velocity at Point 3, Figure 119, is considerably higher than when higher baffles are used with the same discharge. This is because a larger volume of water passes over the tops of the low baffles and the decelerating effect of the baffles on the entire volume of flow is less, Figure 120.

Although the velocity at Point 3, for 60 c.f.s. per foot and the 6-foot baffles, was considerably less than for the 3-foot baffles, the erosion was more severe. When the 6-foot baffles were used, erosion was to elevation 900, exposing the end of the chute. When the 3-foot baffles were used, erosion was only to elevation 905 and the extent of the erosion was also less. Appearance of the flow on the chute and in the downstream channel for the 5-foot baffles, Figure 121B, was better than for the 6-foot baffles, but the appearance for the 4-foot baffles was still better, Figure 121A. The erosion patterns for the 4- and 5-foot baffles were better than for the 3- or 6-foot baffles. The least splash occurred with the 3- and 4-foot baffles.

The same relative performance was evident for the 50 c.f.s. per foot flow. The 4- and 5-foot baffles produced the best flow appearance and the 5-foot baffles produced the most favorable scour and splash patterns. Figure 121 shows the flow for 50 c.f.s. per foot with the 4- and 5-foot baffles.

At 35 c.f.s. per foot, the flow patterns were all satisfactory in appearance. The most favorable erosion patterns occurred with the 3- and 4-foot baffles, the deepest erosion being to elevation 906. The deepest erosion hole with the 5-foot

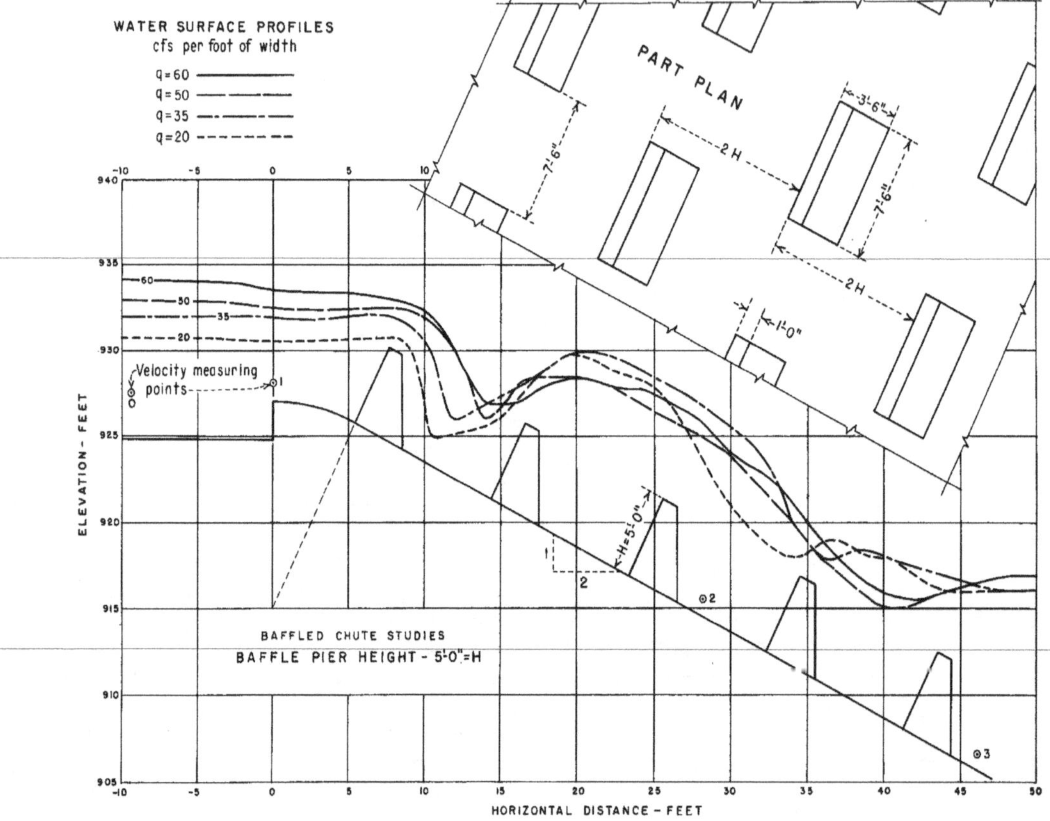

FIGURE 117.—*Baffled chute studies. Baffle pier height, H=5'0''.*

baffles was to elevation 905. Splash was minimum with the 4-foot baffles but was not much greater with the 3-foot baffles. Figure 122A shows the flow pattern and erosion for the 3-foot baffles and 35 c.f.s. per foot of width..

For 20 c.f.s. per foot, flow appearances were all good but the 3-foot baffles showed a slightly better flow pattern. The scour pattern was also most favorable with the 3-foot baffles. The deepest erosion hole was to elevation 908. Erosion with the 4-foot baffles was to elevation 907; with the 5-foot baffles to elevation 905, and with the 6-foot baffles to elevation 906. The 4-foot baffles produced the least erosion near the wing wall at the end of the chute. The splash patterns for 3-, 4-, and 5-foot baffles were almost identical, but the splash for the 6-foot baffles was somewhat greater. Figure 122B shows the flow pattern and erosion for the 3-foot baffles and 20 c.f.s. per foot of width.

After partial analysis of the test data, it was apparent that baffles 2 feet high might provide ample scour protection for a design discharge of 20 c.f.s. per foot of width. Scour tests showed this to be true, although scour depths were about the same as found for the 3-foot-high baffles. For a discharge of 35 c.f.s. per foot, the scour depth exceeded that for the 3-foot baffles and the flow appearance was not good; too much high velocity flow passed over the tops of the piers.

A summary of scour test data is given in Table 21. Listed are the lowest scour-hole elevations (1) at the wing wall visible in the photographs, (2) downstream from chute, and (3) the average of the elevations in (1) and (2). Scour along the wall of symmetry was not con-

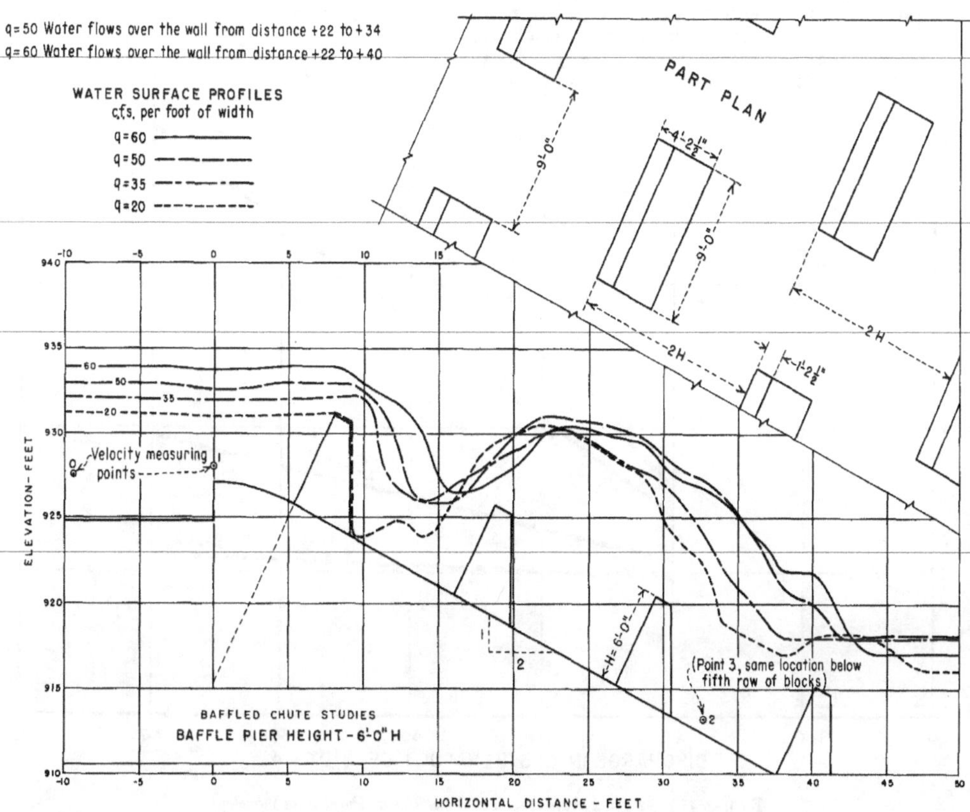

FIGURE 118.—*Baffled chute studies. Baffle pier height, H=6'0''.*

sidered because the adjacent wall affected the scour depth adversely.

Figures 123 and 124 show three groups of curves, A, B, and C, plotted from the data in Table 21, one group, D, plotted from the velocity curves of Figure 119 and one group, E, plotted from the splash tests. In Group A, the scour depth at the wing wall is a minimum for the 2- and 3-foot-high piers for a discharge of 20 c.f.s. At 35 c.f.s., the 3- and 4-foot piers provided the minimum scour depth, and at 60 c.f.s., the 4- and 5-foot piers provided minimum scour depth. In Groups B and C, the depth of scour at the end of the chute and the average of the maximum depths show the same general trend, except that the 3- and 4-foot piers show minimum scour for the maximum design discharge of 60 c.f.s.

If envelope curves were drawn in A, B, and C to determine the height of baffle pier which produces the least scour, the pier heights would vary from 2 feet for 20 c.f.s. in all cases to 3, 4 or 5 feet in the other cases for 60 c.f.s. An envelope curve drawn on the velocity curves to determine the height of pier to produce the lowest velocity on the chute would indicate baffle piers 6 feet high for all discharges. Since 6-foot piers produce maximum scour depth, a compromise must be made. Scour depth is more important than the velocity on the chute, and since the water surface profiles of Figures 115 to 118 favor the lower baffle piers, the most practical height for the baffle piers is indicated by the circles in Figures 123 and 124. The circles have been plotted on each set of curves and represent baffle piers 2 feet high for design discharge 20 c.f.s.; 3 feet high for design discharge 35 c.f.s.; 3.8 feet high for

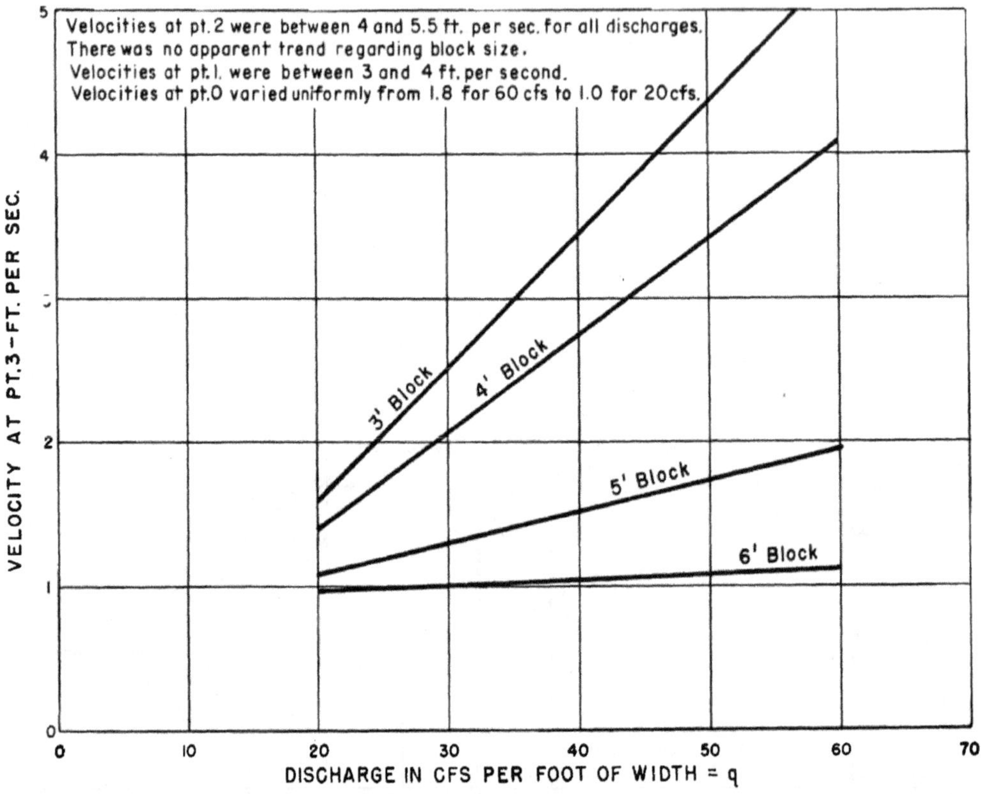

FIGURE 119.—*Baffled chute studies. Velocities at Point 3 on model.*

Baffle piers 6′0″ high.

Baffle piers 3′0″ high.

FIGURE 120.—*Baffled chute studies—discharge 60 c.f.s. per foot of width.*

50 c.f.s. per foot of width.

60 c.f.s. per foot of width.

Baffle piers 4′0″ high.

50 c.f.s. per foot of width.

60 c.f.s. per foot of width.

Baffle piers 5′0″ high.

FIGURE 121.—*Baffled chute studies—discharges 50 and 60 c.f.s. per foot of width.*

design discharge 50 c.f.s. and 4.3 feet high for design discharge 60 c.f.s.

Piers of this height produce near minimum depths of scour for all design discharges and near minimum velocity on the chute. In addition, they produce near minimum splash for all discharges as shown by Curves E of Figure 124. Finally, an inspection of the photographs made of each test (only a few representative photographs are reproduced in this report) show that the flow appearance is satisfactory for each of the recommended piers.

The height of baffle piers shown by the circles in Figures 123 and 124 may be expressed as 0.8 D_c where $D_c=\sqrt[3]{\frac{q^2}{g}}$=critical depth on the chute. Curve B, Figure 125, shows the recommended height of baffle piers.

Generalization of the Hydraulic Design

The general rules for the design of baffled overchutes have been derived from tests on individual models, prototype experiences, and on the verifi-

Discharge 35 c.f.s. per foot of width.

Discharge 20 c.f.s. per foot of width.

NOTE: Bed at elevation 914 at start of 30-minute test.

FIGURE 122.—*Baffled chute studies—baffle piers 3′0″ high.*

cation tests described in detail in this section. Since many of the rules are flexible to a certain degree, an attempt has been made in the following discussion to indicate how rigidly each rule applies.

The rules apply to chute slopes in the range 2:1 (steep) to 4:1 (flat). For slopes flatter than 2:1, the baffle pier row spacing should be modified as discussed on page 175.

Design discharge. The chute should be designed for the full capacity expected to be passed through the structure. The maximum unit discharge may be as high as 60 c.f.s. Generally

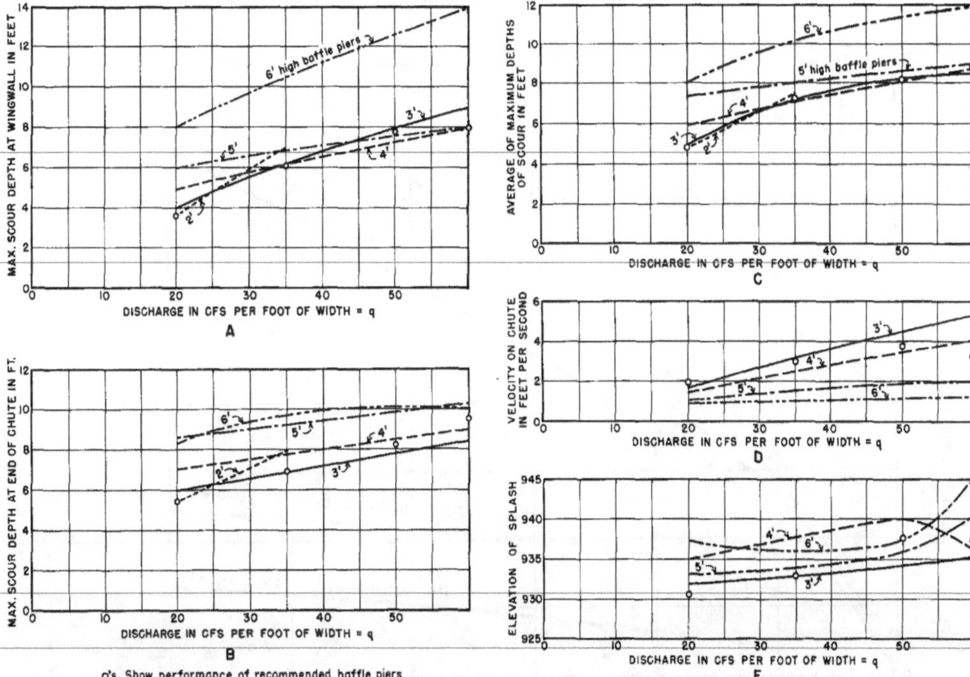

FIGURE 123.—*Baffled chute studies—scour test results.*

FIGURE 124.—*Baffled chute studies—scour, velocity, and splash test results.*

TABLE 21.—*Scour test results*

Baffle pier height ft.	Discharge per foot of width q in c.f.s.	Elevation of deepest erosion		
		(1) At wing-wall	(2) End of chute	Average 1 and 2
2	20	910−	908−	909−
	35	907−	906	906. 5
3	20	910	908	909
	35	908	907	907. 5
	50	906	906+	906. 1
	60	905	906−	905. 5
4	20	909	907	908
	35	908	906	907
	50	907	906−	906. 5
	60	906	905	905. 5
5	20	908	905	906. 5
	35	907	905	906
	50	907	904	905. 5
	60	906	904	905
6	20	906	906	906
	35	903	904	903. 5
	50	902	904	903
	60	900	904	902

speaking, however, unit discharges in the range of 35 c.f.s. provide less severe conditions on the chute and in the downstream channel, and a unit discharge of 20 c.f.s. provides a relatively mild condition.

In installations where downstream degradation is not a problem and an energy dissipating pool can be expected to form at the base of the chute, more acceptable operation for a unit discharge of 60 c.f.s. will occur than will be the case in steeper channels where no energy dissipation occurs. The design maximum unit discharge may be limited by the economics of baffle pier sizes or chute training wall heights. A wider chute with a correspondingly reduced unit discharge may provide a more economical structure.

Reports have been received from the field that baffled aprons designed for a unit discharge of 60 c.f.s. have operated at estimated values up to 120 c.f.s. for short periods without excessive

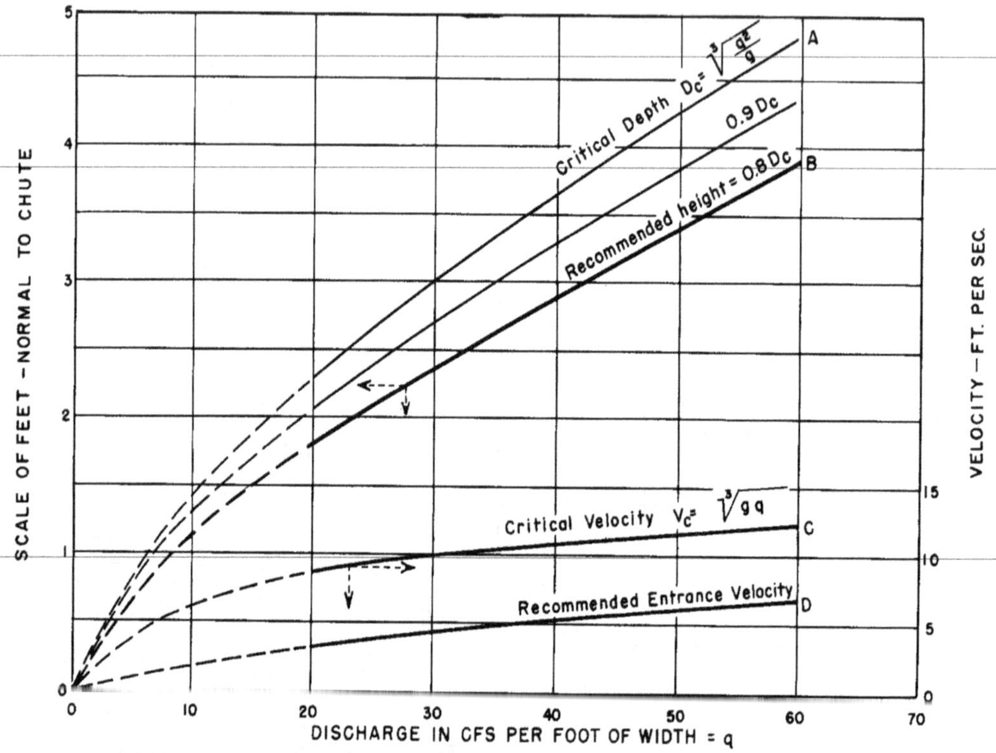

FIGURE 125.—*Baffled chute studies—recommended baffle pier heights and allowable velocities.*

erosion and spillage over the walls. This is mentioned only to indicate that a baffled apron can discharge more than the design flow without immediate disaster; it is not intended to suggest that baffled aprons should be underdesigned as a matter of general practice.

Chute entrance. Flow entering the chute should be well distributed laterally across the width of the chute. The velocity should be well below the critical velocity, preferably the values shown in Curve D of Figure 125. The critical velocity in a rectangular channel is $V_c = \sqrt[3]{gq}$. Velocities near critical or above cause the flow to be thrown vertically into the air after striking the first baffle pier. When the initial velocity is high, the flow has been observed to pass completely over the next row or two of baffle piers in a model. The baffled apron is not a device to reduce the velocity of the incoming flow; rather, it is intended only to prevent excessive acceleration of the flow passing down the chute.

To insure low velocities at the upstream end of the chute, it may be necessary to provide a short energy dissipating pool similar to the ones shown in Figures 103, 105, 107, and 109. A hydraulic jump stilling basin may be suitable if the flow is discharged under a gate as shown in Figure 109. The sequent or conjugate depth in the basin should be maintained to prevent jump sweepout, but the basin length may be considerably less than a conventional hydraulic jump basin, since the primary purpose of this pool is to reduce the average velocity. This is accomplished in the upstream portion of the stilling basin. The downstream third of the basin may therefore be eliminated, since the purpose of this portion of the basin is to complete the jump action and provide a smoother water surface. A basin length of twice the sequent depth will usually provide ample basin length. The end sill of the pool may be used as the crest of the chute, as shown in Figures 103, 105, 107, and 109.

Again, it is very important that proper flow conditions be provided at the entrance to the baffled apron. In fact, satisfactory performance of the entire structure may hinge on whether entrance flow conditions are favorable. If unusual entrance problems are encountered or if any doubt exists, a hydraulic model study is recommended.

Design of chute. The drop section, or chute, is usually constructed on a 2:1 slope. The upstream end of the chute floor should be joined to the horizontal floor by a curve to prevent excessive vertical contraction of the flow. However, the radius should be sufficiently small that the curved surface does not interfere with the placement of the first row of baffles. The upstream face of the first row should be no more than 1 foot (vertically) below the high point of the chute. It is important that the first row of baffles be placed as high on the chute as practicable, since half of the water will not be intercepted until the flow strikes the second row of baffles. To prevent overtopping of the training walls at the second row of baffles, a partial baffle (one-third to two-thirds of the width of a full baffle) should be placed against the training walls in the top or first row. This will place a space of the same width adjacent to the walls in the second row. Alternate rows are then made identical. (Rows 1, 3, 5, 7, etc., are identical; Rows 2, 4, 6, 8, etc., are identical.) Four rows of baffles are necessary to establish the expected flow pattern at the base of the chute.

The height of the training walls on the chute should be three or more times the baffle height, measured normal to the chute floor. Walls of this height will contain the main flow and most of the splash. The greatest tendency to overtop the walls occurs in the vicinity of the second and third rows of baffles, as indicated in the profiles and photographs. If it is important to keep the adjacent area entirely dry, it may be desirable to increase the wall height near the top of the chute.

Several rows of baffle piers are usually constructed below the channel grade and backfill is placed over the piers to restore original bottom topography. To determine the depth below channel grade to which the chute should be constructed, the following methods have been used. When the downstream channel has a control, the slope of a stable channel from the control upstream to the structure should be used to determine the elevation of the end of the chute. Usually, data are not available or sufficient to compute a stable channel grade. In these instances, a slope of 0.0018 is then used. Experience has shown that a slope of 0.015 is much too steep. If a stable downstream control does not exist, the probable stable channel must be determined by estimating the amount of material which will be moved during the maximum design flood.

Baffle pier heights and spacing. Curve A of Figure 125 shows the critical depth in a rectangular channel. The curve was plotted from the equation

$$D_c = \sqrt[3]{\frac{q^2}{g}}$$

Curve B gives values for 0.8 D_c; a curve for 0.9 D_c is also shown. Baffle pier heights for unit design discharges up to 60 c.f.s. may be obtained from Curve B. As indicated by the tests, the baffle pier heights are not critical and the height may be varied by several inches to provide a convenient dimension.

The width of the baffle piers should equal the width of the spaces between baffles in the same horizontal row and may vary between one and one and one-half times the block height—preferred width is one and one-half times the block height. Greater baffle widths may result in too few baffles to break up the flow thoroughly; narrower widths do not intercept enough of the flow at one place and also may result in slots too narrow for easy passage of trash.

As a general rule, the slope distance between rows of baffles (measured face to face on the 2:1 slope) should be twice the baffle height. When baffles less than 3 feet in height are used, the row spacing may be increased but should not exceed 6 feet. Greater spacing with small baffles allows the shallower flows to accelerate excessively before being intercepted by a baffle pier. Alternate rows should be staggered to provide a space below a block and vice versa.

Extensive tests made to determine the baffle pier sizes, spacing, etc., for chutes flatter than 2:1 indicated that the only modification required to produce optimum performance was in the baffle pier row spacing. It was found that a chute on a slope flatter than 2:1 should contain the same number of rows of piers as a 2:1 chute constructed between the identical top and bottom elevations.

In other words, the vertical fall distance between rows should be the same for all chutes, whether on slopes of 2:1 or flatter.

It was also determined that there is a disadvantage in supplying a greater number of rows than specified. Too many rows reduce the efficiency of the stilling action which occurs in the spaces between rows. For example, if a sufficient number of extra rows were added, a smooth floor consisting of the tops of the piers would result, and no energy dissipation could be expected.

The baffles may be constructed with their upstream faces normal to the chute or truly vertical; the difference in performance is hardly measurable in a model. There is a tendency, however, for the vertical faces to produce more splash and less scour than the normal faces, Figure 112. Other dimensions of the blocks are not important except from the structural standpoint. The proportions shown in Figure 115 have been found acceptable for both structural and hydraulic requirements and are recommended for general use. The forces on a baffle pier may be estimated from the baffle pier pressure measurements shown in Figures 103 and 105.

Prototype Performance

Field performance of baffled chutes, designed and constructed according to the suggestions given in this section, has been excellent at most installations. This has been verified by inspection teams working out of design offices and by field personnel responsible for operating the structures. Where deficiencies in performance have been noted, the cause was as obvious as the deficiency and simple remedial measures have resulted in satisfactory performance. The only difficulties reported have been associated with unstable channel banks, lack of riprap, or both. Proper bank protection has resulted in a satisfactory structure in all cases.

Figures 126 through 138 show a variety of installations in the field and indicate construction techniques. Also shown are completed baffled aprons which have operated for several years and structures performing for various fractions of the design flow. Each structure shown has been reported as satisfactory, either at the outset of operation or after bank stabilization had been accomplished. Each structure was built according to the general rules given in this section.

Baffle pier dimensions, spacing and arrangement, wall heights, and other rules for baffled chutes on a 2:1 slope were followed precisely. Table 22 contains data on other structures which have been built following the general rules. Although no reports on the performance of the tabulated aprons have been received, it is believed that they are operating as expected. No adverse comments on their performance have been forthcoming.

TABLE 22.—*Baffled chute structures in use*

Spec. No.	Drawing No.	Location	Station	Chute width, feet	Design discharge, c.f.s.
		Franklin Canal			
DC-3720	271-D-549	Drain F-1.5-D	0+50	8 Trap	85
DC-3720	271-D-549	Drain F-10. 1-U	1+10	8 Trap	80
DC-3720	271-D-550	Drain F-1. 9-D	1+25	6 Trap	64
DC-3720	271-D-550	Drain F-10. 1-D	2+00	6 Trap	51
DC-3720	271-D-551	Drain F-10. 1	84+68+	18 Rect	625
DC-3891	271-D-648	Drain F-14. 1-D	1+44	10 Trap	100
DC-3891	271-D-649	Drain F-14. 9	5+20	32 Rect	1, 100
DC-3891	271-D-650	Drain F-14. 9-D	23+20	14 Rect	280
DC-3891	271-D-651	Drain F-15. 8	5+00	23 Rect	800
DC-3891	271-D-653	Drain F-23. 5-U	2+80	10 Trap	100
		Courtland Canal			
DC-4501	271-D-1031	Drain C-42. 3-U	2+80	10 Trap	120

TABLE 22.—*Baffled chute structures in use*—Continued

Spec. No.	Drawing No.	Location	Station	Chute width, feet	Design discharge, c.f.s.
		Courtland West Canal			
DC-4874	271-D-1344	Drain CW-0. 7-D	3+00	10 Rect	123
DC-4874	271-D-1344	Drain CW-1. 4-U	2+00	6 Rect	123
DC-4874	271-D-1344	Drain CW-10. 5	8+00	6 Rect	46
		Sargent Canal			
DC-4681	499-D-263	Airport Wasteway	16+00	11. 5 Rect	130
DC-4681	499-D-263	Airport Wasteway	36+50	11. 5 Rect	130
DC-4681	499-D-264	Airport Wasteway	51+20	17 Rect	300
DC-4681	499-D-264	Airport Wasteway	98+00	17 Rect	300
DC-4681	499-D-229	Big Oak Drain	11+25	11 Rect	220
DC-4681	499-D-230	Big Oak Drain	13+00	12. 5 Rect	150
DC-4681	499-D-248	Drain S-21. 9	4+60	25. 5 Rect	800
DC-4681	499-D-249	Drain S-22. 6	4+00	19. 5 Rect	650
DC-4681	499-D-250	Drain S-22. 6-U	0+60	14. 5 Rect	165
DC-4681	499-D-260	Drain S-38. 0	29+35	15 Rect	180
		Gila Project			
DC-2688	50-D-2417	Wellton-Mohawk Canal	7+14. 48	84 Rect	35 c.f.s.
DC-2688	50-D-2432	Wellton-Mohawk Canal	151+39. 25	52 Rect	per foot of
DC-2688	50-D-2438	Wellton-Mohawk Canal	234+60	36 Rect	width.
DC-2972	50-D-2668	Mohawk Dike No. 1	0+00	140 Rect	
DC-2972	50-D-2679	Mohawk Dike No. 1	12+30	25 Rect	
DC-2972	50-D-2646	Mohawk Canal	1125+95. 74	180 Rect	
DC-2972	50-D-2654	Mohawk Canal	1406+22. 25	124 Rect	
DC-2972	50-D-2659	Mohawk Canal	1479+78. 47	46 Rect	
DC-2972	50-D-2661	Mohawk Canal	1546+90	8 Rect	35 c.f.s
DC-3683	50-D-2982	Radium Hot Springs	179+84. 91	18 Rect	per foot
DC-4983	50-D-3359	Wellton-Mohawk Canal	661+16	90 Rect	of width.
DC-2822	50-D-2446	Wellton-Mohawk Canal	489+21. 71	65 Rect	
DC-2822	50-D-2453	Wellton-Mohawk Canal	563+50	39 Rect	
DC-2822	50-D-2456	Wellton-Mohawk Canal	614+21. 71	65 Rect	
DC-2822	50-D-2459	Wellton-Mohawk Canal	660+00	62 Rect	
DC-2822	50-D-2470	Wellton-Mohawk Canal	822+17. 17	200 Rect	
DC-2822	50-D-2473	Wellton-Mohawk Canal	938+00	36 Rect	
DC-5019	50-D-3366	Texas Hill Floodway	113+00	11 Rect	200
DC-5019	50-D-3368	Texas Hill Floodway	133+00	28. 5 Rect	1, 000
		Eden Project			
DC-3558	153-D-152	Means Canal	7+30. 77	18 Rect	630
		Columbia Basin Project			
DC-4888	222-D-19589	WB5WW1	36+90	18 Rect	226
DC-4888	222-D-19596	WB5WW1	564+95	7 Rect	85
DC-4888	222-D-19596	WB5WW1	280+10	7 Rect	85
DC-4888	222-D-19596	WB5WW1	286+60	11 Rect	127

TABLE 22.—*Baffled chute structures in use*—Continued

Spec. No.	Drawing No.	Location	Station	Chute width, feet	Design discharge, c.f.s.
		Columbia Basin Project—Continued			
DC-4888	222-D-19596	WB5WW1	303+10	11 Rect	127
DC-4888	222-D-19596	WB5WW1	329+10	11 Rect	127
DC-4888	222-D-19596	WB5WW1	346+25	11 Rect	127
DC-4888	222-D-19596	WB5WW1	363+10	11 Rect	127
DC-4888	222-D-19596	WB5WW1	396+60	11 Rect	127
DC-4888	222-D-19597	WB5WW1	410+10	14 Rect	172
DC-4888	222-D-19597	WB5WW1	420+60	14 Rect	172
DC-4888	222-D-19597	WB5WW1	432+10	14 Rect	172
DC-4888	222-D-19597	WB5WW1	441+45	14 Rect	172
DC-4888	222-D-19597	WB5WW1	456+75	14 Rect	172
DC-4888	222-D-19597	WB5WW1	465+70	14 Rect	172
DC-4888	222-D-19597	WB5WW1	472+90	14 Rect	172
DC-4888	222-D-19598	WB5WW1	481+85	14 Rect	172
DC-4888	222-D-19598	WB5WW1	489+60	14 Rect	172
DC-4888	222-D-19598	WB5WW1	497+10	14 Rect	172
DC-4888	222-D-19598	WB5WW1	505+10	14 Rect	172
DC-4888	222-D-19598	WB5WW1	513+40	14 Rect	172
DC-4888	222-D-19598	WB5WW1	520+40	14 Rect	172
DC-4888	222-D-19598	WB5WW1	527+60	14 Rect	172
DC-4696	222-D-18763	EL68DWW	321+55.70	14 Rect	146
DC-4696	222-D-18817	EL68DWW	551+07AH	22 Rect	450
DC-4696	222-D-18775	EL81WW	202+02	18 Rect	365
DC-4696	222-D-18770	EL68DWW	Dike No. 1	9 Rect	96
DC-4696	222-D-18776	EL68DWW	Dike No. 4	14 Rect	198
DC-4696	222-D-18776	EL68DWW	Dike No. 5	14 Rect	198
DC-4696	222-D-18776	EL68DWW	Dike No. 6	14 Rect	198
DC-4696	222-D-18776	EL68DWW	Dike No. 7	14 Rect	198
DC-4696	222-D-18776	EL68DWW	Dike No. 9	18 Rect	313
DC-4696	222-D-18776	EL68DWW	Dike No. 10	20 Rect	363
DC-4696	222-D-18776	EL68DWW	Dike No. 11	22 Rect	414
DC-4696	222-D-18776	EL83WW	Dike No. 12	11 Rect	220
DC-4696	222-D-18776	EL83WW	Dike No. 13	11 Rect	220
DC-4696	222-D-18776	EL83WW	Dike No. 14	11 Rect	220
DC-4696	222-D-19601	WB5WW1	531+17.53	14 Rect	172
DC-4696	222-D-19601	WB5WW1	535+80	14 Rect	172
DC-4571	222-D-18422	PE16.4WW	1594+63	22 Rect	770
DC-4749	222-D-19090	Potholes East Canal	1369+11	46.5 Rect	3,900
		Colorado Big Thompson Project			
DC-3657	245-D-6645	St. Vrain Supply	513+86	18 Rect	575
DC-4150	245-D-7137	Boulder Creek Supply	667+78	10 Rect	200
		Solano Project			
DC-4881	413-D-513	Putah South Canal	1099+79	13 Rect	156
DC-4555	413-D-317	Putah South Canal	263+50	6 Rect	48

(Above) Setting forms for baffled chute at Sta. 3+35 of Wasteway 10.7, and (upper right) compacting backfill at Sta. 2+85 of Wasteway 11.1, Culbertson Canal, Missouri River Basin project.

(Lower right). A discharge of 63 c.f.s. flowing into Helena Valley Regulating Reservoir, Missouri River Basin project, from Helena Canal. Soft earth bank is eroded, otherwise, performance is excellent.

FIGURE 126.—*Construction and performance of baffled chutes.*

Figure 126 shows construction techniques used on two baffled chutes and operation of another at partial capacity. In the latter photograph, a small quantity of riprap on the earth bank would have prevented undermining and sloughing of the soft earth at the downstream end of the right training wall.

The baffled chute shown in Figure 127 is on the Boulder Creek Supply Canal and has operated many times over a range of discharges approaching the design discharge. As a result, a shallow pool has been scoured at the base of the structure. This is desirable, since the pool tends to reduce surface waves and make bank protection downstream from the structure unnecessary. A relatively small quantity of riprap has been placed to achieve the maximum benefit. Also, the wetted area (darker color) adjacent to the training walls starts at about the second row of baffles. This is caused by a small amount of splash which rises above the walls and is carried by air currents. No reports have ever been received that this splash or water loss is of any consequence.

Figure 128 shows a low-drop baffled chute on the Bostwick Courtland Canal. It appears that grass has stabilized the banks sufficiently for the

Dark rock area adjacent to training walls is wet from spray.

Baffled chute at Sta. 667+78, Extension Boulder Creek Supply Canal, Colorado-Big Thompson project, designed for 200 c.f.s. Discharges of 150 c.f.s. (upper left) and 100 c.f.s. (lower left) show excellent performance in both instances.

FIGURE 127.—*Prototype installation of baffled chute.*

No flow through structure on Bostwick Courtland Canal, Drain A, Sta. 6+08, designed for 924 c.f.s.

With a discharge of about 5 c.f.s., the structure performs well. It is reported that larger flows are handled satisfactorily.

FIGURE 128.—*Prototype installation of baffled chute.*

No flow in structure on Bostwick Courtland Canal, Drain A, Sta. 67+93. Trash has accumulated at foot of chute. Design discharge 277 c.f.s.

Discharge about 3 c.f.s. Reports received indicate that structure performs well for large discharges.

FIGURE 129. *Prototype installation of baffled chute.*

height of fall indicated. Little, if any, riprap is evident and the structure has performed satisfactorily for a number of years with little maintenance. There is a shallow scoured pool at the base of the apron.

Figure 129 shows another structure on the Bostwick Courtland Canal. Trash has accumulated near the base of the structure. Field reports indicate that trash tends to collect during a falling stage and is removed by the water during the rising stage. Generally speaking, trash is not a problem on baffled chutes and does not contribute materially to maintenance costs. Well-placed riprap at the base of the structure contributes to bank stability.

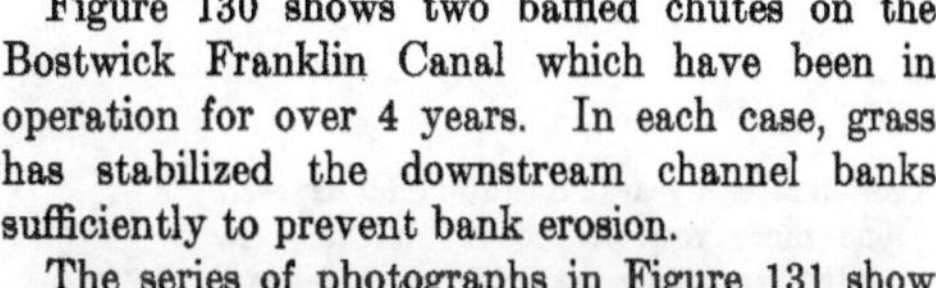

Figure 130 shows two baffled chutes on the Bostwick Franklin Canal which have been in operation for over 4 years. In each case, grass has stabilized the downstream channel banks sufficiently to prevent bank erosion.

The series of photographs in Figure 131 show the progress of downstream scour from October 1956 to the spring of 1959 at a drain on the Bostwick Division, Missouri River Basin project. It may be noted that between October 1956 and September 1957, scouring occurred which exposed one row of the buried blocks. The bed material which was carried away consisted of fines; the coarse material that resembles riprap was left in place as shown in the photographs.

Structure after 4 years of operation. Performance has been satisfactory. Design discharge 625 c.f.s.

Bostwick Franklin Canal, Drain F-10.1, Sta. 84+88.

Structure after 5 years of operation. Performance has been satisfactory. Design discharge 1,100 c.f.s.

Bostwick Franklin Canal, Drain F-14.9, Sta. 5+20.

FIGURE 130.—*Prototype installations of baffled chutes.*

No flow in October 1956.

Erosion after a year of operation has exposed one more row of blocks. Rocks were sorted from finer material which moved. Rubbish has collected by September 1957.

Erosion did not continue at original rate, and is no more severe after 2½ years of operation in April 1959.

FIGURE 131.—*Progress of erosion in Bostwick Crow Creek Drain, Sta. 28+90. Design discharge 2,000 c.f.s.*

Figure 132 shows the Bostwick Superior Canal Drain after only a few months of operation. The soft earth banks were badly eroded, both upstream and downstream from the structure. The small amount of riprap placed downstream did much to protect the structure from complete failure. Stabilization of the banks with a grass cover eliminated sloughing of the banks. Figure 133 shows the same structure 6 years later, operating satisfactorily for a fraction of the design discharge. Now that the banks are stable, there is no maintenance problem.

The left photograph in Figure 134 shows Frenchman-Cambridge Drain 8C after 4½ years of operation. Performance has been excellent. Riprap originally placed at the base of the walls is covered by weed and grass cover. The shallow energy-dissipating pool has helped to reduce bank maintenance downstream. In the right photograph, the Culbertson Canal Wasteway 3.3 is

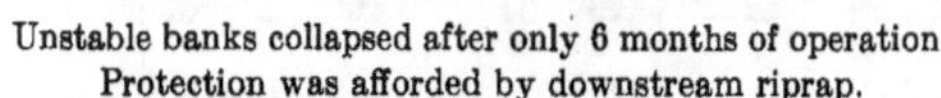

Unstable banks collapsed after only 6 months of operation. Protection was afforded by downstream riprap.

Upstream banks were badly eroded.

FIGURE 132.—*Unstable banks create an erosion problem on Bostwick Superior Canal, Drain 2A, Sta. 36+82.4.*

shown in operation shortly after construction was completed. The need for riprap at the waterline near the base of the baffled apron is beginning to become apparent. Figure 135 shows closer views of this same structure and indicates how energy dissipation is accomplished on the chute. Action in hydraulic models of baffled chutes is identical to that shown here. The left photograph in Figure 136 shows the wasteway after the discharge was stopped. It appears that additional riprap protection would be desirable, particularly if the discharge is greater than 75 c.f.s.

Figure 136, right photograph, shows the Robles-Casitas Canal discharging 500 c.f.s. into a baffled chute. The riprap affords adequate protection to the structure. Operation is excellent.

Figure 137 shows a drop on the Frenchman-Cambridge Wasteway. The right photograph

Stabilized banks in April 1959 show no evidence of erosion.

Performance of structure during rainstorm. Discharge 81 c.f.s. in May 1959. Design discharge 400 c.f.s.

FIGURE 133.—*S abilized banks present no erosion problem after the work was done on Bostwick Superior Canal, Drain 2A, Sta. 36+82.4. (See Fig. 132.)*

Frenchman-Cambridge Drain 8–C after 4½ years of operation. Excellent performance. Design discharge 1,000 c.f.s.

Baffled apron of Culbertson Canal Wasteway 3.3 discharging 75 c.f.s. Good performance. Design discharge 400 c.f.s.

FIGURE 134.—*Performance of prototype structures.*

shows how wingwalls can be used to protect the structure and how a small amount of riprap can be used to protect the wingwalls from undercutting. The left photograph shows the action of the water on the baffled chute for a very small discharge. There is practically no turbulence at the base of the apron (see right photograph also).

Figure 138, left photograph, shows the North Branch Wasteway—Picacho Arroyo System discharging at about half capacity after a violent rainstorm. The water is carrying a high concentration of sediment. After the flood, right photograph, it was found that the downstream channel had aggraded rather than scoured, partially covering one row of blocks which had been more exposed before the runoff. In this case, the reduction in velocity at the base of the apron caused sediment to settle out of the wasteway water.

★ ★ ★

Recapitulation

Baffled aprons or chutes are used in flow ways where water is to be lowered from one level to

The second row of baffles is completely covered because of acceleration of flow between the first and second rows with a flow of 75 c.f.s. at top of baffled chute.

Same flow in midportion of chute. See right-hand photograph in Figure 134 for general view.

FIGURE 135.—*Performance of baffled chute on Culbertson Canal Wasteway 3.3.*

Culbertson Canal Wasteway 3.3 after a discharge of 75 c.f.s. in May 1959.

Robles-Casitas Canal between Sta. 294 and Sta. 298 with 500 c.f.s. discharging into Santa Ana Creek. Waves in canal section occasionally splash over top of canal concrete lining.

FIGURE 136.—*Performance of prototype structures.*

another. The baffle piers prevent undue acceleration of the flow as it passes down the chute. Since the flow velocities entering the downstream channel are relatively low, no stilling basin is required. The chute, on a 2:1 slope or flatter, may be designed to discharge up to 60 c.f.s. per foot of width, and the drop may be as high as structurally feasible. The lower end of the chute is constructed to below stream-bed level and backfilled as necessary. Degradation or scour of the stream bed, therefore, does not adversely affect the performance of the structure. The simplified hydraulic design procedure given in the numbered steps refers to Figure 140. More detailed explanations have been given in the text.

Simplified Design Procedure

1. The baffled apron should be designed for the maximum expected discharge, Q.

2. The unit design discharge $q=\frac{Q}{W}$ may be as high as 60 c.f.s. per foot of chute width, W. Less severe flow conditions at the base of the chute exist for 35 c.f.s. and a relatively mild condition occurs for unit discharges of 20 c.f.s. and less.

3. Entrance velocity, V_1, should be as low as practical. Ideal conditions exist when $V_1=\sqrt[3]{gq}-5$, Curve D, Figure 125. Flow conditions are not acceptable when $V_1=\sqrt[3]{gq}$, Curve C, Figure 125.

4. The vertical offset between the approach channel floor and the chute is used to create a stilling pool or desirable V_1 and will vary in individual installations; Figures 103, 105, 107, and 109 show various types of approach pools. Use a short radius curve to provide a crest on the sloping chute. Place the first row of baffle piers close to the top of the chute no more than 12 inches in elevation below the crest.

▲ Stilling action of blocks is most effective for small discharges.

A small amount of riprap provides excellent protection to foot of chute. ▶

FIGURE 137.—*Frenchman-Cambridge Meeker Extension Canal Wasteway, Sta. 1777+18. Discharge about 5 c.f.s., design discharge 269 c.f.s.*

5. The baffle pier height, H, should be about 0.8 D_c, Curve B, Figure 125. The critical depth on the rectangular chute is $D_c=\sqrt[3]{\frac{q^2}{g}}$, Curve A. Baffle pier height is not a critical dimension but should not be less than recommended. The height may be increased to 0.9 D_c, Figure 125.

6. Baffle pier widths and spaces should be equal, preferably about 3/2 H, but not less than H. Other baffle pier dimensions are not critical;

Estimated discharge 15 c.f.s. per foot width (half capacity).

Channel after flood—material was deposited rather than scoured.

North Branch Wasteway Channel, Picacho Arroyo System, Rio Grande project.

FIGURE 138.—*Baffled chute may produce channel aggradation rather than scour.*

Baffle piers 18″ high and 18″ wide—18″ spaces. Row spacing, 6′0″.

Chute 9′ wide and 90′ long—2 : 1 slope. Training walls 5′ high.

FIGURE 139.—*Kopp Wasteway on the Main East Canal, Michaud Flats project, Idaho, discharging 25 c.f.s. (one-third capacity).*

suggested cross section is shown. Partial blocks, width 1/3 H to 2/3 H, should be placed against the training walls in Rows 1, 3, 5, 7, etc., alternating with spaces of the same width in Rows 2, 4, 6, etc.

7. The slope distance (along a 2:1 slope) between rows of baffle piers should be 2 H, twice the baffle height H. When the baffle height is less than 3 feet, the row spacing may be greater than 2 H but should not exceed 6 feet. For slopes flatter than 2:1, the row spacing may be increased to provide the same vertical differential between rows as expressed by the spacing for a 2:1 slope.

8. The baffle piers are usually constructed with their upstream faces normal to the chute surface; however, piers with vertical faces may be used. Vertical face piers tend to produce more splash and less bed scour, but differences are not significant.

9. Four rows of baffle piers are required to establish full control of the flow, although fewer rows have operated successfully. Additional rows beyond the fourth maintain the control established upstream, and as many rows may be constructed as is necessary. The chute should be extended to below the normal downstream channel elevation as explained in the text of this section, and at least one row of baffles should be buried in the backfill.

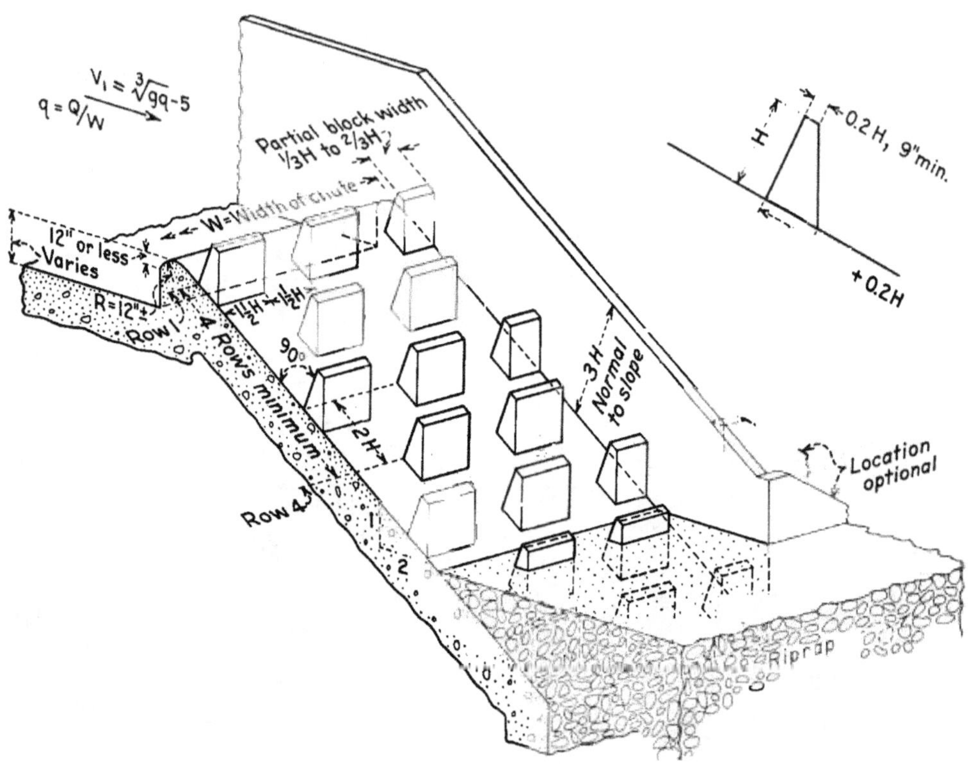

FIGURE 140. *Basic proportions of a baffled chute.*

10. The chute training walls should be three times as high as the baffle piers (measured normal to the chute floor) to contain the main flow of water and splash. It is impractical to increase the wall heights to contain all the splash.

11. Riprap consisting of 6- to 12-inch stones should be placed at the downstream ends of the training walls to prevent eddies from working behind the chute. The riprap should not extend appreciably into the flow area. Figures 126 to 139 show effective and ineffective methods of placement on field structures.

Section 10

Improved tunnel spillway flip buckets (Basin X)

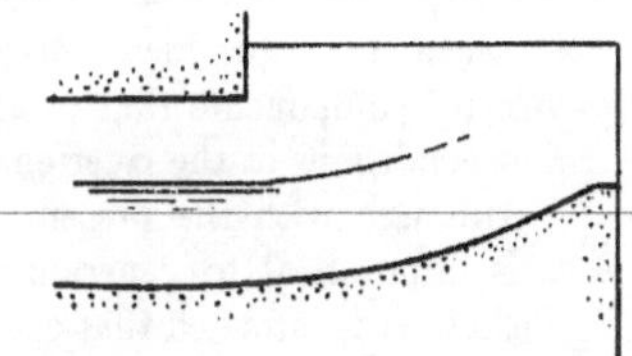

THE two basic parts of a tunnel spillway are an upstream spillway crest, free or controlled, and a downstream tunnel, part of which is sloping and part near horizontal. From the standpoint of economy the tunnel diameter must be kept to a minimum, but the tunnel is never allowed to flow full because of the possibility of siphonic action producing dangerous flow conditions. It is therefore necessary to keep flow velocities high and to prevent turbulent areas in the tunnel. Spillway tunnels are usually designed to flow from ¾ to ⅞ full at maximum discharge, making the outflow at the tunnel portal relatively deep. The combination of depth and velocity produces the highest possible concentration of energy and increases the difficulty of obtaining satisfactory flow conditions where the flow spills into the river. As an example, on the Glen Canyon tunnel spillways, the maximum discharge of 276,000 c.f.s. produces 159,000 hp. per foot of width at the tunnel portals. On Grand Coulee, an overfall spillway, where the maximum discharge is 1,000,000 c.f.s., the energy per foot of width is only 15,650 hp., or one-tenth that on Glen Canyon (Table 23).

If it were feasible to construct an efficient hydraulic jump stilling basin at the end of one of the Glen Canyon tunnels, the basin depth, from apron to tail water elevation, would need to be 170 feet. The hydraulic jump length would be more than 1,000 feet and would require a basin 700 ft. to 800 feet long or more. Basin appurtenances, such as baffle piers, could not be used effectively because the high entrance velocity, 165 f.p.s., would produce cavitation problems. The cost of a structure this size would be prohibitive, and it is readily seen why other types of structures are used at the end of tunnel spillways. Buckets have been the most common of these structures and were probably derived from the slight upturns placed at the base of early overfall spillways. It is not clear whether the designers intended that these buckets operate free or submerged. In some cases, the upturn was too slight to produce a measurable effect on a thick jet, but probably the intended purpose was to deflect the jet downstream to prevent undermining of the spillway structures. Buckets of this type are referred to variously as ski-jump, deflector, diffuser, trajectory, or flip buckets. For uniformity, the term flip bucket will be used in this section.

Flip buckets are not a substitute for energy dissipators because such a bucket is inherently incapable of dissipating energy within itself. The purpose of a flip bucket is to throw the water downstream where the riverbed damage, which is usually certain to occur, does not endanger the safety of the dam, powerplant, or other structures, including the flip bucket itself. In accomplishing this primary function, buckets are also designed to spread the flow across as much of the downstream channel as is considered desirable in order to reduce riverbed damage as much as possible. The jet trajectory is modified as necessary to cause the jet to impinge on the tail water surface at the desired location, and when possible, the steepness of the jet trajectory at the point of impingement is selected to produce horizontal and vertical velocity components that produce most favorable flow conditions in the river channel.

Although with the present state of knowledge it is impractical to generalize the design of flip buckets, it is intended that certain basic facts that have been found to be true as a result of extensive hydraulic model testing and prototype observation be presented.

Bucket Design Problems

It is usually difficult or impossible to predict the flow pattern to be expected from a particular bucket by mere inspection of the bucket shape. Because of variations in velocity and depth, the spreading and trajectory characteristics of a given bucket can be determined only by testing in a hydraulic model. Because of the opportunity to test various types of buckets and to observe first hand their performance in the field, the findings of these tests should be of interest to designers who must often select a bucket type before the hydraulic-model tests are made.

In the course of developing and improving bucket designs, a number of difficulties have been found and overcome. The following examples indicate the problems that may be encountered in bucket design and that may not be generally known.

The flip buckets on the tunnel spillways at Hungry Horse Dam and Yellowtail Dam of the Bureau of Reclamation projects, and Bhumiphol Dam and Wu-Sheh Dam being built in Thailand and Formosa, respectively, are similar (Table 23) and are what may be called a "standard" type. The buckets are placed downstream from a transition that changes the circular or horseshoe shaped tunnel to a flat bottom in order to correspond to the flat bottom of the bucket. High-velocity flow in the tunnel makes it difficult to design a short transition, and long transitions are usually costly. If the transition is not carefully designed, and checked by model studies, there is the possibility of dangerous subatmospheric pressures occurring in the corners. The transition, therefore, becomes as much of a design problem as the bucket.

The Fontana Dam spillway buckets do not have an upstream transition (Table 23). The bucket inverts are circular, the same as the tunnel inverts, Figure 141. The buckets were shaped by trial in a 1:100 scale model tested in the Tennessee Valley Authority Hydraulic Laboratory. The curved surfaces of the finally developed buckets could not be defined by ordinary dimensioning or even by mathematical equations. That the buckets were well designed has been proved by subsequent operation of the structure, but the methods necessary to convert the model dimensions at a scale of 1:100 to prototype dimensions were quite laborious. Because of the high-velocity flow in the bucket, dimensions taken from the model could not be scaled up directly. Any small irregularity or misalinement when multiplied by 100 could have been sufficiently large to produce cavitation in the prototype bucket. It was, therefore, necessary to convert the dimensions to a 1:10 scale bucket, and after smoothing these, to convert the corrected dimensions to a 1:1 scale.

On some buckets, particularly those on dams outside of the United States, a serrated or toothed edge has been placed at the downstream end of the bucket. The teeth are to provide greater dispersion of the jet before it strikes the tail water surface. High velocity flow passing over the sharp edges may produce cavitation damage on the concrete surfaces.

The problem of draining a tunnel that has a flip bucket at the downstream end provides a challenge in design. The drain must be placed in a surface exposed to high-velocity flow. Even though it is

TABLE 23.—*Description of spillway tunnels on various projects*

Name and location	Agency	Maximum discharge in cubic feet per second	Fall-max headwater to bucket invert in feet	Tunnel dimensions
(1)	(2)	(3)	(4)	(5)
Glen Canyon Dam, Colorado River Storage project, Arizona.	Bureau of Reclamation	276,000	588	2 tunnels each 41-ft. diameter.
Grand Coulee Dam, Columbia Basin project, Washington.	do	1,000,000	420	Overfall spillway 1,650 ft. wide.
Hungry Horse Dam, Hungry Horse Dam project, Montana.	do	50,000	488	31-ft. diameter.
Yellowtail Dam, Missouri River Basin project, Montana.	do	173,000	512	20.5-ft. diameter conduit.
Bhumiphol Dam, Thailand	Kingdom of Thailand Ministry of Agriculture Royal Irrigation Department.	212,000	402	2 horseshoe tunnels 37.08-ft. diameter
Wu-Sheh Dam, Taiwan, China	Taiwan Power Co.	66,000	387	27-ft. diameter.
Fontana Dam, North Carolina	Tennessee Valley Authority	180,000	423	2 tunnels each 34-ft diameter.
Trinity Dam, Central Valley project, California.	Bureau of Reclamation	24,000	475	20-ft. diameter.

possible to design or develop a drain opening in the laboratory that will not produce cavitation pressures, it is difficult to obtain field construction to the necessary tolerances to prevent cavitation from occurring. An ideal bucket design would be self-draining and would not present a cavitation problem at the drain structure.

Improved Bucket Designs

A number of tunnel-spillway flip buckets have been developed in the Bureau Hydraulic Laboratory that seem to offer simple but effective methods of directing the flow away from the structure and which also overcome, in part, the

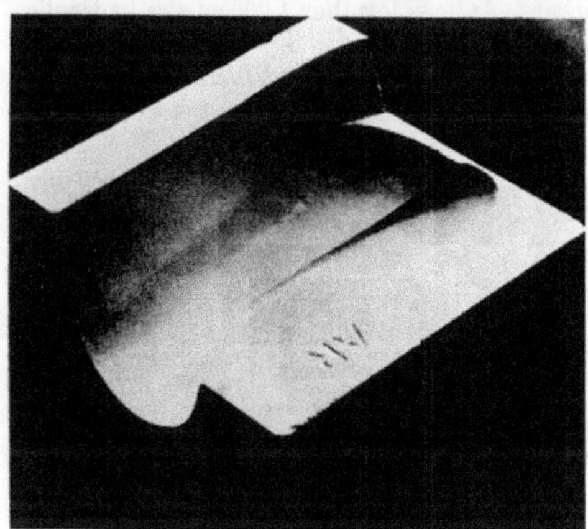

Bucket used at Tunnel 1 outlet.

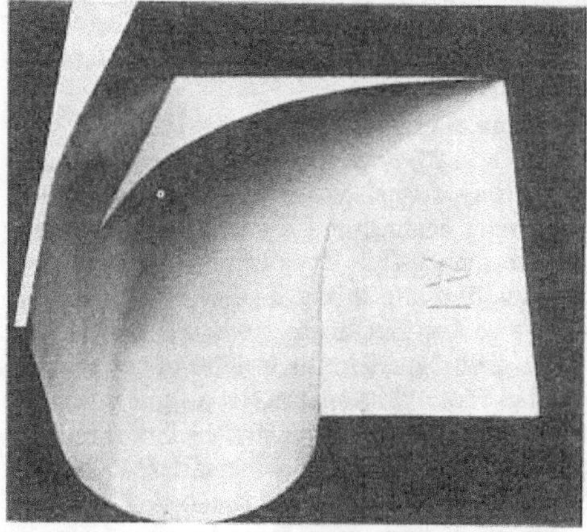

Bucket used at Tunnel 2 outlet.

FIGURE 141.—*Fontana Dam spillway flip bucket models.*

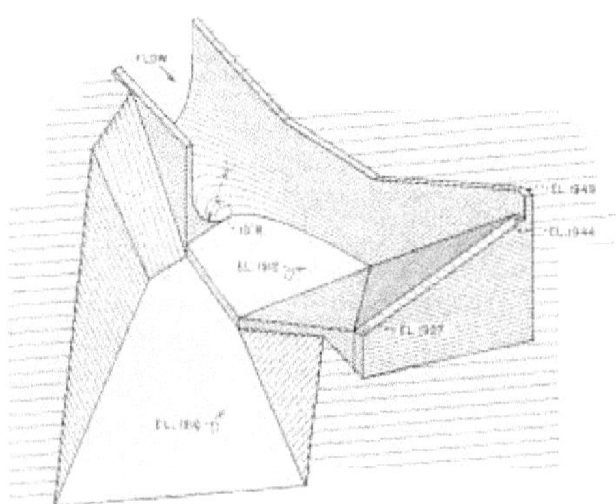

FIGURE 142. ·*Dispersion flip bucket.*

difficulties previously described. Although no single bucket eliminates all of the undesirable features, the use of the principles to be described will help the designer to provide an improved bucket on a particular structure. Thus, an ideal bucket should provide (1) easy drainage of the tunnel, (2) a bucket shape that can be defined and expressed in prototype size by ordinary dimensioning on ordinary drawings, (3) no need for an upstream transition, and (4) an impingement area that may be shaped, by simple additions to a basic bucket, to fit the existing topographic conditions. Some of the buckets described are unique and probably cannot be generally used without some adaptation. However, the others are basic in type and need only minor additions to accomplish some specific function.

A unique design was the Trinity Dam spillway bucket (Table 23) developed on a 1:80 scale model. The spillway tunnel enters one side of a wide, shallow river channel and the flow tends to cross the river diagonally. It was necessary to discharge the flow into this channel without creating excessive eddies that might erode the riverbanks or cause disturbances in the vicinity of the powerhouse tail race. The spillway is an uncontrolled morning-glory, and consequently the flow can vary from a few second-feet to a maximum of 24,000 c.f.s. The velocity at the bucket is 122 f.p.s. Because small flows may occur for days, it was desirable for low flows to leave the bucket as close to the riverbed elevation as possible to prevent excessive erosion near the base of the structure. On the other hand, large flows should be flipped downstream away from the structure with as much dispersion as possible to prevent erosion and induced eddies from damaging the structure. In the usual flip bucket, a hydraulic jump forms in the bucket for small flows and the water dribbles over the bucket end and falls onto the riverbed. This could cause erosion that would undermine the structure. When the jump is first swept out of the bucket, the jet usually lands near the structure and erosion and undermining of the structure may still occur.

At Trinity Dam, the foundation conditions at the end of the tunnel were such that it was deemed necessary to protect against the possibility of erosion and undermining. In order to place the bucket near riverbed level, the semicircular channel constructed downstream from the tunnel portal was curved downward in a trajectory curve, and the flip-bucket structure was placed at the end, Figure 142. The flip-bucket surface consists of three plane surfaces so placed that they spread and shape the jet to fit the surrounding topography. Large flows are spread into a thin sheet having a contact line with the tail water surface a considerable distance downstream, Figure 143. However, even small flows are thrown downstream well away from the base of the bucket.

A training wall was used to prevent spreading of the jet on the high, or land side, of the bucket. There was no wall on the low or river side of the bucket. At flows less than 1,000 c.f.s., a hydraulic jump formed over the horizontal surface and part way up the slope of the bucket and the flow spilled out of the low side of the bucket into

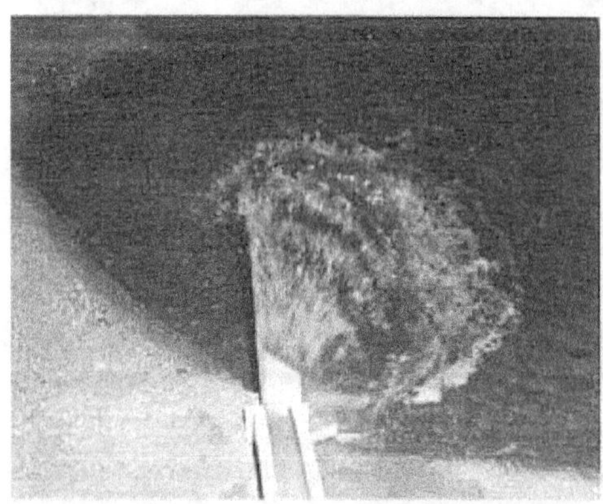

FIGURE 143. ·*Dispersion-type flip bucket—Q=24,000 c.f.s.*

the river channel. The open side of the bucket was only 4 feet or 5 feet above the river. Had the flow been confined on both sides and forced to spill out the end, the drop would have been more than 40 feet and additional protection of the bucket foundation would have been required. At discharges greater than 1,000 c.f.s., the jump swept out of the bucket without hesitation and with sufficient velocity that the flow was carried well downstream away from the structure. As the discharge increased, the jet was flipped farther downstream and became increasingly dispersed. The long contact line between the jet and the tail water reduced the unit forces on the tail water and the eddies induced at the ends of the contact line were thereby found to be a minimum. Since one side of the bucket is entirely open, the bucket is self-draining. Other advantages of this design are that the bucket may be defined for prototype construction with a few simple dimensions, and no curved or warped forms are necessary for prototype construction.

Another unusual type of flip bucket was developed for the Wu-Sheh Dam tunnel spillway. Construction schedules and geologic conditions in the field made it necessary to modify this bucket from the standard type previously described. After the line of the tunnel had been established and construction of the tunnel started, it was found necessary, as a result of model tests, to change the direction of the flow entering the river channel. Earth and rock slides during the diversion period made it necessary to construct retaining walls in the tunnel portal area which restricted the length of the flip bucket. Hydraulic model studies were made to determine how much turning of the jet was required and whether the turning could be accomplished in the tunnel. The tests showed that it was undesirable to turn the tunnel and that all turning should be accomplished in the bucket. The final bucket, as determined from model studies, used curved walls to turn the flow, a batter in the left wall to prevent congestion in the bucket and reduce hydraulic loads at the larger discharges, and a fillet at the junction of the left wall and floor to smooth up or control the jet undernappe, Figure 144. The resulting bucket was "tailormade" to direct the flow to impinge

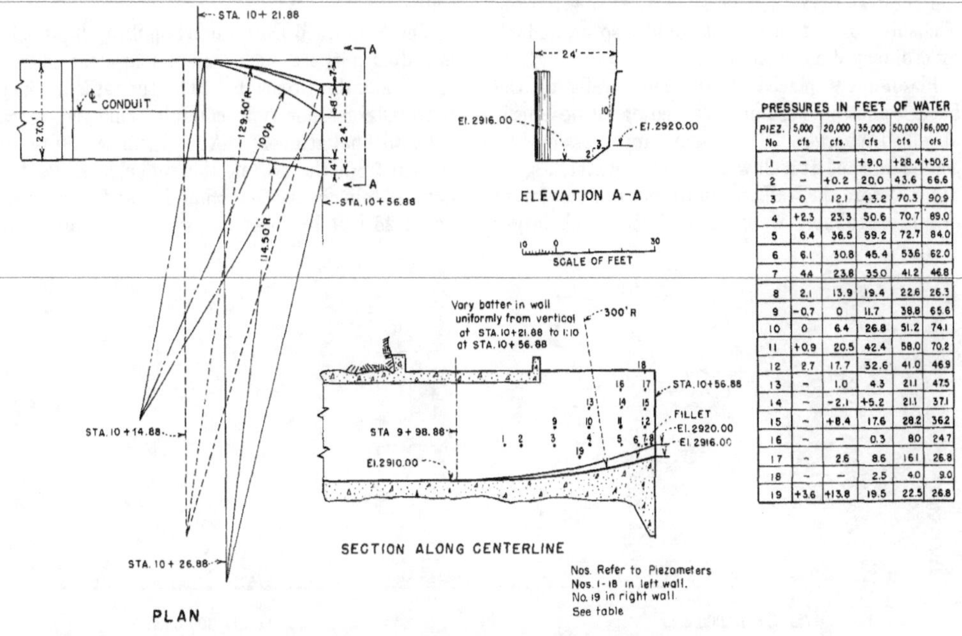

PRESSURES IN FEET OF WATER

PIEZ. No	5,000 cfs	20,000 cfs.	35,000 cfs	50,000 cfs	66,000 cfs
1	–	–	+9.0	+28.4	+50.2
2	–	+0.2	20.0	43.6	66.6
3	0	12.1	43.2	70.3	90.9
4	+2.3	23.3	50.6	70.7	89.0
5	6.4	36.5	59.2	72.7	84.0
6	6.1	30.8	45.4	53.6	62.0
7	4.4	23.8	35.0	41.2	46.8
8	2.1	13.9	19.4	22.6	26.3
9	-0.7	0	11.7	38.8	65.6
10	0	6.4	26.8	51.2	74.1
11	+0.9	20.5	42.4	58.0	70.2
12	2.7	17.7	32.6	41.0	46.9
13	–	1.0	4.3	21.1	47.5
14	–	-2.1	+5.2	21.1	37.1
15	–	+8.4	17.6	28.2	36.2
16	–	–	0.3	8.0	24.7
17	–	2.6	8.6	16.1	26.8
18	–	–	2.5	4.0	9.0
19	+3.6	+13.8	19.5	22.5	26.8

FIGURE 144.—*Recommended bucket, Wu-Sheh Dam.*

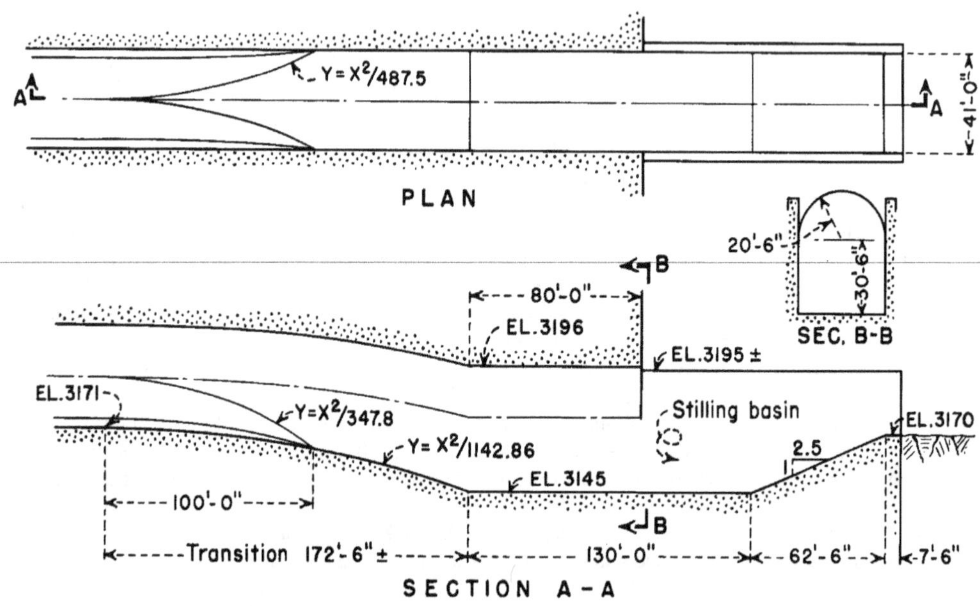

FIGURE 145.—*Yellowtail Dam stilling basin (preliminary design).*

near the middle of the river channel and to obtain the greatest dispersion possible at all discharges. The surfaces in this bucket could also be defined by ordinary dimensioning.

Piezometers placed in the side walls of the bucket showed pressures exceeding atmospheric at all discharges. The maximum pressure recorded on the left wall was 91 feet of water, Figure 144. Before the wall was battered, the maximum pressure probably would have been much larger because of a more direct impact on the converging wall.

The Yellowtail Dam tunnel-spillway flip bucket is a dual purpose bucket similar in some respects to the standard buckets. The tunnel is a curved bottom horseshoe-type conduit (changed to circular in final studies). At a distance 250 ft. upstream from the portal, the tunnel changes to a flat bottom horseshoe conduit, and the invert drops 26 feet by means of a combination transi-

(a) Q=12,000 c.f.s.

(b) Q=13,000 c.f.s.

FIGURE 146.—*Combination hydraulic jump basin flip bucket.*

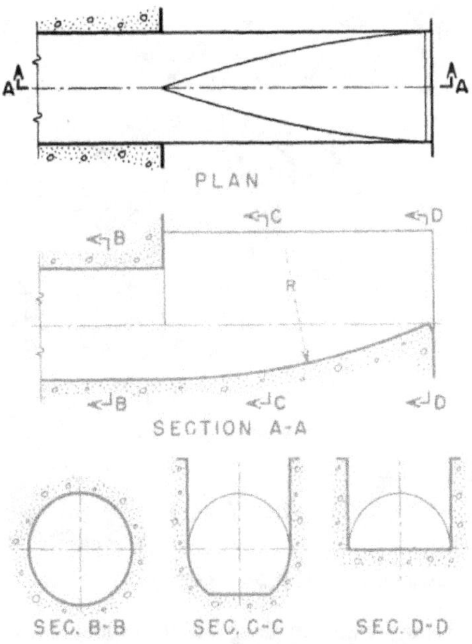

FIGURE 147. *Transition flip bucket.*

tion-trajectory curve 172.5 feet long. The bucket has a flat horizontal floor 130 ft. long and a 62.5-foot-long upward sloping sill, Figure 145. At spillway flows up to 12,000 c.f.s., a hydraulic jump forms in the bucket, Figure 146(a), and relatively quiet water is discharged into the downstream channel. As the spillway discharge increases, the jump moves downstream and at 13,000 c.f.s. sweeps out of the basin, Figure 146(b). For greater discharges and up to the maximum, 173,000 c.f.s., the basin acts as a flip bucket. The basin or bucket is placed low in solid rock so that discharges in the unstable zone, 12,000 to 13,000 c.f.s., cannot undermine the structure. This basin was developed in the laboratory to serve the specific purpose of acting as a hydraulic jump basin for the most prevalent spillway discharges, discharges expected to be exceeded only every 100 years, and acting as a flip bucket to prevent damage to the structures during large floods. The hydraulic jump was used for part of the discharge range in order to protect the river channel against clogging with talus, which was present in the canyon in large quantities and was expected to move if a high-velocity stream struck it. It was expected that reopening of the channel after large discharges would be necessary to reestablish full power head.

The flip buckets for the Glen Canyon Dam tunnel spillways are an example of buckets developed to eliminate the tunnel transition and the need for a flat-bucket invert. The buckets at the portals of the 41-foot diameter tunnels are on opposite sides of the river and are aimed to discharge at

FIGURE 148.—*Standard flat-bottom flip bucket. Glen Canyon Dam studies* ($F=7.59$) $Q=138,000$ *c.f.s.*

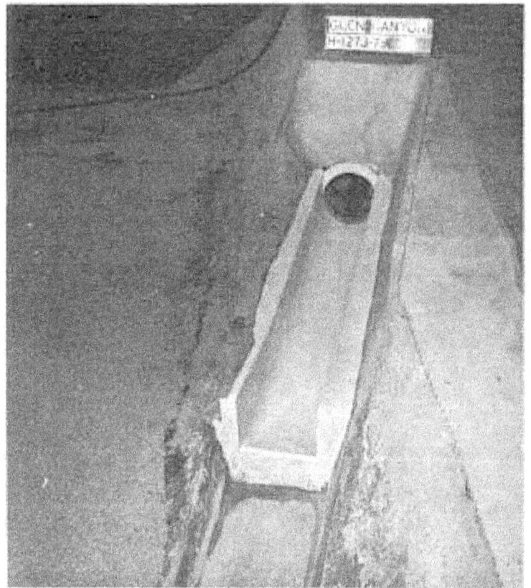

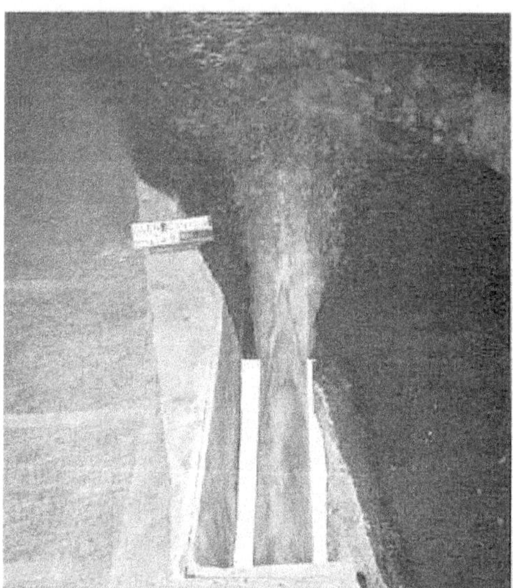

FIGURE 149. *Transition flip bucket, Glen Canyon Dam studies (F=5.64) Q=138,000 c.f.s.*

acute angles with the center of the river. The left bucket is farther downstream than the right. Each bucket is designed to handle the maximum discharge of 138,000 c.f.s. at a velocity of approximately 165 f.p.s. This represents more than 13,000,000 hp. in energy released into the river during maximum discharge.

In the preliminary design, a 70-foot-long transition was placed between the circular tunnel and the rectangular channel containing the flip bucket. Hydraulic-model studies indicated that the transition was too short, and that subatmospheric pressures would be sufficiently low to produce cavitation and damage to the structure. Two alterna-

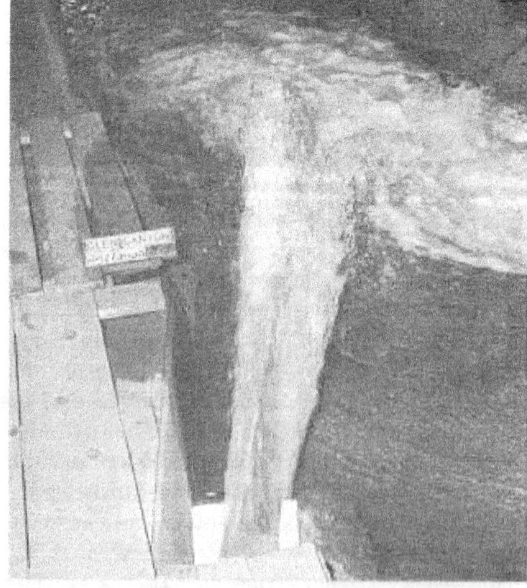

FIGURE 150. *Transition flip bucket with sidewall deflectors, Glen Canyon Dam studies (F=5.64) Q=138,000 c.f.s.*

FIGURE 151. *Typical jet profile for 35° transition flip bucket, Glen Canyon Dam studies—Q=75,000 c.f.s.*

tives were developed during the model studies. One was to use a 100-foot-long transition in which the change in cross section was accomplished without dangerous pressures occurring, and the other was to eliminate the transition by continuing the circular tunnel invert downstream to intersect the upward curve of the flip bucket, Figure 147. The latter scheme was developed and is used in the prototype structure; identical buckets are placed on the twin spillways. In effect, the transition and the bucket are combined into the bucket structure without complicating the design of the bucket.

Because the flat-bottom portion of these buckets diverges in plan, small flows are spread laterally more than for the flat-bottomed bucket. As the discharge increases, the rate of spreading decreases so that it is easier to accommodate the jet for flood flows in a relatively narrow channel. Figures 148 and 149 show a comparison of the flow from the two types of buckets. In the flat-bottomed bucket, Figure 148, which is preceded by a transition, the flip curve extends across the full width of the bucket for its entire length. All of the flow elements at a given elevation are turned simultaneously. In the alternative bucket, Figure 149, the flip curve turns the lower-flow elements in the center of the stream first and gradually widens its zone of influence as the flow moves downstream, resulting in greater dispersion of the jet. In effect, the flow along the center line of the bucket is turned upward while the flow elements on either side of the center are turned upward and laterally. Training walls may be used to limit the lateral spreading. In subsequent testing, deflectors were added to the bucket training walls to make the jets conform to the shape of the river channel and surrounding topography, Figures 150 and 151.

The flip bucket used on the Flaming Gorge Dam tunnel spillway was of the same type as that used on the Glen Canyon spillways. The maximum design flow for Flaming Gorge spillway is 28,800 c.f.s., the velocity of the flow at the portal of the 18-foot-diameter tunnel is approximately 140 f.p.s. The energy in the jet at the flip bucket is equivalent to 1 million hp. In operation, the flow appearance of the Flaming Gorge bucket was entirely different than that of the Glen Canyon buckets. The Flaming Gorge jet was well dispersed at the lower discharges, Figure 152(a), and became more compact as the discharge increased, Figure 152(b). The Glen Canyon jets were well dispersed for

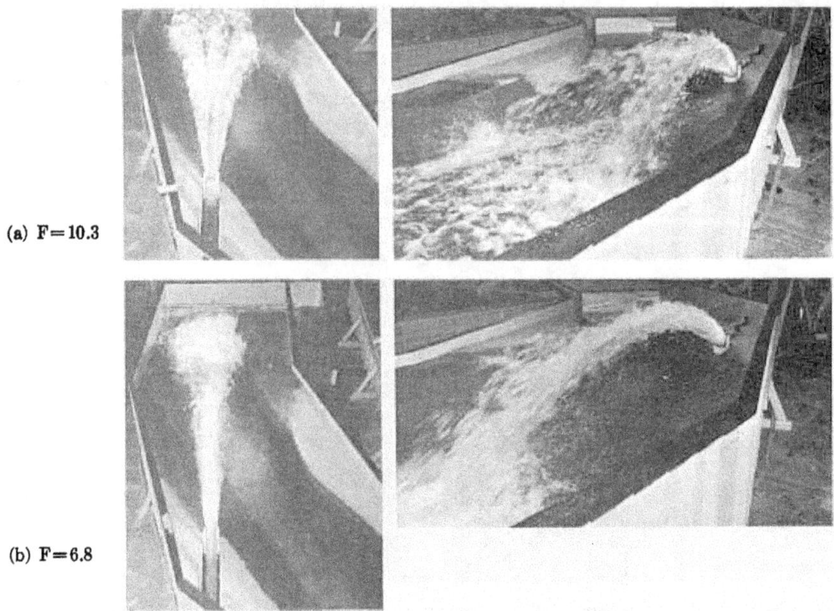

(a) F=10.3

(b) F=6.8

FIGURE 152. *Flip bucket studies for 35° transition bucket, Flaming Gorge Dam studies (a) Q=7,200 c.f.s.; (b) Q=28,800 c.f.s.*

all flows, and the change in lateral spreading with discharge was not so apparent. In the Flaming Gorge bucket, the water rose on the sides of the bucket at low flows, forming in effect a U-shaped sheet of water in which the bottom and sides were of equal thickness. The vertical sides of the U followed the line of the bucket side walls after leaving the bucket, while the bottom sheet of water had a tendency to diverge to either side. The vertical fins had a shorter trajectory than the lower sheet and on falling would penetrate the lower jet, tending to spread or disperse it. This can be seen in Figure 152(a). As the discharge increased, the size of the fins relative to the thickness of the lower sheet became insignificant and no longer had this spreading effect. The differences in the Glen Canyon and Flaming Gorge jets might be explained by the fact that the flow depth for maximum discharge was approximately 61 percent of the diameter of the Glen Canyon tunnel and 81 percent of the diameter of the Flaming Gorge tunnel. For a flow 0.61D in Flaming Gorge, the jet was still well dispersed.

Both the Flaming Gorge and Glen Canyon buckets were modified by reducing the height of the river sidewall. The Flaming Gorge bucket is located well above the maximum tail water elevation so that the wall could be cut down to the spring line of the tunnel invert curve without tail water interference. The effect was to eliminate the fin that formerly rose along the wall. The jet spread out evenly to the right and was better dispersed than before. The Glen Canyon buckets are located closer to the maximum tail water elevation, and in order to prevent the tail water from interfering with the jet, the river wall could be cut down to only 5 feet above the spring line of the tunnel invert. Sufficient wall remained to train the jet and little difference in the flow pattern could be detected.

The flip bucket used on the Whiskeytown Dam tunnel spillway differs from both the flat bottom and transition flip buckets. Instead of changing the bucket invert to a flat bottom, the semicircular invert is curved upward radially, forming in effect, a turned-up tube or elbow, Figure 153. The sidewalls above the spring line of the invert

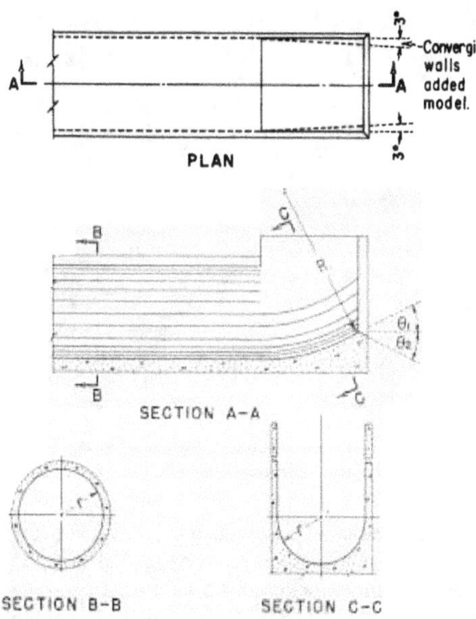

FIGURE 153. *Tube elbow flip bucket.*

are vertical. In the Whiskeytown bucket, 3° wedges 25 feet long were placed along both sidewalls to converge the flow lines and to reduce the spreading of the jet, Figure 154.

Hydraulic model studies determined that the jet from a transition-type bucket did not "fit" the downstream river channel because of excessive jet divergence. The tube-elbow type of flip bucket was developed specifically to provide a narrow jet to conform to the topographic features in the discharge channel, Figures 155 and 156.

Design Considerations

Tunnel spillways usually make use of part of the river diversion tunnel. The downstream portion of the diversion tunnel becomes the horizontal portion of the spillway tunnel. The bucket is added after diversion needs have been satisfied. Because the diversion tunnel is one of the first items of construction, and because of time limitations and construction schedules, it is often necessary to determine line and grade for the diversion tunnels before the details of the spillway are known. Care in selecting the exact position and elevation of the diversion tunnel, while keeping in mind its ultimate use as a spillway tunnel, will help to provide a dual-purpose tunnel that will satisfy the temporary as well as the final demands with the least amount of modification when the bucket is added.

Items that should be considered during design and that will help to provide a simple bucket structure having desirable performance characteristics will now be examined.

Elevation of bucket invert. It is desirable to construct the bucket and tunnel inverts at the same elevation. Because diversion requirements make it necessary to keep the diversion tunnel low in order to provide the diversion capacity, the greatest danger is that the tunnel will be set too low for ideal spillway operation. This will require building up the bucket lip to prevent the tail water from submerging the bucket. As a general rule, maximum tail water should be no higher than the elevation of the center line of the tunnel. If the bucket is set lower, difficulty may be experienced in obtaining free flow at low spillway discharges. The shape of the tail water curve will determine the exact requirements. The drawdown in tail water elevation at the bucket

FIGURE 154.—*Tube-elbow-type flip bucket used on Whiskeytown Dam spillway tunnel has 3° converging walls to limit spreading of jet.*

caused by the ejector action of the jet may also affect the vertical placement.

Flow direction. The bucket center line should be a continuation of the tunnel center line, and the portion of the diversion tunnel used for the spillway tunnel should be straight. Therefore, the objective is to aim the diversion tunnel so that it may be used without change for the spillway tunnel. The tunnel direction should be set so that spillway flows will be aimed downriver and so that the design discharge impinges in the center of the discharge channel. The flow should be directed to minimize the diameter of induced eddies at the sides of the jet because these can be damaging to channel banks. In an ideal arrangement, the jet will be exactly as wide as the channel so that there will be little return flow from the downstream tail water.

Figure 157 shows the angle of divergence of one side of the jet leaving the bucket for two types of buckets, the flat-bottom type and the transition bucket used on Glen Canyon and Flaming Gorge spillways. In both cases, the angle of divergence is plotted against the angle of inclination θ for a range of Froude numbers (of the flow entering the bucket) The flat-bottom bucket produced little change in angle of divergence for a range of Froude numbers or inclination angles. The transition bucket showed considerable change in divergence angle (from 4° to 12°) for a Froude number range of 6 to 11. Because the higher Froude numbers occur at low discharges, the transition-bucket jet divergence is greatest at low flows. As the discharge increases, the Froude number becomes smaller and the divergence angle decreases. In most designs this is a favorable characteristic and results in improved river flow conditions for all discharges.

FIGURE 155.—*Tube elbow bucket produces a narrow jet for the narrow channel below Whiskeytown Dam. Discharge 28,650 c.f.s. maximum.*

FIGURE 156.—*Tube elbow bucket produces clear-cut stable jet with little spray. Discharge 28,650 c.f.s. maximum.*

Drawdown. For the conditions previously described, the jet will act as an ejector to lower the tail water upstream from the jet impingement area. From the Hungry Horse Dam model tests, 26 feet of drawdown was predicted for 35,000 c.f.s. discharge, and it was recommended that a weir be constructed in the powerplant tailrace to prevent unwatering of the turbines. Prototype tests made for 30,000 c.f.s. showed 25 feet of drawdown and demonstrated that the weir was indeed necessary. At Hungry Horse the flow leaves the bucket at a 15° angle, making the trajectory relatively flat, Figure 158. The jet is as wide as the downstream channel. The drawdown is maximum under these conditions. At Glen Canyon the spillway jets do not occupy the entire width of channel, but the jet trajectory is steeper, and the discharge is considerably greater. Hydraulic model tests have indicated that up to 25 feet of drawdown may be expected.

Other hydraulic model bucket tests have shown the drawdown to be appreciable, particularly when the jet occupies a large proportion of the channel width. No means have been found to compute the amount of drawdown to be expected except by making careful measurements on a hydraulic model. However, by using measurements ob-

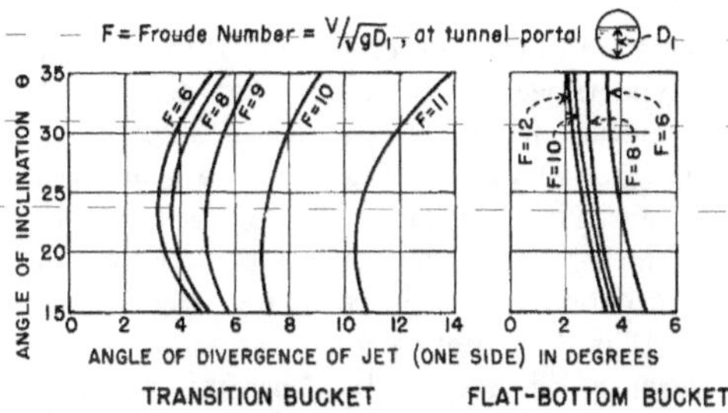

FIGURE 157.—*Spreading of jet.*

tained on several model studies and from limited prototype observations, the curve in Figure 159 was derived. It is presented herein as a means of estimating the drawdown that can be expected with a tunnel spillway and flip bucket.

The intensity of the ejector action and the resultant lowering of the tail water at the bucket have been found to be a function of the energy in the jet and the amount of resistance encountered when the jet strikes the tail water. In the curve of Figure 159, the abscissa is the cross-sectional area of the river flow near the point of impact of the jet divided by the cross-sectional area of the flow at the tunnel portal. The river flow area is the product of the difference between the no-flow tail water elevation and the tail water elevation, for the discharge being investigated, and the average width of the river near the point of impact. The area of the flow at the portal is obtained by dividing the spillway-discharge quantity by the average velocity. The ordinate is the ratio of the amount of drawdown to the depth of tail water. The depth of tail water is the same depth used to determine the river cross-sectional area.

(a) Prototype: Q=30,000 c.f.s.

(b) Model: Q=35,000 c.f.s.

FIGURE 158.—*Model-prototype comparison, Hungry Horse spillway flip buckets.*

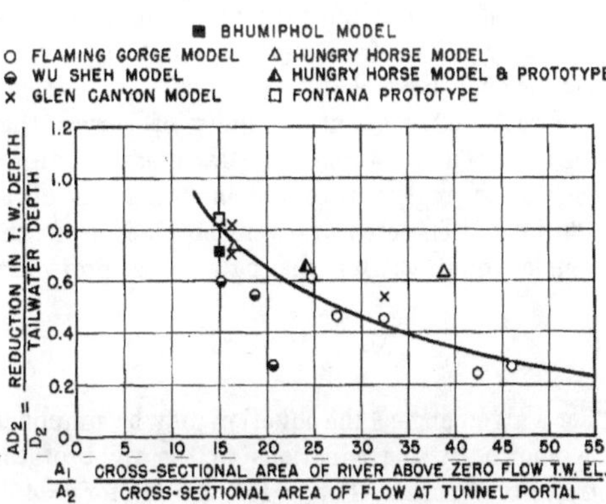

FIGURE 159.—*Tail water drawdown.*

The curve, defined by the test points shown, indicates with reasonable accuracy the drawdown at each dam site for which data were available. The test points include various shapes and depths of channel and various types of bucket jets. Furthermore, the two prototype tests—on Fontana and Hungry Horse Dams—showed good agreement between model and prototype test results. However, in predicting drawdown at future sites the curve should be used with caution until more data are available.

Effect of trajectory shape. In addition to the effects of drawdown that were previously explained, the jet trajectory is important in other ways. The angle of the bucket lip with respect to horizontal determines the distance the water will be thrown downstream. However, the steeper the angle, the more the jet will be broken up and slowed down by air resistance. Both of these effects cause the jet to enter the tail water at a steeper angle. With a steep entry, the vertical component of velocity will be greater, and the jet will tend to dig into the channel bottom. With flatter trajectories, the horizontal component will be greater, and the forward velocity will be higher. High-velocity channel flow may persist downstream from the impingement area for a considerable distance if the channel bottom does not erode to produce a deep pool. High-velocity flow along the channel banks will then occur. If the bottom erodes, an energy dissipating pool will be formed, and flow downstream will be smoother. Bucket flip angles are usually constructed from between 15° and 35°. Angles less than 15° do not give enough lift to clear the bucket structure, and little is usually gained, from any viewpoint, by increasing the angle beyond 35°.

Figure 160 contains a family of curves that may be used to estimate the trajectory length for inclination angles up to 45° and velocities up to 160 f.p.s. These curves were obtained from the simple equations for the path of a projectile,

$$X=\frac{V^2}{g}\sin 2\theta$$

For a given angle θ the equation may be simplified as shown by the equations to the right of the trajectory curves. For $\theta=15°$, $X=H'$; for $\theta=45°$, $X=2H'$; and so on, in which H' is the velocity head at the bucket entrance. To estimate H', the curve in the lower right of Figure 160 may be used. Here, H', expressed as a percentage of the total head, H, is plotted against the percentage of maximum tunnel discharge. The term H' is seen to vary from approximately 61 percent for 20 percent of maximum discharge, to about 75 percent for maximum discharge. Maximum discharge is considered to occur when the tunnel is about three-fourths full at the outlet portal. The points that determine the curve have ratios of vertical drop to horizontal tunnel length, H/L, from 0.15 to 1.9.

Trajectory lengths taken from these curves have been found to be reasonably accurate when checked by hydraulic models. Some difference between model- and prototype-trajectory lengths may be expected to occur, however. Little is known regarding model- and prototype-trajectory-length agreement, but from measurements estimated or scaled from photographs, and from actual measurements,[1] it appears that the differences are usually not critical in nature. The prototype trajectory is shorter than the model or theoretical jet and has a steeper angle of entry into the tail water. The difference is believed to be caused by the greater air resistance encountered by the high-velocity prototype jet. From sketchy information on a few structures, the trajectory length in the prototype for 20 percent of maximum discharge is believed to be 15 percent to 20 percent shorter than in the model, an assumption inferred from comparing the photographs in Figure 161. There also are indications that the difference becomes less as the prototype discharge increases.

A bucket radius at least four times as great as the maximum depth of flow is needed. This provides an incline sufficiently long to turn most of the water before it leaves the bucket and provides assurance that the jet will be thrown into the desired area downstream.

Pressures in the transition bucket. Because of the simplicity and effectiveness of the transition bucket, it will probably be used on many future tunnel spillways. Extensive pressure measurements were therefore made on several buckets having two different inclination angles, 15° and 35°, to indicate that the buckets were safe against cavitation pressures and to provide data for structural design. The results of these tests have been summarized in Figures 162, 163, and 164, and may be helpful in making preliminary designs.

Figure 162 shows pressures along the center line of the transition-bucket floor. The envelope curve includes inclination angles from 15° to 35° and flows in the Froude number range of 6.8 to 10.3, the usual range of operating conditions. The maximum pressure was found to be slightly greater than given by D. B. Gumensky[2] from theoretical considerations. The theoretical pressure P_t, is expressed as

$$P_t=(1.94\ w^2\ R+62.5)\ D_I$$

in which

$$w=\frac{V}{R}$$

This maximum pressure occurred approximately 0.6 of the bucket length from the upstream end. Pressures rapidly became less toward the downstream end of the bucket and reached atmospheric at the bucket lip.

[1] Peterka, A. J., "Model and Prototype Studies on Unique Spillway: Part III of Symposium on Fontana Dam Spillway," *Civil Engineering*, Vol. 16, No. 6, June 1946.

[2] Gumensky, D. B., "Design of Sidewalls in Chutes and Spillways," *Transactions*, ASCE, Vol. 119, 1954.

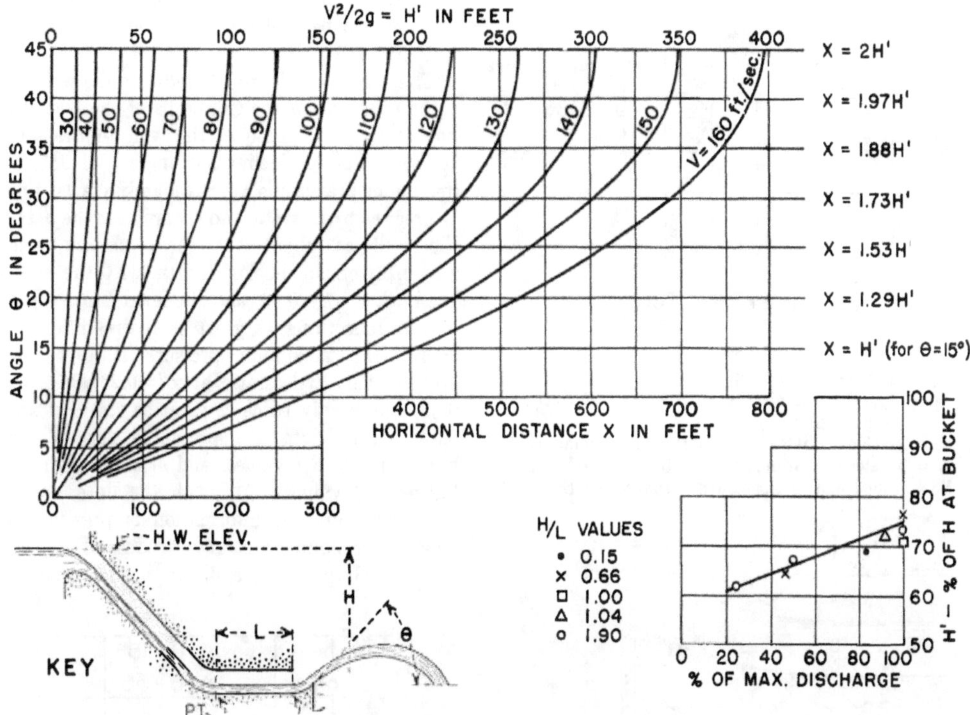

FIGURE 160. *Trajectory lengths and head loss.*

(a) Both prototype tunnels: Q=10,000 c.f.s. each.

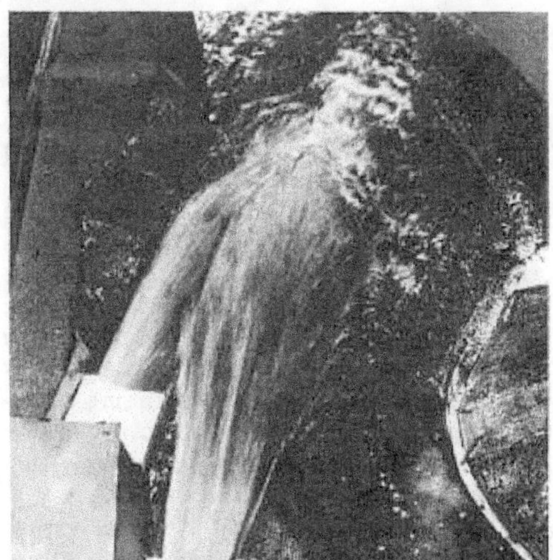

(b) Both model tunnels: Q=12,500 c.f.s. each.

FIGURE 161 *Model-prototype comparison, Fontana Dam spillway flip buckets.*

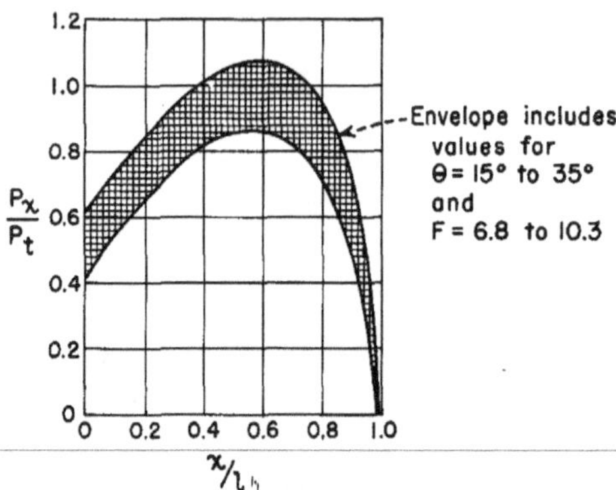

P_x = Measured pressure
P_t = Theoretical pressure; $(1.94\omega^2 R + 62.5) D_1$; where $\omega = V/R$
x = Developed distance from P.C. to piezometer
l = Developed distance from P.C. to end of bucket
F = Froude number, computed from V and D_1 at P.C.

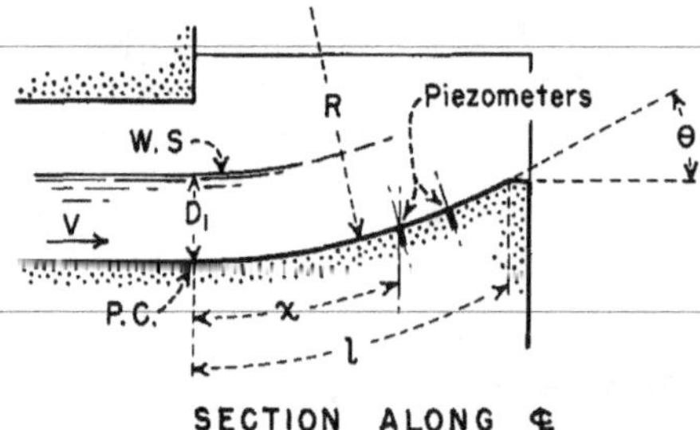

FIGURE 162.—*Pressures on transition bucket floor.*

For some tests a piezometer placed just upstream from the bucket lip, Figure 163, indicated pressures below atmospheric, a phenomenon which has not been satisfactorily explained. Experiments on model buckets showed that the pressure on this piezometer was affected by the shape or angle of the downstream portion of the bucket lip. The curve of Figure 163 shows the relation between pressure and the angle β. The curve indicates that for a given angle of inclination θ, β should be 35° or more to insure atmospheric pressures or above at the lip piezometer. The curve also indicates that if β is 0° the pressure will be atmospheric. This is not a practical solution, however, since if β=0° the piezometer will then be upstream from the lip and a new problem will be created at the end of the extended bucket. It should be noted that the bucket side walls extend beyond the lip piezometer as shown in Figure 163.

The curves of Figure 164 indicate the pressures to be expected on the side walls of the transition bucket from the base of the wall to the water surface. For an inclination angle θ of 35°, the maximum pressure is approximately eleven times as great as hydrostatic and occurs near the base of the wall at about the three-quarters point, x/l=0.75, of the bucket length. At the end of the bucket, x/l=0.99, and the maximum pressure is only four times as great as hydrostatic. For θ=15° the maximum pressure is four times greater than hydrostatic at x/l=0.26, 0.55, and 0.80, and is only twice as great as static at x/l= 0.99. Other data are presented for different bucket radii, R/l values, and stations along the bucket, x/l values. Although the data are not complete, sufficient information is presented to make a preliminary structural design. On the Flaming Gorge Dam spillway bucket, one side

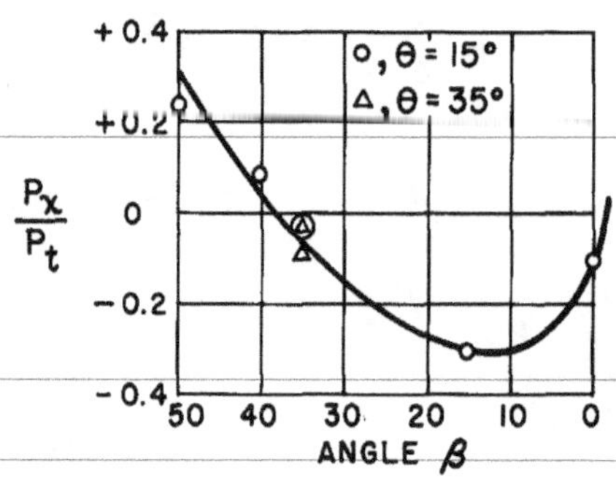

P_x = Measured pressure at end of bucket
P_t = Theoretical pressure (See figure 158)

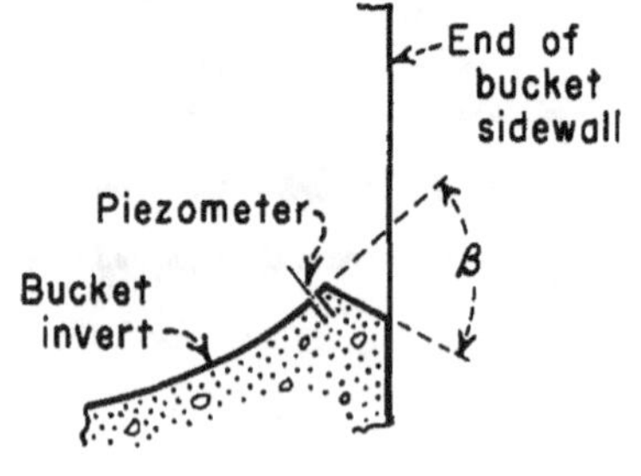

FIGURE 163.—*Pressures at end of bucket.*

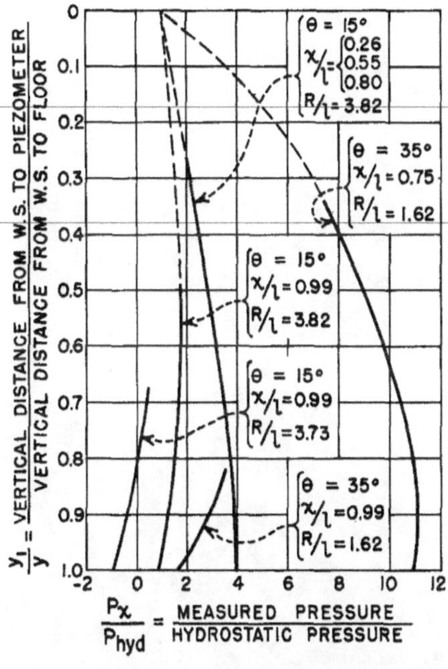

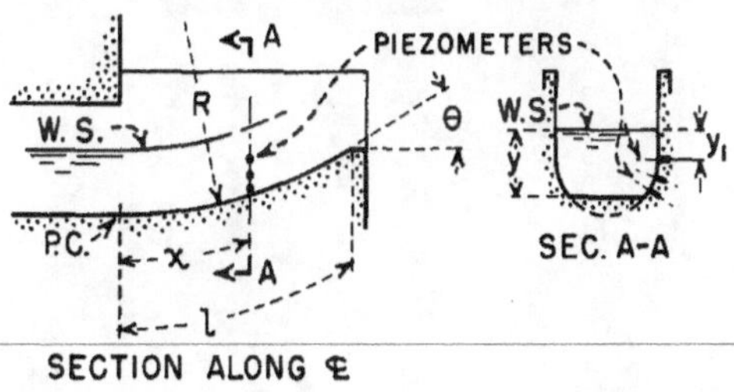

FIGURE 164.—*Pressures on sidewall of transition bucket.*

wall was cut down to the spring line of the tunnel without objectionable spreading of the jet occurring when the flow depth exceeded the height of the wall. This procedure simplified the structural design of the bucket by reducing the overall load on a wall which had no rock behind it.

Conclusions

1. Flip-bucket designs need not be as complicated in concept as some which have been used in the past. Simplified buckets formed by plane or simple curved surfaces can be made to be as effective as those using warped or compound surfaces.

2. A simplified bucket, geometrically formed by three planes, was developed to reduce the possibility of low flows dribbling over the lip; to flip larger flows into the river channel in an unsymmetrical pattern; and to be self-draining after cessation of spillway discharges.

3. The "transition bucket," formed geometrically by the intersection of two cylinders and developed for use on a circular tunnel spillway, eliminated the need for a transition to change the invert to a rectangular cross section. Hydraulic model studies on a group of these buckets provided information to generalize the design of this type of bucket both hydraulically and structurally. Available data will allow the designer to establish the following:

a. The spreading angle of the jet, which is greatest at low flows and decreases as the discharge increases.

b. The jet trajectory geometry.

c. The dynamic pressures on the sidewalls and floor of the bucket.

d. The amount of tail water drawdown to be expected. These data are important in determining the proper vertical placement of the bucket structure.

4. In the present state of knowledge, new "transition bucket" designs will still require hydraulic model testing if it is thought necessary to protect the downstream channel banks against damage from high-velocity flows. More tests and prototype observations are needed to establish confidence in the performance of buckets used in critical locations.

Section 11

Size of riprap to be used downstream from stilling basins*

PREVENTING bank damage caused by surges from a stilling basin and forestalling possible undermining of the structure caused by erosive currents passing over the end sill usually requires placing riprap on the channel bottom and banks downstream. Many factors affect the stone size required to resist the forces tending to move the riprap. In terms of the riprap itself, these include the size or weight of the individual stones, the shape of the large stones, the gradation of the entire mass of riprap, the thickness of the layer, the type of filter or bedding material placed beneath the riprap, the slope of the riprap layer, and perhaps others. In terms of the flow leaving the basin, the factors known to affect riprap stability are velocity, current direction, eddy action and waves.

Extensive tests, both in the laboratory and in the field, would be required to evaluate all of these factors, and only a few of the investigations needed to evaluate riprap stability characteristics have been accomplished. However, using the means at hand, including laboratory tests, prototype observations, photographs of riprap failures combined with known facts which produced the failure, and theoretical considerations, the designer or investigator may predict within practical limits the stone sizes required for the protection of channel bottoms and banks for the usual conditions which prevail downstream from a stilling basin. In order for the predictions to be valid, the investigator can only deal with a channel relatively free of large waves and surges and having a reasonably well distributed velocity pattern.

Experience has shown that the primary reason for riprap failure is undersized individual stones in the maximum size range. Failure has occurred because of (1) underestimation of the required stone size, and (2) a general tendency for the riprap in place to be smaller than specified, despite attempts at rigid control procedures.

When the investigation described in this section was started, the purpose was to determine the

*Includes prototype tests on Basin VI.

individual stone size necessary to resist a range of velocities which usually exist in open channels downstream from stilling structures. Using published material, a tentative curve was selected and the lower end was modified from laboratory observations and field experience, Figure 165. The curve was then used with caution to predict the size of the large stones required in laboratory and field riprap installations. The gradation of the smaller stones in the riprap layer was based on judgment and experience, but no exact method of specifying the gradation was agreed upon. Field tests on riprap installations that had been based on the data from Figure 165 showed the riprap to be stable and satisfactory. Thus, it was established that a well-graded riprap layer containing about 40 percent of the rock pieces smaller than the required size was as stable, or more stable, than a single stone of the required size. This may be due to compensating factors provided by the interlocking of the stones and the boundary layer velocity reduction produced by the rough riprap surface in contact with the flow. No attempt was made to specify stone shapes except to say that they should not be "flats."

The curve in Figure 165 gives the individual minimum stone size (diameter and weight of a spherical specimen) for a range of bottom velocities up to 17 feet per second. The points shown adjacent to the curve indicate riprap failures, "F," and successful, "S," installations observed in the field. Thus, six points indicate that when the maximum stone size was less than the curve value, the riprap failed; five points indicate that when the maximum stone size was equal to or greater than the curve value, the riprap remained stable. The observed field data, although sometimes sketchy and incomplete, tended to confirm the derived curve of Figure 165 and helped to provide a basis for the selection of the maximum stone size in a graded riprap mixture. Lacking more specific information, it may only be stated that most of the mixture should consist of stones having length, width, and thickness dimensions as nearly equal as practical, and of curve size or larger; or the stones should be of curve weight or more (weight is computed on the basis of 165 pounds per cubic foot) and should not be flat slabs.

General field experience has shown that the riprap layer should be 1½ times, or more, as thick as the dimension of the large stones (curve size) and should be placed over a gravel or reverse filter layer.

Prototype and model tests on Stilling Basin VI are described and indicate how prototype observations were used to help establish the validity of the stone-size curve. The tests also showed that the field performance of Stilling Basin VI agrees with the predictions made during the basin development tests. Other prototype observations are cited and used to help confirm the stone-size curve.

Stone-size determination. The lower portion of the curve of Figure 165 is an average of data reported by Du Buat in 1786, Bouniceau in 1845, Blackwell in 1857, Sainjon in 1871, Suchier in 1874, and Gilbert in 1914. It checks well with results of tests made at the State University of Iowa by Chitty Ho, Yun-Cheng Tu, Te Yun Liu, and Edward Soucek. The data were assembled and discussed in a paper, "A Reappraisal of the Beginnings of Bed Movement-Competent Velocity" by F. T. Mavis and L. M. Laushey, for the International Association for Hydraulic Structures Research, 1948, Stockholm, Sweden. In a thesis by N. K. Berry (*35*), University of Colorado, 1948, a similar curve was determined and an equation for it presented:

$$V_b=2.57\sqrt{d}$$

where

V_b=bottom velocity in channel in feet per second
d=diameter of particle in inches (specific gravity 2.65)

Mavis and Laushey (*34*) proposed an equation for use with particles of any specific gravity:

$$V_b=1/2\sqrt{d_1}\sqrt{s-1}$$

where

s=specific gravity of the particle
d=diameter of particle in millimeters

Tests made in the Bureau of Reclamation hydraulic laboratory on sands, gravels, and selected stone sizes up to 2½ inches in maximum dimension indicate the lower middle portion of the curve to be essentially correct. Field observations of riprap up to 18-inch size also indicated the curve to be accurate within wide practical limits. Ration-

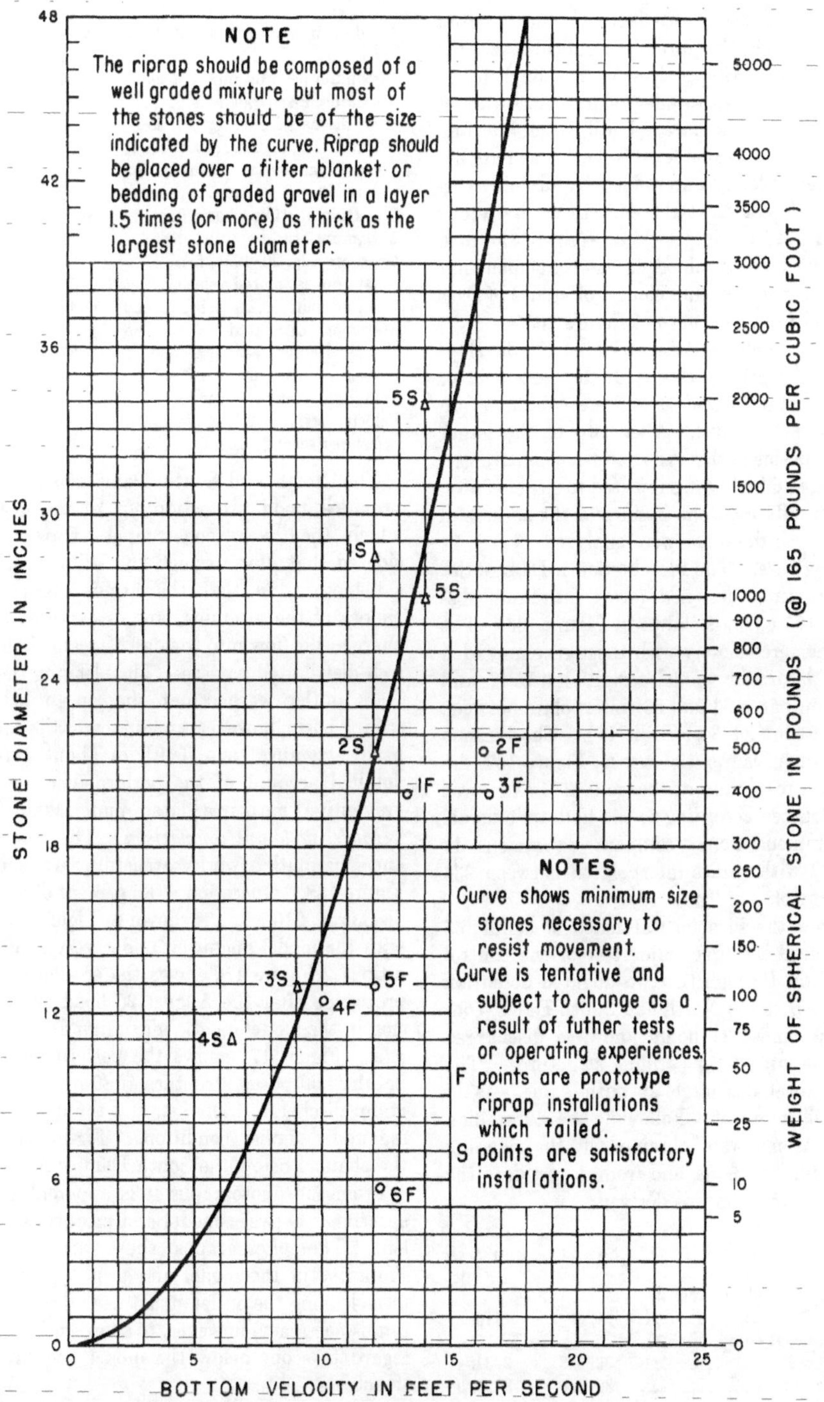

FIGURE 165.—*Curve to determine maximum stone size in riprap mixture.*

alization of all the known factors indicated that the curve might be directly applicable to the determination of riprap sizes, even though many of the factors known to affect riprap sizes are not accounted for in the data. Until more data and experience were available, it was decided to use the average velocity determined by dividing discharge by flow area at the end sill to find stone sizes; and until the interlocking effect of the rock pieces could be determined, it was concluded that most of the riprap should consist of stones of sizes determined from the curve of Figure 165.

Model and prototype tests. During the development tests for Stilling Basin VI, the placement of riprap was studied and tests were conducted on several sizes of gravel. After two of the larger prototype basins of this type had been constructed in the field and had been subjected to sizable flows, the riprap failure on one basin and the successful installation on the other were analyzed.

Prototype tests. The Picacho South Dam outlet works structure, designed for a maximum discharge of 165 c.f.s., is shown in Figure 166. The dimensions agree closely with those recommended from the hydraulic model tests, Figure 42 and Table 11 on pages 83 and 86 (interpolate between 151 and 191 in Col. 3 of Table 11). The Picacho North Branch Dam outlet works, Figure 167, designed for a maximum discharge of 275 c.f.s., was also constructed according to the hydraulic model test recommendations (compare dimensions in Figure 167 with values interpolated between 236 and 339 in Col. 3 of Table 11).

Rain over the Picacho watershed of about 0.5 inch produced the first major test of the control works. Flow through the two ungated detention dams, known as the North and South Dams, continued for almost 24 hours and was discharged through the impact-type stilling structures. The combined total discharge at both dams was in excess of 400 acre-feet. Following the storm, high water elevations were obtained in the ponding basins behind the dams, and from design data the following information was derived:

	North Dam	*South Dam*
Maximum water surface elevations	3938. 0	3941. 0
Acre-feet impounded on Aug. 20, 1954	125	110
High water elevation on Aug. 20, 1954	3920. 3	3931. 4
Intake elevation	3911. 0	3921. 0
Head on intake on Aug. 20, 1954, feet	9. 3	10. 4
Maximum head on intake, feet	27. 0	20. 0
Maximum discharge, c.f.s	275	165
High discharge on Aug. 20, 1954, c.f.s	210	130
Percent maximum discharge on Aug. 20	80	80
Elevation of stilling basin floor	3895. 71	3912. 92
Maximum head on outlet, feet	42. 29	28. 08
Head on Aug. 20, 1954, feet	24. 59	18. 48
Maximum estimated velocity, feet per second	48. 0	39. 0
Maximum estimated velocity on Aug. 20, 1954, feet per second	37. 0	31. 8
Critical velocity over end sill on Aug. 20, feet per second	7. 6	6. 9
Velocity striking riprap Aug. 20, feet per second	12±	5±

The North and South Dams control facilities provided flood protection up to the degree for which they were constructed. Observers considered that flow through the stilling basins was satisfactory, in that the basins dissipated the energy of the incoming flow as expected and discharged the flow into the downstream channel in a well-distributed pattern. Flow leaving the North Dam outlet washed out the riprap below the stilling basin, however, and undercut the end of the basin structure to a depth of about 2 feet. A detailed account of the performance and scour-preventive measures later undertaken follows.

Model-prototype comparison. The North Dam and the outlet works structure are shown in Figure 168. Operation at 80 percent of maximum discharge, 210 c.f.s., is shown in Figure 169, along with the model operating under very similar conditions. Figure 170 shows the erosion below the prototype after the August 20 flood and the erosion in the model for the maximum discharge, 130 c.f.s. Figure 171 shows the performance of the South Dam outlet structure at 80 percent of maximum discharge, 130 c.f.s., and the model operating under similar conditions. Figure 172 shows the channel below the South Dam outlet.

From the photographs it is apparent that the agreement between model and prototype is excellent. The photographs show the remarkable similarity in the model and prototype flow patterns leaving the outlet structures. Closer inspection is necessary, however, to show similarity with regard to scour below the model and prototype structures.

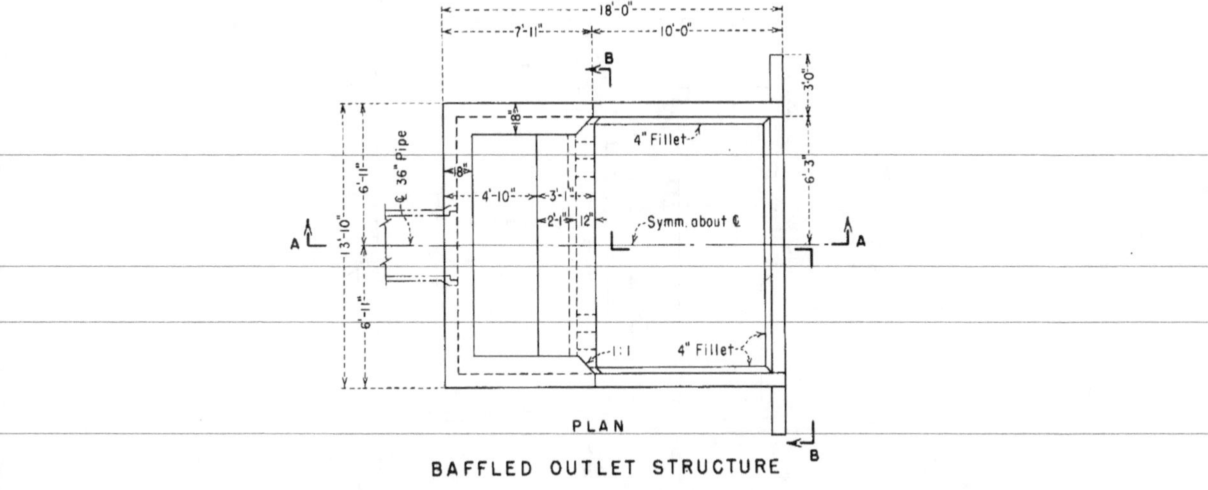

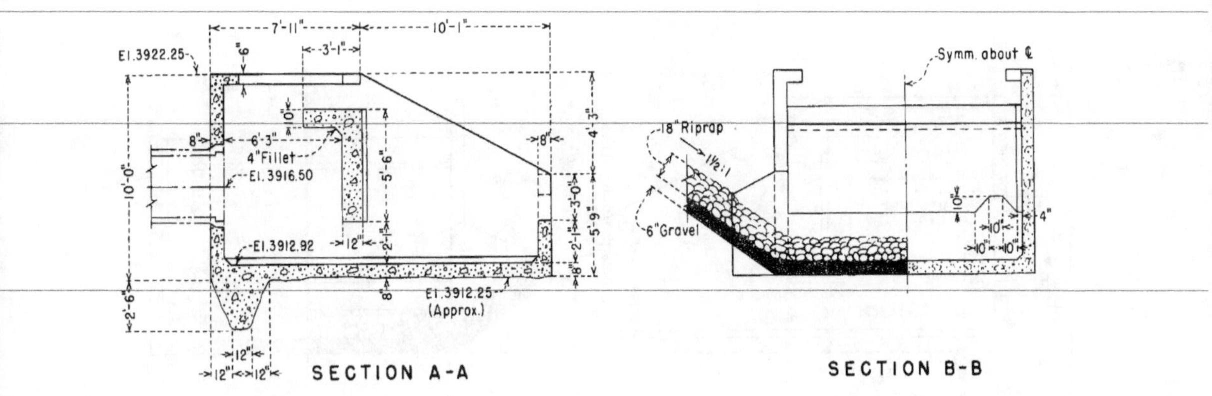

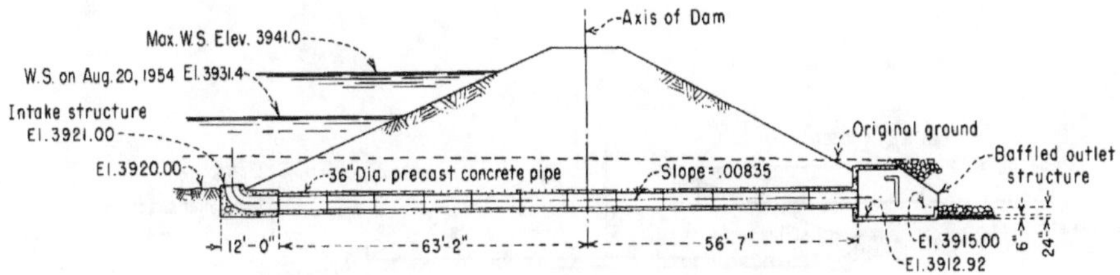

FIGURE 166.—*Outlet works of Picacho South Dam, Las Cruces Division, Rio Grande project.*

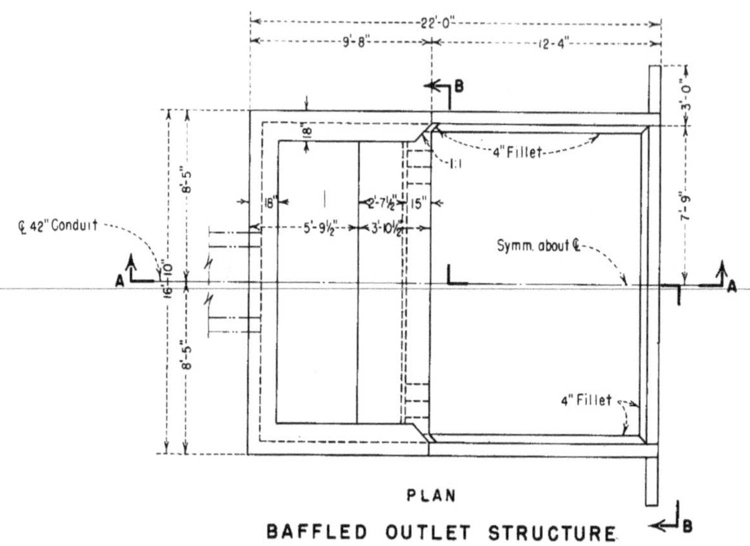

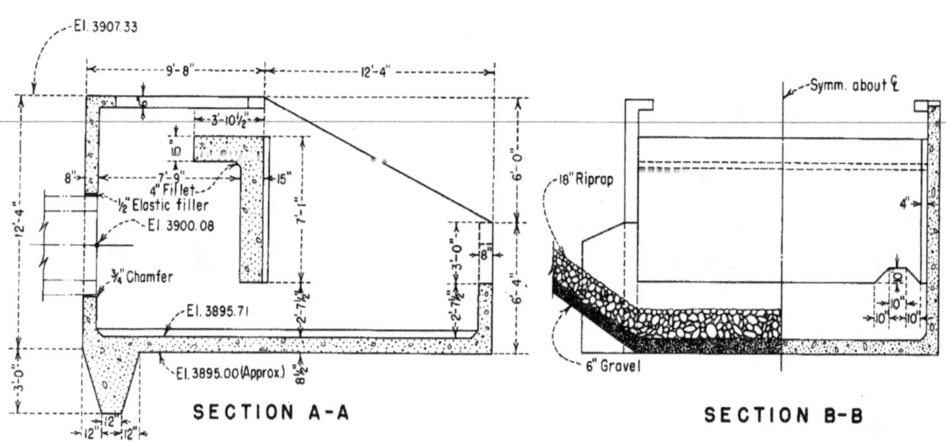

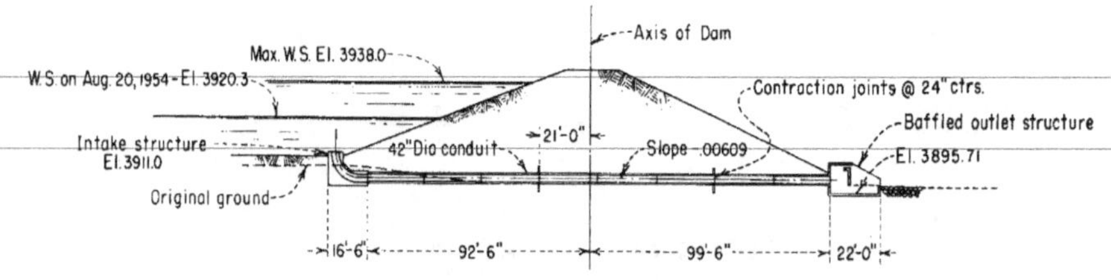

FIGURE 167.—*Outlet works of Picacho North Dam, Las Cruces Division, Rio Grande project.*

FIGURE 168.—*Impact-type stilling basin structure, Picacho North Dam, following flood of August 20, 1954.*

Picacho North Dam outlet works structure discharging 210 c.f.s. (80 percent of maximum capacity).

Hydraulic model discharging maximum capacity under similar conditions of head and tailwater.

FIGURE 169.—*Model-prototype comparison, Picacho North Dam.*

The photograph of the model in Figure 170 shows the extent and depth of scour when there was no riprap protection provided in the channel. The pea-gravel used in the model was considered to be an erodible bed. The contours, visible as white lines, show that the erosion depth was 19/26 of the sill height below apron elevation. Since the prototype sill height is 31.5 inches, scour depth in the prototype without riprap protection should be about 23 inches below apron elevation. This compares very favorably with the 2 feet measured in the prototype. The more general erosion which occurred in the prototype is probably attributable to the higher velocity entering the prototype stilling basin. The estimated velocity (based on calculations) of 37 feet per second is greater than the upper velocity limit, 30 feet per second, used in the model tests and recommended for the upper limit in prototype structures of this type. Larger riprap would have prevented the erosion.

According to construction specifications, the riprap below the outlet was to ". . . consist of durable rock fragments reasonably graded in size . . ." from 1/8 cubic yard to 1/10 cubic foot. The individual rocks, therefore, would vary from about 18- to 5½-inch cubes, or in weight from about 500 to 15 pounds. Although it is impossible from the photograph of the prototype in Figure 170 to determine the size of riprap in the channel at the start of the run, the bank riprap indicates that there were very few rockpieces of the 500-pound size. The few remaining pieces near the man at the right seem to be in the upper size range and apparently these did not move. In the hydraulic model test made to develop this basin, shown in Figure 173, riprap corresponding to 9- to 18-inch stones did not show excessive movement of the rock mass, but did show some erosion downstream from the end sill.

To further analyze the stone size necessary to withstand the erosive forces, the curve of Figure 165 is used. Using the curve for the case at hand, the critical stone size is about 20 inches. This checks the equivalent 9- to 18-inch stone size used in the model tests to a reasonable degree, since some of the model riprap did move.

It appears that 18- to 20-inch minimum stones would have been required to prevent movement of the riprap below the North Dam outlet. To withstand the maximum velocity to be expected when the structure is subjected to full head and

Scour below Picacho North Dam outlet works following flood of August 20, 1954. Evidence points to undersized riprap.

Hydraulic model indicates erosion similar to prototype when riprap size is inadequate.

FIGURE 170.—*Model-prototype comparison, Picacho North Dam.*

discharge conditions, larger stones would be required, perhaps 24-inch minimum.

In contrast to the riprap failure at the North Dam the riprap at the outlet of the South Dam was relatively undisturbed, Figure 172. No damage was found after inspection of the dry channel. Flow conditions below the South Dam outlet are shown in Figure 171.

The velocity over the end sill of the South Dam Basin is much lower than at the North Dam, being only about 5 feet per second. According to the curve of Figure 165, the stone size required would be about 4 inches. Since the riprap sizes given in specifications for the South Dam outlet are the same as for the North Dam outlet (5 to 18 inches) the stones were sufficiently large to resist movement.

It should be noted in Figure 169 that the tail water is low with respect to the elevation of the stilling basin. Therefore, the velocity over the end sill is considerably lower than the velocity striking the riprap. For the North Dam basin which has a sill length of 15.5 feet, a critical depth of 1.5 feet over the end sill and a discharge of 210 c.f.s., the critical velocity would be 7.6 feet per second. According to the curve in Figure 165, riprap about 9 inches in diameter is required. Further acceleration of the flow by a drop over the

South Dam outlet works structure discharging 130 c.f.s. (80 percent of maximum).

Flow appearance in model for the same conditions. Note similarity both upstream and downstream from vertical baffle.

FIGURE 171.—*Model-prototype comparison, Picacho South Dam.*

FIGURE 172.—*Flow conditions downstream from Picacho South Dam outlet works are entirely satisfactory. There was no disturbance or loss of riprap under a discharge of 130 c.f.s.*

end sill to the tail water surface of 2 feet, Figure 169, would result in a velocity of about 14 feet per second, requiring riprap about 30 inches in diameter. Thus, the importance of matching the basin elevation to the probable tail water elevation is evident. In the case of the North Dam basin, however, the tail water was, no doubt, higher before the riprap was lost.

The analysis indicates that, according to the curve of Figure 165, the riprap below the North Dam outlet would be expected to fail and did; at the South Dam outlet the riprap would be expected to remain in place and did. The curve of Figure 165, therefore, appears to have merit in the determination of riprap sizes. Other prototype observations, none as conclusive as the example above, tended to indicate the same degree of confirmation.

Riprap stability observations. Riprap stability observations made in the field by various personnel were analyzed, interpreted and plotted on Figure 165 as "F" (failure) and "S" (satisfactory) points to help in determining the validity of the riprap size curve. Data were not always complete and in some cases rock sizes had to be scaled from photographs taken prior to the riprap failure. Even when ample data were available, it was not always clear where a test point should be plotted since the riprap size analysis was difficult to interpret, rock sizes varied in a single reach of channel, or the riprap had been exposed to a range of velocities with failure occurring at less than maximum velocity. However, since all 11 plotted points indicate that the stone-size curve of Figure 165 is correct, it is relatively unimportant that there is some doubt connected with the plotting of each individual point. Each point on Figure 165 is discussed in terms of the known factors; unknown factors are not mentioned. "F" points indicate failure of the riprap installation to a degree sufficient to require extensive repairs or total loss. "S" points indicate a satisfactory installation that required only routine maintenance or none.

Point 1F

Riprap on banks with variable slope and flat bottom; failure occurred near water line on 1:1 slope. Riprap size analysis: 100 percent finer than

FIGURE 173.—*Hydraulic model tests using 9- to 18-inch diameter (equivalent) stones show some movement of riprap.*

700 pounds, 90 percent finer than 400 pounds, 80 percent finer than 200 pounds, 50 percent finer than 80 pounds, 40 percent finer than 35 pounds, 24 percent finer than 9 pounds. Subjected to velocity of from 10 to 14 feet per second. Failure at 13.3 feet per second, based on discharge divided by flow area.

Point 2F

Riprap on banks and bottom; failure was general on banks and bottom—it is not known which occurred first. Riprap size analysis: 100 percent finer than 500 pounds, 90 percent finer than 350 pounds, 80 percent finer than 260 pounds, 50 percent finer than 80 pounds, 40 percent finer than 45 pounds, 20 percent finer than 10 pounds. Subjected to velocities up to 23 feet per second. Failure occurred after sustained average velocity of 16.3 feet per second.

Point 3F

Riprap failure on banks of channel, slope 1:1 approximately and flatter. Riprap size 90 percent finer than 410 pounds. Failure was rapid at a velocity of 16.5 feet per second.

Point 4F

Riprap failure on flat sloping banks. Riprap was 12-inch-thick layer, 50 percent of rock was 20 to 90 pounds (remainder spalls). Continual maintenance required above 10 feet per second and believed to be marginal at just less than 10 feet per second.

Point 5F

Riprap failure on bottom of channel. Material consisted of 18-inch-thick layer of rocks averaging 100 pounds (few larger pieces). Resisted 12 feet per second for a time but eventually failed.

Point 6F

Riprap failure on steep bank projecting into current. Six-inch cobbles placed about two stones deep. Complete failure at 12 feet per second.

Point 1S

Riprap on 1½:1 banks, 10 feet high, stood up well with only occasional maintenance. Mixture of 1,000- and 2,000-pound stones; smaller pieces often on surface. Flow velocity 14 feet per second in center of channel; velocity near riprap estimated at 12 feet per second. No filter or bedding but riprap was placed on natural sand.

Point 2S

Hand-placed stone about 500 pounds each. Resisted 12 feet per second, required maintenance at greater unknown velocity.

Point 3S

Dumped riprap fill, 100 pounds maximum weight, resisted 6 to 9 feet per second for indefinite period.

Point 4S

Flat paving of 10- to 12-inch stones below a canal headgate resisted a velocity of 6 to 7 feet per second for many years.

Point 5S

This is actually two points, as shown on graph. In a natural mountain stream in flood, 1,000-pound stones moved under 14 feet per second velocity; 2,000-pound stones did not.

Conclusions

The passage of the flood of August 20 through the two outlet works structures of the Picacho Arroyo control indicates that the prototype performance was as predicted by the hydraulic model tests described in Section 6. Despite the fact that the general design rules presented limit the incoming velocity to 30 feet per second, the North and South Dams structures performed very well for velocities computed to be about 37 and 32 feet per second, respectively, with discharges equal to 80 percent of design capacity. The only adverse performance of these structures was the loss of the riprap below the North Dam outlet works. This loss was shown to be consistent with the curve in Figure 165.

The outlet works structures at the North and South Dams appear, offhand, to be of about the same general size, both in physical dimensions and in the quantity of water to be handled. On this basis, riprap sizes for both structures were specified to consist of material from ⅛ cubic yard to ⅒ cubic foot. On the North Dam outlet works this material was entirely removed from the channel bottom by outflow having a velocity of about 12 feet per second. Below the South Dam outlet works the same material remained in place with an outflow velocity of about 5 feet per second, considerably lower than at the North Dam. Therefore, it is evident that the minimum stone sizes are critical with respect to the velocity below the structure. Stone size in a riprap layer used for channel bank or bottom protection is indicated by Figure 165. It is felt that this curve, even though only partially proven by the "F" and "S" points, will provide a starting point for the development of a more accurate method of determining stone sizes and specifying riprap mixtures. The curve indicates over most of its range that doubling the flow velocity leaving a structure makes it necessary to provide riprap about 4 times larger in nominal diameter or 16 times larger in volume or weight. These factors alone provide a basis for thought in specifying riprap material.

Recommendations

The riprap sizes given in Table 11, Column 19, are based on the data and discussions presented here. For other types of stilling basins use the bottom velocity if known, or the average velocity based on discharge divided by cross-sectional area at the end sill of the stilling basin, to find the maximum stone size in Figure 165. Specify riprap so that most of the graded mixture consists of this size. Place the riprap in a layer at least 1½ times as thick as the maximum stone size. It is a fairly well established fact that better performance of the riprap results when it is placed over a filter, or bedding, composed of gravel or graded gravel having the larger particles on the surface.

FIGURE 174.—*Surge-type waves extracted fine earth material from behind coarse riprap, causing entire mass to settle away from top of bank. High water line was below elevation where man stands.*

Figure 174 shows an installation of oversized riprap laid directly on fine soil. The riprap has partially failed because waves removed material from beneath the riprap layer. The top of the riprap was originally at the top of the bank. A filter layer would have prevented settlement.

Following this text are:

1. Bibliography
2. Nomograph
3. Pictorial Summary

Works listed in the Bibliography supplied both source and reference material for this monograph, although most of the material contained herein is original in nature.

The Nomograph will be found to be extremely useful in solving hydraulic jump problems, particularly on a first-trial basis. The rate of change of the variables to be seen by manipulating a straightedge can be of definite help to both student and design engineer.

The Pictorial Summary is particularly useful in locating a particular item in the monograph or for suggesting the proper structure for a given set of conditions.

Bibliography

1. Bakhmeteff, B. A. and Matzke, A. E., "The Hydraulic Jump in Terms of Dynamic Similarity," *Transactions* ASCE, Vol. 101, p. 630, 1936.
2. Puls, L. G., "Mechanics of the Hydraulic Jump," Bureau of Reclamation Technical Memorandum No. 623, Denver, Colo., October 1941.
3. Safranez, Kurt, "Untersuchungen uber den Wechselsprung" (Research Relating to the Hydraulic Jump), Bausinginieur, 1929, Heft. 37, 38. Translation by D. P. Barnes, Bureau of Reclamation files, Denver, Colo. Also Civil Engineer, Vol. 4, p. 262, 1934.
4. Woycicki, K., "The Hydraulic Jump and its Top Roll and the Discharge of Sluice Gates," a translation from German by I. B. Hosig, Bureau of Reclamation Technical Memorandum No. 435, Denver, Colo., January 1934.
5. Kindsvater, Carl E., "The Hydraulic Jump in Sloping Channels," *Transactions* ASCE, Vol. 109, p. 1107, 1944, with discussions by G. H. Hickox, et al.
6. Bakhmeteff, B. A. and Matzke, A. E., "The Hydraulic Jump in Sloped Channels," *Transactions* ASME, Vol. 60, p. 111, 1938.
7. Rindlaub, B. D., "The Hydraulic Jump in Sloping Channels," Thesis for Master of Science degree in Civil Engineering, University of California, Berkeley, Calif.
8. Blaisdell, F. W., "The SAF Stilling Basin," U.S. Department of Agriculture, Soil Conservation Service, St. Anthony Falls Hydraulic Laboratory, Minneapolis, Minn., December 1943.
9. Bakhmeteff, B. A., "The Hydraulic Jump and Related Phenomena," *Transactions* ASME, Vol. 54, 1932. Paper APM-54-1.
10. Chertonosov, M. D. (Some Considerations Regarding the Length of the Hydraulic Jump), *Transactions* Scientific Research Institute of Hydrotechnics, Vol. 17, 1935, Leningrad. Translation from Russian in files of University of Minnesota.
11. Einwachter, J., "Wossersprung und Deckwalzenlange" (The Hydraulic Jump and Length of the Surface Roller) Wasserkraft und Wasserwirtschaft, Vol. 30, April 17, 1935.
12. Ellms, R. W., "Computation of Tail-water Depth of the Hydraulic Jump in Sloping Flumes," *Transactions* ASME, Vol. 50, 1928.
13. Ellms, R. W., "Hydraulic Jump in Sloping and Horizontal Flumes," *Transactions* ASME, Vol. 54, Paper Hyd. 54-6, 1932.
14. Hinds, Julian, "The Hydraulic Jump and Critical Depth in the Design of Hydraulic Structures," *Engineering News-Record*, Vol. 85, No. 22, p. 1034, Nov. 25, 1920.
15. King, H. W., "Handbook of Hydraulics," McGraw-Hill Book Co., Second Edition, p. 334, 1929.
16. Blaisdell, F. W., "Development and Hydraulic Design, Saint Anthony Falls Stilling Basin," *Transactions* ASCE, Vol. 113, p. 483, 1948.
17. Lancaster, D. M., "Field Measurements to Evaluate the Characteristics of Flow Down a Spillway Face with Special Reference to Grand Coulee and Shasta Dams," Bureau of Reclamation Hydraulic Laboratory Report Hyd. 368.
18. Rouse, Hunter, "Engineering Hydraulics," John Wiley & Sons, p. 571, 1950.

19. Newman, J. B. and LaBoon, F. A., "Effect of Baffle Piers on the Hydraulic Jump," Master of Science Thesis, Massachusetts Institute of Technology, 1953.
20. Higgins, D. J., "The Direct Measure of Forces on Baffle Piers in the Hydraulic Jump," Master of Science Thesis, Massachusetts Institute of Technology, 1953.
21. Forchheimer, "Ueber den Wechselsprung," (The Hydraulic Jump), Die Wasserkraft, Vol. 20, p. 238, 1925.
22. Kennison, K. R., "The Hydraulic Jump in Open Channel Flow," *Transactions* ASCE, 1916.
23. Kozeny, J., "Der Wassersprung," (The Hydraulic Jump), Die Wasserwirtschaft, Vol. 22, p. 537, 1929.
24. Rehbock, T., "Die Wasserwalze als Regler des Energie—Haushaltes der Wasserlaufe," (The Hydraulic Roller as a Regulator of the Energy Content of a Stream). *Proceedings* 1st International Congress for Applied Mechanics, Delft, 1924.
25. Schoklitsch, A., "Wasserkraft und Wasserwirtschaft," Vol. 21, p. 108, 1926.
26. Woodward, S. M., "Hydraulic Jump and Backwater Curve," *Engineering News-Record*, Vol. 80, p. 574, 1918, Vol. 86, p. 185, 1921.
27. Moore, W. L., "Energy Loss at the Base of a Free Overfall," *Transactions* ASCE, Vol. 109, p. 1343, 1943.
28. "Hydraulic Model Studies, Fontana Project," Technical Monograph 68, Tennessee Valley Authority, p. 99.
29. Rhone, T. J., "Hydraulic Model Studies on the Wave Suppressor Device at Friant-Kern Canal Headworks," Bureau of Reclamation Hydraulic Laboratory Report Hyd. 395.
30. Beichley, G. L., "Hydraulic Model Studies of the Outlet Works at Carter Lake Reservoir Dam No. 1 Joining the St. Vrain Canal," Bureau of Reclamation Hydraulic Laboratory Report Hyd. 394.
31. Peterka, A. J., "Impact Type Energy Dissipators for Flow at Pipe Outlets, Franklin Canal," Bureau of Reclamation Hydraulic Laboratory Report Hyd. 398.
32. Simmons, W. P., "Hydraulic Model Studies of Outlet Works and Wasteway for Lovewell Dam," Bureau of Reclamation Hydraulic Laboratory Report Hyd. 400.
33. Schuster, J. C., "Model Studies of Davis Aqueduct Turnouts 15.4 and 11.7, Weber Basin, Utah," Bureau of Reclamation Hydraulic Laboratory Paper No. 62.
34. Mavis, F. T., and Laushey, L. M., "A Reappraisal of the Beginnings of Bed Movement-Competent Velocity," *Proceedings* of the International Association for Hydraulic Structures Research, 1948, Stockholm, Sweden.
35. Berry, N. K., "The Start of Bed Load Movement," Thesis, University of Colorado, 1948.
36. Rhone, T. J., "Hydraulic Model Studies of the Trinity Dam Spillway Flip Bucket, Central Valley Project, California," Bureau of Reclamation Hydraulic Laboratory Report Hyd. 467, August 10, 1960.
37. Beichley, G. L., and Peterka, A. J., "Progress Report IV—Research Study on Stilling Basin for High Head Outlet Works Utilizing Hollow-Jet Valve Control," Bureau of Reclamation Hydraulic Laboratory Report Hyd. 446.
38. Peterka, A. J., "Progress Report V—Research Study on Stilling Basins, Energy Dissipators, and Associated Appurtenances—Section 9—Baffled Apron on 2:1 Slope for Canal or Spillway Drops," Bureau of Reclamation Hydraulic Laboratory Report Hyd. 445.
39. Beichley, G. L., "Hydraulic Model Studies of Trinity Dam Outlet Works," Bureau of Reclamation Hydraulic Laboratory Report Hyd. 439.
40. Rhone, T. J., "Hydraulic Model Studies of Wu-Sheh Dam Tunnel Spillway," Bureau of Reclamation Hydraulic Laboratory Report Hyd. 430.
41. Beichley, G. L., and Peterka, A. J., "Progress Report III—Section 7—Slotted and Solid Buckets for High, Medium, and Low Dam Spillways," Bureau of Reclamation Hydraulic Laboratory Report Hyd. 415.
42. Peterka, A. J., "Stilling Basin Performance Studies—An Aid in Determining Riprap Sizes," Bureau of Reclamation Hydraulic Laboratory Report Hyd. 409.
43. Bradley, J. N., and Peterka, A. J., "Progress Report—Research Study on Stilling Basins, Energy Dissipators, and Associated Appurtenances," Bureau of Reclamation Hydraulic Laboratory Report Hyd. 399.
44. "Bibliography on the Hydraulic Design of Spillways," *Journal* of the Hydraulics Division, ASCE, Vol. 89, HY 4, Paper No. 3573, 1963.
45. Wagner, W. E., "Hydraulic Model Studies of the Check Intake Structure—Potholes East Canal," Bureau of Reclamation Hydraulic Laboratory Report Hyd. 411.
46. Lancaster, D. M., and Dexter, R. B., "Hydraulic Characteristics of Hollow-Jet Valves," Bureau of Reclamation Hydraulic Laboratory Paper No. 101.
47. Bradley, J. N., and Peterka, A. J., "The Hydraulic Design of Stilling Basins," ASCE Journal of Hydraulics Division, Vol. 83, No. HY 5, October 1957, Papers 1401 through 1406.
48. Rhone, T. J., and Peterka, A. J., "Improved Tunnel Spillway Flip Buckets," *Transactions* ASCE, Vol. 126, Part 1, 1961, p. 1270.
49. Wagner, W. E., "Hydraulic Model Studies of the Outlet Control Structure; Culvert Under Dike; and Wash Overchute at Station 938+00—Wellton-Mohawk Division, Gila Project, Arizona," Bureau of Reclamation Hydraulic Laboratory Report Hyd. 359.
50. Beichley, G. L., "Hydraulic Model Studies of Yellowtail Dam Spillway and Outlet Works (Preliminary Studies)," Bureau of Reclamation Hydraulic Laboratory Report Hyd. 414.
51. Rhone, T. J., "Hydraulic Model Studies of Yellowtail Dam Outlet Works (Final Studies)," Bureau of

Reclamation Hydraulic Laboratory Report Hyd. 482, February 1962.

52. Rhone, T. J., "Hydraulic Model Studies of Baffled Apron Drops—Willard Canal Pumping Plant No. 1, Weber Basin Project, Utah," Bureau of Reclamation Hydraulic Laboratory Report Hyd. 490, July 1962.
53. Beichley, G. L., "Hydraulic Model Studies of Whiskeytown Dam Spillway Diversion Structure and Outlet Works," Bureau of Reclamation Hydraulic Laboratory Report Hyd. 498.
54. Reinhart, A. S., "Hydraulic Model Studies of Falcon Dam—International Boundary and Water Commission Report," Bureau of Reclamation Hydraulic Laboratory Report Hyd. 276, July 1950.
55. Rusho, E. J., "Hydraulic Model Studies of the Outlet Works—Boysen Dam, Missouri River Basin Project," Bureau of Reclamation Hydraulic Laboratory Report Hyd. 283, December 1950.
56. Peterka, A. J., and Tabor, H. W., "Progress in New Designs for Outlet Works Stilling Basins," Bureau of Reclamation Hydraulic Laboratory Report Hyd. 302, December 1950.
57. Rusho, E. J., "Hydraulic Model Studies of the Overflow Spillway and the Hale Ditch Irrigation Outlet—Bonny Dam," Bureau of Reclamation Hydraulic Laboratory Report Hyd. 331, January 1952.
58. Scobey, Fred C., "Notes on the Hydraulic Jump," *Civil Engineering*, August 1939.
59. Posey, C. J., "Why Bridges Fail in Floods," *Civil Engineering*, February 1949.
60. Einstein, H. A., "Bed Load Transportation in Mountain Creek," Soil Conservation Service, U.S. Department of Agriculture, Washington, D.C., August 1944, Vol. SCS–TP–55, 54 pp.
61. Meyer-Peter, E., and Muller, R., "Formulas for Bed Load Transport," Appendix 2. Second Meeting International Association for Hydraulic Research, Stockholm, Sweden, June 1948, pp. 39-64.
62. Johnson, J. W., "Scale Effects in Hydraulic Models Involving Wave Motion," *Transactions* American Geophysical Union, Vol. 30, No. 4, August 1949.
63. Rouse, Hunter and Ince, Simon, "History of Hydraulics," Iowa Institute of Hydraulic Research, State University of Iowa.
64. Moore, W. L., and Morgan, C. W., "Hydraulic Jump at an Abrupt Drop," *Proceedings* ASCE, 83, HY 6, No. 1449, December 1957, and Discussion *Proceedings* ASCE, 84, HY 3, No. 1690, June 1958.
65. McPherson, M. B., and Karr, M. H., "Study of Bucket-type Energy Dissipator Characteristics," *Proceedings* ASCE, 83, HY 3, No. 1266, June 1957, and Discussion, HY 5, No. 1417, October 1957, and *Proceedings* 84, HY 1, No. 1558, February 1958.
66. Ippen, A. T., "Chapter VIII, Engineering Hydraulics," edited by Hunter Rouse, John Wiley & Sons, Inc., New York, 1950.
67. Albertson, M., Dai, Y. B., Jensen, R. A., and Rouse, H., "Diffusion of Submerged Jets," *Transactions* ASCE, Paper 2409, Vol. 115, 1950.
68. Fiola, G. R., and Albertson, M. L., "The Manifold Stilling Basin," Colorado State University Research Foundation, CE.
69. Round, W., "Flow Geometry at Straight Drop Spillways," *Proceedings* ASCE, Vol. 81, Separate No. 791, September 1955.
70. Clark, R. R., "Bonneville Dam Stilling Basin Repaired After 17 Years' Service," *Journal* ACI, Vol. 27, No. 8, April 1956.
71. Warnock, J. E., "Spillways and Energy Dissipators," *Proceedings* of Hydraulic Conference, State University of Iowa Engineering Bulletin 20, 1940.
72. Houk, Ivan E., "Irrigation Engineering," Vol. 2, John Wiley & Sons, Inc., New York, 1956.
73. Keener, K. B., "Spillway Erosion at Grand Coulee Dam," *Engineering News-Record*, Vol. 133, No. 2, July 13, 1944.
74. Blaisdell, F. W., and Donnelly, C. A., "Straight Drop Spillway Stilling Basins," St. Anthony Falls Hydraulic Laboratory, Paper 15, Series B, October 1954.
75. Ahmad, Mushtaq, "Mechanism of Erosion Below Hydraulic Works," *Proceedings* International Association for Hydraulic Research and ASCE, University of Minnesota, Minneapolis, Minn.

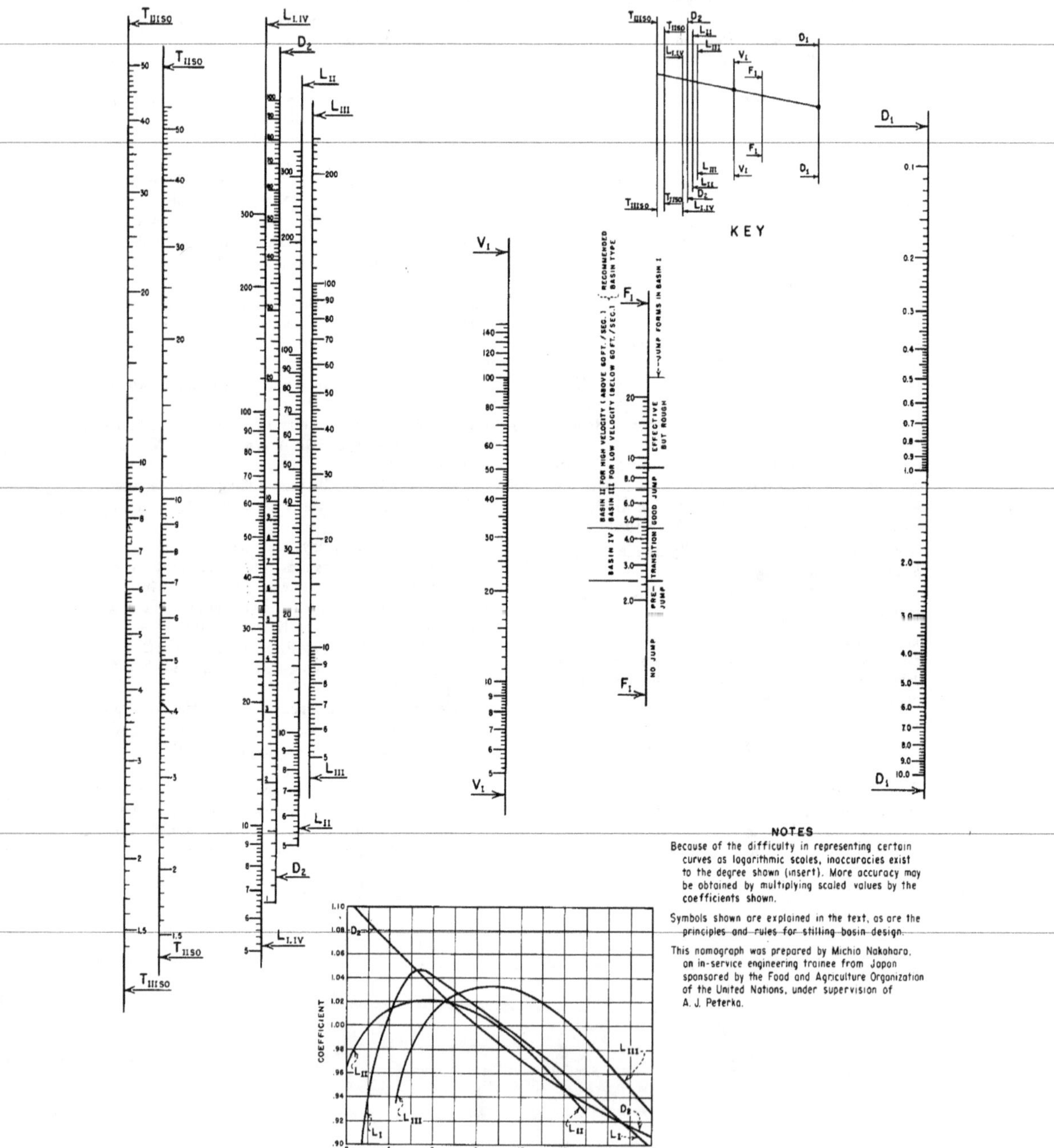

Nomograph to determine hydraulic jump stilling basin characteristics and dimensions.

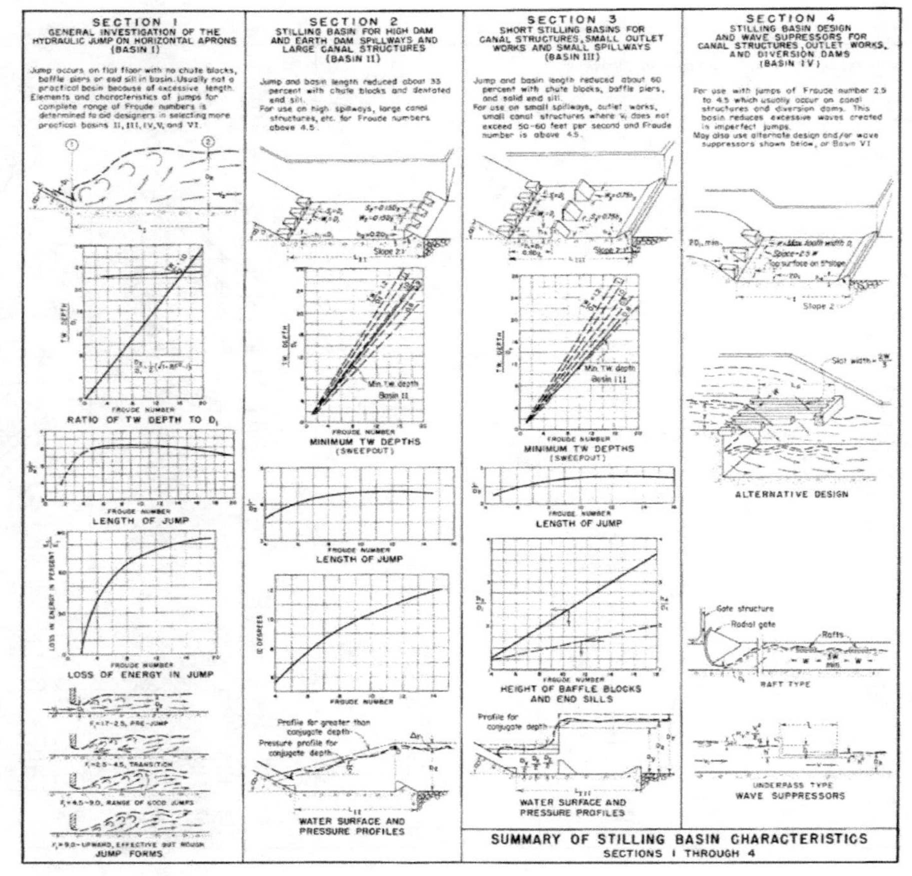
SECTION I
GENERAL INVESTIGATION OF THE HYDRAULIC JUMP ON HORIZONTAL APRONS (BASIN I)
Jump occurs on flat floor with no chute blocks, baffle piers or end sill in basin. Usually not a practical basin because of excessive length.
Elements and characteristics of jumps for complete range of Froude numbers is determined to aid designers in selecting more practical basins II, III, IV, V, and VI.
FROUDE NUMBER
RATIO OF TW DEPTH TO D1
FROUDE NUMBER
LENGTH OF JUMP
FROUDE NUMBER
LOSS OF ENERGY IN JUMP
LOSS IN ENERGY IN PERCENT
F1=1.7-2.5, PRE-JUMP
F1=2.5-4.5, TRANSITION
F1=4.5-9.0, RANGE OF GOOD JUMPS
F1=9.0-UPWARD, EFFECTIVE BUT ROUGH
JUMP FORMS
SECTION 2
STILLING BASIN FOR HIGH DAM AND EARTH DAM SPILLWAYS AND LARGE CANAL STRUCTURES (BASIN II)
Jump and basin length reduced about 33 percent with chute blocks and dentated end sill.
For use on high spillways, large canal structures, etc. for Froude numbers above 4.5.
Slope 2:1
Min. TW depth Basin II
FROUDE NUMBER
MINIMUM TW DEPTHS (SWEEPOUT)
FROUDE NUMBER
LENGTH OF JUMP
FROUDE NUMBER
Profile for greater than conjugate depth
Pressure profile for conjugate depth
WATER SURFACE AND PRESSURE PROFILES
SECTION 3
SHORT STILLING BASINS FOR CANAL STRUCTURES, SMALL OUTLET WORKS AND SMALL SPILLWAYS (BASIN III)
Jump and basin length reduced about 60 percent with chute blocks, baffle piers, and solid end sill.
For use on small spillways, outlet works, small canal structures where V1 does not exceed 50-60 feet per second and Froude number is above 4.5.
Slope 2:1
Min. TW depth Basin III
FROUDE NUMBER
MINIMUM TW DEPTHS (SWEEPOUT)
FROUDE NUMBER
LENGTH OF JUMP
FROUDE NUMBER
HEIGHT OF BAFFLE BLOCKS AND END SILLS
Profile for conjugate depth
WATER SURFACE AND PRESSURE PROFILES
SECTION 4
STILLING BASIN DESIGN AND WAVE SUPRESSORS FOR CANAL STRUCTURES, OUTLET WORKS, AND DIVERSION DAMS (BASIN IV)
For use with jumps of Froude number 2.5 to 4.5 which usually occur on canal structures and diversion dams. This basin reduces excessive waves created in imperfect jumps.
May also use alternate design and/or wave suppressors shown below, or Basin VI
Slope 2
ALTERNATIVE DESIGN
Gate structure
Radial gate
Rafts
RAFT TYPE
UNDERPASS TYPE
WAVE SUPPRESSORS
SUMMARY OF STILLING BASIN CHARACTERISTICS
SECTIONS I THROUGH 4

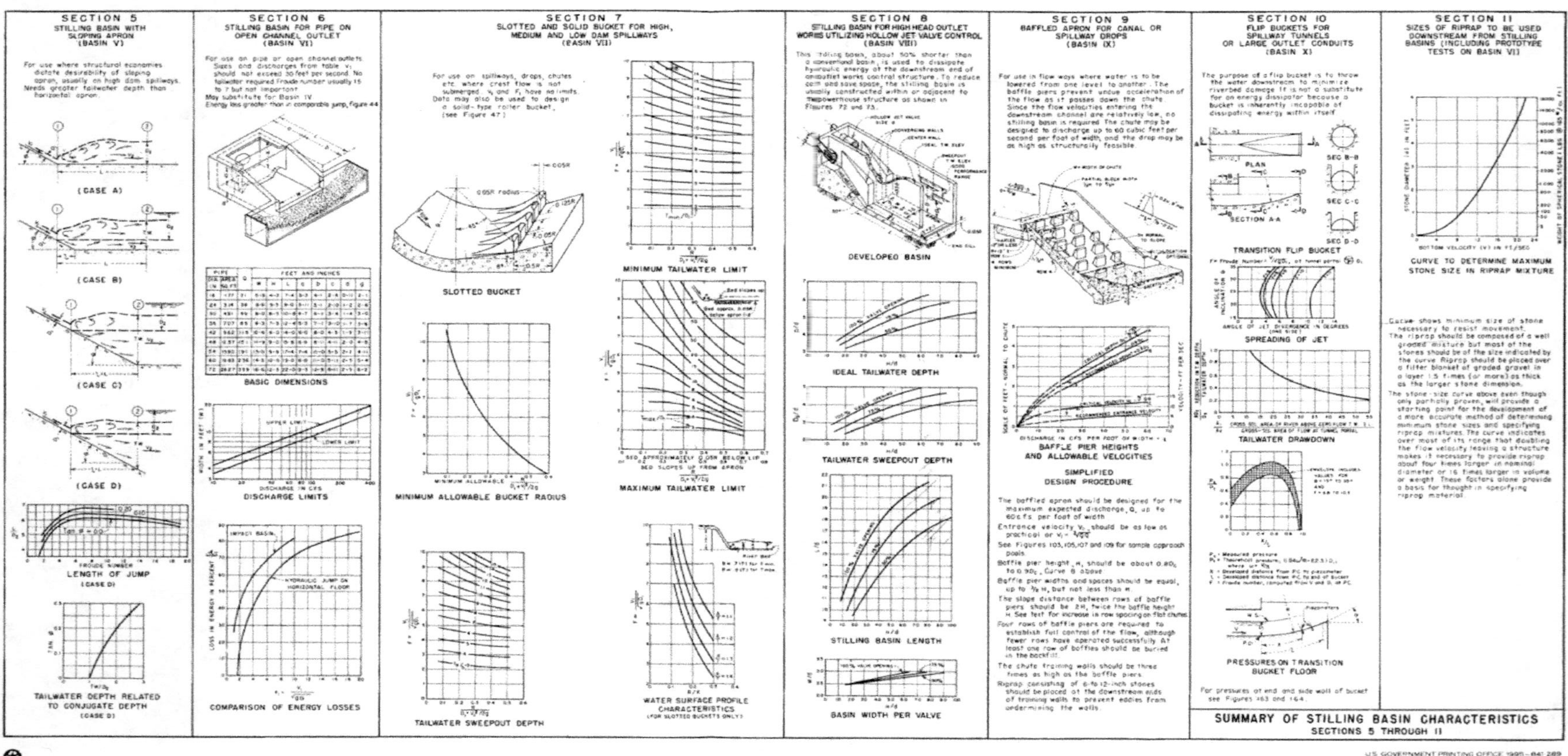

Printed on Recycled Paper

U.S. GOVERNMENT PRINTING OFFICE 1995—841-289

Mission of the Bureau of Reclamation

The Bureau of Reclamation of the U.S. Department of the Interior is responsible for the development and conservation of the Nation's water resources in the Western United States.

The Bureau's original purpose "to provide for the reclamation of arid and semiarid lands in the West" today covers a wide range of interrelated functions. These include providing municipal and industrial water supplies; hydroelectric power generation; irrigation water for agriculture; water quality improvement; flood control; river navigation; river regulation and control; fish and wildlife enhancement; outdoor recreation; and research on water-related design, construction, materials, atmospheric management, and wind and solar power.

Bureau programs most frequently are the result of close cooperation with the U.S. Congress, other Federal agencies, States, local governments, academic institutions, water-user organizations, and other concerned groups.

www.ingramcontent.com/pod-product-compliance
Lightning Source LLC
Chambersburg PA
CBHW080636180526
45168CB00008B/3191

* 9 7 8 1 5 0 8 7 2 2 8 1 6 *